TELLING OUR WAY
TO THE SEA

TELLING OUR WAY
TO THE SEA

A Voyage of Discovery in the Sea of Cortez

AARON HIRSH

Farrar, Straus and Giroux   New York

Farrar, Straus and Giroux
18 West 18th Street, New York 10011

The illustrations on pages 182 and 183 are based on N. R. Lovejoy,
"Reinterpreting Recapitulation: Systematics of Needlefishes and Their Allies
(Teleostei: Beloniformes), *Evolution* 54, no. 4 (August 2000): 1349–62.

Library of Congress Cataloging-in-Publication Data
Hirsh, Aaron, 1971–
    Telling our way to the sea : a voyage of discovery in the Sea of Cortez / Aaron
Hirsh. — First edition.
        pages   cm
    ISBN 978-0-374-27284-5 (hardcover)
    1. Ecology—Study and teaching—Mexico—California, Gulf of. 2. Evolution
(Biology)—Study and teaching—Mexico—California, Gulf of. 3. California,
Gulf of (Mexico)—Description and travel. 4. Experiential learning.   I. Title.
II. Title: Voyage of discovery in the Sea of Cortez.

QH541.13.H57 2013
577—dc23

                                                                    2013005478

Designed by Jonathan D. Lippincott

www.fsgbooks.com
www.twitter.com/fsgbooks • www.facebook.com/fsgbooks

1   3   5   7   9   10   8   6   4   2

For Veronica and Henry

We can understand, too, that natural species are chosen not because they are "good to eat" but because they are "good to think with."

—Claude Lévi-Strauss, *Totemism*

Stories, stories, stories. A world and a land and even a river full of the damn slippery things.   —Richard Flanagan, *Death of a River Guide*

We notice that it required a separate intention of the eye, a more free and abstracted vision, to see the reflected trees and the sky, than to see the river bottom merely; and so are there manifold visions in the direction of every object, and even the most opaque reflect the heavens from their surface.   —Henry David Thoreau, *A Week on the Concord and Merrimack Rivers*

CONTENTS

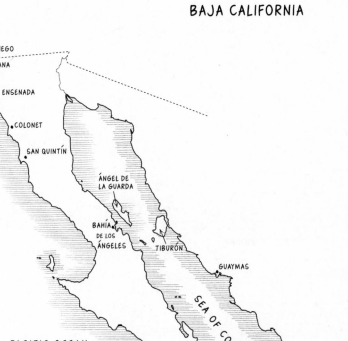
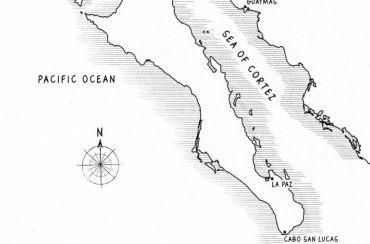

BAJA CALIFORNIA

SAN DIEGO
TIJUANA

ENSENADA

COLONET

SAN QUINTÍN

ÁNGEL DE
LA GUARDA

BAHÍA
DE LOS
ÁNGELES

TIBURÓN

GUAYMAS

SEA OF CORTEZ

PACIFIC OCEAN

N

LA PAZ

CABO SAN LUCAS

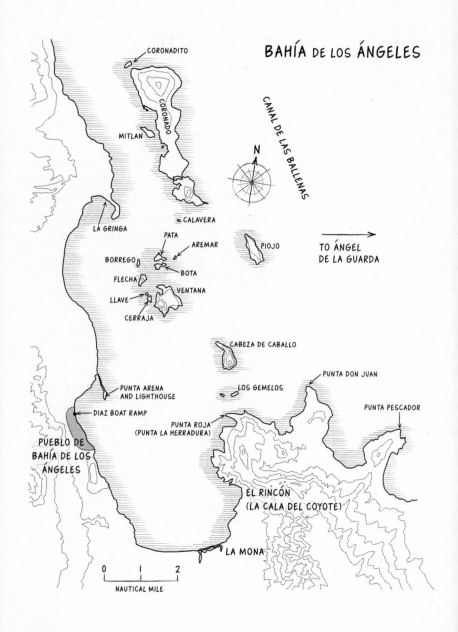

AUTHOR'S NOTE

Each summer for a decade, my wife and I took a group of about twelve college students into the Sea of Cortez for an intensive field course in ecology and evolutionary biology. This book focuses on one special class, but I have also drawn liberally on experiences from a number of other summers. The narrative is therefore a composite. Such compression makes certain deviations from facticity unavoidable. In addition, I have changed many names and some identifying attributes. In general, though, the events that I describe here did actually happen, and the characters are based on real people. The sections on science and history are—to the best of my knowledge and ability—straightforwardly true. References and further technical notes for those sections are available on the web at www.telling ourwaytothesea.com.

I've adopted certain textual conventions that are atypical and therefore merit a note. This book is unusual in having both a lot of dialogue and many quotations from texts. To distinguish between the two, I've used quotation marks only for words spoken aloud; excerpts from texts are in italics.

PART I

Isostichopus fuscus

LEARNING TO SEE

1

A school of needlefish parts to stream around me, and I find myself momentarily among the silver traces of a comet shower. I move to join them, but they accelerate and dissolve into open water, leaving me to stare at the luminous, molten mirror that is the underside of the ocean's surface. Veronica taps my arm—a signal that says both *look at that* and *be right back*—as she slips from the roiled layer of silver and descends swiftly, like a being born underwater. Her skindiver's fins form a single broad fluke, which propels her neoprene form sinuously toward the rocky bottom. Bright bubbles, escaping her snorkel, wobble urgently back to the air above. A thousand times I have seen her descend like this, yet still I find myself wondering if, this time, she might go too deep, or stay too long.

Here, mercifully, the seafloor is only twenty feet down—a depth at which the plunging chutes of sunlight are just converging to their vanishing point. As she approaches the rocks, Veronica twists, glides to a supine and weightless pause, and gazes up at the students who float beside me here at the surface. She seems to be pointing at something on the seafloor.

Allie, the student to my right, turns to look at me. Inside the partial shade of her dive mask, her eyes are hard to read: They look puzzled, a little concerned. She is probably just wondering why Veronica wants them to notice what appears to be a mud-brown lump of sea muck. Though it also seems possible that Allie has already perceived Veronica's tendencies underwater—the strange private gravity that seems to draw her to depth—and she is now asking, in her gentle way, whether something should perhaps be done to bring Veronica back to the surface. I take several long breaths, saturating my blood with oxygen and preparing to dive, but just as I draw my last, deep dose of air, Veronica finally relents. She places her hand gently around the nondescript mound and pulls it from the rocks, holding it as one might hold a soft loaf of bread.

Arriving among us, Veronica holds out her hand, upon which rests her inert quarry. What was mud-colored below is now—in this bright, shallow water—more of a yellow ocher, and it is studded with pale tubercles that are almost the color of lemon drops. The skin, stretched taut over the knobby body, appears thin and mucosal, making the thing look terribly exposed, like a bodily organ drawn by the hand of a surgeon into the sudden brightness of the operating theater. The students—there are five of them here—draw in around Veronica's palm, peering intently through their panes of tempered glass. They seem transfixed, certain that Veronica's plunge must have been for something thrilling, and yet I know their patience can be short, especially this early in the trip, when everything around them feels new. And so, as the thing on Veronica's palm waits us out, stolid as a piece of earth, I begin to worry that the students will soon lose interest, and miss what Veronica wants them to see.

Just when I think they may be eyeing one another through their personal portholes—wondering, perhaps, if it would be rude to resume their search for colorful fish—the lump trembles, inches forward along Veronica's palm. Suddenly it is less vegetable than animal, and the students pull back apprehensively. But as the circle of masks starts to widen, Veronica's free hand catches Allie by the wrist.

Veronica is wise, I think, to choose Allie, because there are others who might not be so trusting. Carefully, she opens Allie's palm and holds it beside her own. As the knobby creature slides from one hand to the other, Allie's eyes widen and she speaks into her snorkel—an incomprehensible but richly expressive string of syllables. For a moment, she seems frozen. But even in her astonishment, she looks to the other students. She takes the hand of the young man floating beside her, opens his palm, and holds it next to her own. The animal slides over obligingly, and as it does, Cameron explores the creature's back with his other hand.

Cameron's hands look muscular, well-worn, and they sometimes move in unusual ways: the fingers seem to explore independently, executing many minor adjustments, as if they were navigating the neck of a string instrument. These hands have learned to perceive more than other hands, because Cameron cannot see. He is blind. And as his fingers creep across the animal's back, investigating, it becomes clear that they are following a pattern: the yellow warts, which at first seemed to be scattered more or less randomly, are in fact arranged—loosely, but nonetheless perceptibly—into two rows.

I have never noticed this rough regularity, but now that I see it, I suspect it might be meaningful: I suspect, in fact, that those two haphazard rows are clues to a deep connection—an invisible but very real thread that links the ugly animal on Cameron's hand to far more beautiful creatures we've seen this morning. Just moments before her plunge, Veronica pointed us to a sun star, *Heliaster kubiniji*, a pink-and-green starfish in the unmistakable shape of a sunflower. And before that, we all hovered in admiration over the crown sea urchin, *Centrostephanus coronatus*, which is a sphere of long and slender spines, each one perfectly black but for the occasional sharp wave of blue light that races from tip to base. To describe these scattered pulses to Cameron, Allie said it looked "like an alien's brain."

After *Heliaster kubiniji* and *Centrostephanus coronatus*, even the name of the animal now sliding across Cameron's hand rings a little prosaic. It is *Isostichopus fuscus*, the brown sea cucumber, and one would not readily assume that it has much in common with those other, much lovelier animals. Yet that is precisely what Cameron's subtle touch has just revealed: those messy lines of tubercles, I now realize, are among the attributes that place the brown cucumber firmly in the broad alliance of animals known as phylum Echinodermata. And who else should number among the echinoderms but the sea stars and urchins. All of these creatures, from ugly *I. fuscus* to brilliant *C. coronatus*, are the descendants of a single ancient species: an ur-echinoderm that inhabited the ocean 520 million years ago. And because that ur-echinoderm, in its own time on earth, underwent several extraordinary modifications—we could even call them innovations—we now find mementos of those changes in every single one of the creature's descendants. In fact, those vague rows of bumps just now detected by Cameron's hands are but a faint reminder of the ur-echinoderm's most fundamental innovation. But exactly what that innovation was, and how it later became a trace so obscure it took Cameron's touch to disclose—these are questions I should raise later, when we're back at the field station. Because right now, Veronica seems to have a plan of her own; she has just lifted her head from the water, letting her snorkel dangle by her face, and the rest of us now follow her lead.

"Cameron," she says, "do you feel it attaching to your hand?"

"Yeah, totally," he responds. Cameron grew up in a small town outside of Santa Barbara, California, and he talks like a surfer. "It's got those wicked little suckers," he continues, "just like the starfish."

"*Exactly*," Veronica says, clearly pleased with his suggestion. "Sea stars and cucumbers both have tube feet."

As Veronica begins to explain how these ingenious little devices operate, Cameron allows the cucumber to creep back to Allie, who then takes the hand of another student. It's Chris, a quietly confident young man who may already know much of what Veronica is explaining. Chris has spent a semester at a marine biological laboratory, a month on an oceanographic research vessel, and, judging from his comfort in the water, a lot of time diving. As I talked with him during our long drive down the Baja Peninsula, Chris seemed strangely familiar, as if I knew him from somewhere. But it wasn't until we arrived at our destination, the small town of Bahía de los Ángeles, that I finally understood: the only people I'd ever met who possessed this calm composure—not arrogance, really, but equanimity—were the more formidable patriarchs here in Bahía. That Chris shares this trait is perhaps not entirely coincidence. His father is Mexican, and Chris spent part of his childhood in a picturesque village not far from Mexico City.

The student floating beside Chris, Anoop Prakash, appears less comfortable in the water. In fact, he seems to be expending a tremendous amount of energy just to remain upright, and occasionally his fin flails out and whacks Chris in the legs. Despite these occasional assaults, the transfer of the cucumber from Allie to Chris goes smoothly, and Chris is now placing one hand in front of the other to make a kind of treadmill for the animal. Even with our faces above water, and the cucumber just below the surface, we can see that as it slides forward, its front seems to dab back and forth laterally, as if it were exploring the curious new terrain. The end of this treadmill ride, I'm afraid, will be Anoop's hand, and the thought of that unsteady platform makes me more than a little anxious—for the animal and Anoop alike.

I assume Veronica too is monitoring Anoop's buoyancy, because his safety in the Vermilion Sea was the subject of some debate when we first reviewed applications for our field course. Anoop had approached us after the Baja info session, an informal presentation that Veronica and I offer for prospective students. As soon as the presentation had ended, he unfolded himself from his writing desk—he is tall for a South Indian, about six foot one—and in three gangly steps, he arrived at the front of the room. With a black goatee and wire-rimmed glasses that gleamed silver against his chestnut skin, he looked quite scholarly. And because the lenses over his eyes appeared somewhat too small—as though they could possibly clarify only what lay straight ahead, leaving the periphery in a blur—he also looked intensely focused. Fixing his narrowly tunneled view on Veronica, he said, "Dr. Volny, I'd like to ask what endemic species of salt-tolerant plants are found around the research station."

Usually, students ask if we might really see a whale shark, or if the dolphin are there every year, or if it's truly so hot as we say it is. They do not, in general, ask about endemic salt-tolerant plants, and it took Veronica a moment to gather her answer. Later, when Veronica and I were laughing about the question, she seemed charmed by it, but she was also earnestly concerned that Anoop's sheer focus might compromise a more general awareness of his surroundings. "He'll step on a scorpion," she said.

The students' written applications arrived two weeks later, and Anoop's was fabulously impressive. As I read certain sections to Veronica, she said, "You're making this up!" and grabbed the page from my hands. But I hadn't made it up. Besides the independent research in yeast genetics and a concentration in philosophy of science, there were graduate seminars on Marcel Proust and Ludwig Feuerbach, in which Anoop, a sophomore undergraduate, had earned perfect grades. But there were also a few sentences that Veronica referred to as "warning signs." In response to the simple question *Can you swim?* Anoop had written, *Last time I checked, I was competent in breast stroke, Australian crawl, back stroke, and side stroke*—a response that Veronica deemed suspicious for its very thoroughness. "He won't step on a scorpion," she said. "Because he'll sink first."

In the end, I argued we simply had to take Anoop; his academic record—not to mention his devotion to salt-tolerant plants—left us no choice. Veronica consented, but added, "He's on your watch." And now I do watch—quite closely—as Anoop's hands rise and open to form the animal's next platform. Without his arms for paddling, he drops slightly: his chin lowers into the water, and he tilts his head back to keep his mouth above the wavelets.

Tube feet, Veronica's been explaining, are the only outward sign of an internal system possessed by all echinoderms—sea cucumbers, urchins, sea lilies, sand dollars, brittlestars, and, as Cameron sensed, sea stars. The bodies of these animals are piped with a network of tubes, and when an echinoderm decides to move, water flows to the appropriate plumbing, creating hydraulic pressure. Each one of those hundreds of tube feet on an echinoderm's underside is the continuation of an internal pipe. When a foot needs to take a step, a small ampule inside the animal contracts, driving water into the tube and extending it until it touches the surface below—in this case, a student's palm. What feels like a miniature suction cup—or, as Cameron put it, a wicked little sucker—is in fact something stranger: As the little foot makes contact with the surface, it secretes a kind of quick-dry superglue, fastening it to its substrate. And when it is ready to let go, the foot secretes a fast-acting antidote to its own adhesive.

Having crept from Allie to Cameron, back to Allie, and on to Chris and Anoop, the cucumber now seems to have reached something of a cul de sac, and I wonder whether any of the other students are growing impatient with Anoop. I don't think he intends to monopolize the sensation of tube feet on the palm; it's just that he's fully absorbed in keeping his head above water and his hands relatively steady. Rafe, the only student in our floating circle who has not yet handled the animal, seems to be sidling closer to Anoop, awaiting the transfer, though he is held at bay by Anoop's erratic kicks.

Ever since he introduced himself at the Baja info session, Rafe has made me uneasy, because his self-assurance sometimes verges on blithe overconfidence. Veronica had asked everyone at the meeting to say a word about themselves, and most students seemed to blush and hurry through rather self-effacingly. But Rafe spoke slowly, with striking nonchalance. In his thick Australian accent, he said, "My name is Kurtis Rafe, but you can call me Rafe. My major is chemistry, but I'm taking a lot of theoretical physics as well." And when he said, serenely, "I'm applying 'cause I think it'd be ace to swim with sharks," the room broke into laughter, expelling its awkward tension, and Rafe sat back, smiling slightly. Even here in Baja, he keeps his long blond hair nicely combed, pulled back in a tidy ponytail, and he wears a small but thick gold hoop in one ear.

Evidently, Veronica is not as concerned as I am about Rafe's looming encroachment; ignoring him, she says she wants to show the students something very strange about the echinoderms. And just from the way she says "very strange"—as if she were some sort of haruspex, preparing to astonish her small audience—I suddenly know what's coming, though I can hardly believe she's really going to do it with the creature resting on Anoop's hands.

The first time Veronica took me diving in these waters, seven years ago, she found a brown cucumber. She placed the animal on my open palm, and when it began to creep forward, she gently pressed a finger against its knobby back. Almost instantly, the animal contracted, like a biceps balling up in tension. I pressed it with my own finger—it was as hard as a billiard ball—and I widened my eyes and shook my head in underwater communication of astonishment. But Veronica held up a finger, as if to say, "Just wait, there's more." She began to rub the animal's back, as though she were polishing its hardened surface. At first nothing happened; the billiard ball rested hard and compact in my hand. But then, with the abruptness of a dropped egg hitting the kitchen floor, the animal changed—it seemed to melt and began to ooze over the edges of my palm. In alarm I tried to hold the creature together, cupping it now in both palms, but the substance seemed to be seeping between my hands, and all I could do was hold still and hope that whatever was digesting the animal would not start working on me as well. Before I could even know whether my hands felt pain, the sludge had slipped from them and was pouring itself into a crevice in the rocks below.

An hour later, I stood alone in the airless heat of the field station library, reading from a zoology text I had just slid off the shelf: *The degree of change in the rigidity of mutable connective tissue is almost as great as that of ice to water. Indeed, people who have firsthand experience of echinoderm tissues passing from rigid to flexible states liken the transition to liquification.*

Liquefaction—ice to water—seemed exactly right. What was surprising, however, was that this *mutable connective tissue* is made almost entirely of the protein collagen. That seemed odd, almost disturbing, because the very same protein composes our own cartilage—the firm stuff of spinal disks, as well as noses. As I read this strange new fact, registering it slowly in the stifling heat, my fingers rose involuntarily to check on my nose.

What this tells us, first of all, is that collagen is an old molecule. It must have evolved even before the ur-echinoderm branched off from our own phylum's ancestor, the ur-chordate; otherwise, we and the cucumber would not both have it. But once those ancient species had separated, and each was following its own evolutionary path, the ur-echinoderm must have hit upon some extraordinary innovations in the construction of tissue from collagen, because all the echinoderms, from elegant *C. coronatus* to homely *I. fuscus*, are able to go billiard-ball-hard one moment and pudding-soft the next.

But what was it, exactly, that the ur-echinoderm discovered? How does a creature freeze or melt its own tissue at will? Collagen forms a rigid substance when millions of individual molecules gather into a thick fiber: imagine winding together twist-ties until they made something like a suspension bridge cable. Our own collagen is permanently wound into cable, such that our noses do not suddenly melt. But echinoderm collagen, by contrast, can suddenly disassemble, allowing the cable to relax and fall apart into so many millions of twist-ties. More fantastically, the twist-ties can then reassemble, linking up once again into cable. The mysterious substance that triggers this change is yet to be isolated in a lab, but we know it's a protein, and we know it's secreted by cells that are embedded in the mutable tissue.

At the moment, however, the more pressing mystery is why Veronica would take Anoop, of all people, as her volunteer for a magic trick with mutable connective tissue. Given that we will try this only once in our two-week course—Veronica will surely ask the students not to repeat it, because it must take a toll on the cucumber—I don't see why she would make Anoop the one whose nerves will be tested. Does she think his excitability will enhance the shock of transformation? Or has she really put him on my watch, allowing herself to forget her own warnings? In any case, when she places her snorkel in her mouth and puts her face in the water, all of us follow suit, except Anoop: with the animal stranded on his trembling hands, he is unable to replace his snorkel, and is left alone above water while everyone watches his hands below.

Through my half-submerged mask, I keep a watchful eye on Anoop's face. At first he looks somewhat resigned, like a surgical patient separated from the doctors' work by a curtain across his chest. But a second later, he whips his head suddenly to the side, trying to catch his snorkel's mouthpiece with his teeth. He makes three such lunges, each of them unsuccessful, before he gives up, takes a deep breath, and submerges his face—just in time to see Veronica's index finger pressing on the animal's back.

The creature compresses instantly, but Anoop holds steady. Veronica places Cameron's hand on the frozen animal, and the other students, too, feel the hardness of cross-linked collagen. As Anoop watches their hands reaching in to investigate, his cheeks begin to swell with the pressure of unexhaled breath, and just when it appears they might burst open, admitting a gasp of seawater, Allie reaches across our circle and places Anoop's snorkel in his mouth. Through his glinting mask, he seems to look gratefully at her.

Veronica begins to rub the animal's back, and at first, it appears to relax. But then something goes wrong. The creature skips liquefaction and moves straight to more drastic measures—the last line of defense. Because an echinoderm's nervous system speaks directly to those special cells embedded in the collagen matrix, telling them when to release their potent catalyst of change, the animal can freeze or liquefy whichever piece of tissue it needs to. This is how a brittlestar caught by the arm can throw off the entire limb, leaving it behind like the detached tail of a lizard: it simply liquefies the narrow segment of tissue that connects the arm to the central disk. And when a cucumber is under attack, it resorts to an even more radical tactic. It swiftly disassembles the collagen cables that hold its organs, violently contracts its entire body wall, and shoots its viscera out its anus.

One can imagine that even the most menacing predator might be taken aback by such a move, and even if it weren't, it might at least be tricked into pursuing the evacuated innards instead of the now-hollow cucumber. Taking a bite of floating viscera, the predator would quickly learn that the animal is laced with a powerful toxin. The hollow cucumber, meanwhile, would have moved on, and would later regenerate, from stem cells in its empty body cavity, a complete set of internal organs.

But if evisceration might distract a menacing predator, just think what it could do to Anoop. When the dark purple organs explode from the animal's posterior, he startles and flails, attempting to back away quickly. The guts purl and twist in his turbulence, forming a kinetic design of dark ribbons, diffusing colors, and loose round forms at the center of our circle. From the bottom of this turning mobile, the cucumber body sinks toward the seafloor. Rafe, evidently reluctant to let the animal escape, dives for it. He moves less smoothly than Veronica, but nonetheless manages to kick his way downward fairly rapidly until, about halfway to the bottom, he clutches the sides of his head and halts his dive.

As he makes his way back toward the surface, holding his head all the way, Veronica and I look at each other through our masks. Had we suspected that Rafe didn't know what he was doing, we would have tried to catch him on his way down. But he plunged so suddenly, and it occurred to neither of us that he would neglect to equalize the pressure in his ears. The class has often included students who didn't know how to snorkel, but they have always learned the basics in just a day or two—and from the other students, not from us. Equalizing ear pressure is such a simple trick—as simple as holding one's nose and blowing out—and the early tutorials among newly acquainted students have always seemed to bring them together, so Veronica and I have never offered any formal introduction to snorkeling. Rafe, however, must have been watching Veronica, who somehow equalizes without holding her nose, and he must have decided that the graceless plugging of nostrils is unnecessary for talented skindivers. And now, as he rises to the surface holding his head in pain, Veronica and I can only hope that our habit of leaving it to the students to teach one another hasn't cost Rafe an eardrum.

2

The Vermilion Sea Field Station is solidly built. The foundation and the first few feet of wall are stone: a sturdy masonry of massive blocks. Atop the stone, the structure is framed with thick wooden beams, which are exposed around windows and doorways but elsewhere hidden beneath an earthy stucco. Wherever wood is visible, it is traced with the vermiform designs of a century of termites. The station was built in the 1880s by an American mining corporation, and I've often thought that the admirable sturdiness of construction must testify to a mixture of corporate optimism, which in the end proved misplaced, and colonial determination, which in the end proved futile. *We're here to stay*, the construction seems to say. *We build for posterity.* But around the time of the Mexican Revolution, the corporation disappeared. I've read conflicting accounts of the mine's closure. Some say the vein of silver at Las Flores, ten miles from the station, simply ran dry. Other accounts refer darkly to *political circumstances*. But what seems most likely is that the mining operation, like so many ventures before and after, was slowly but surely battered out of existence by the sheer difficulty of life and work in a relentlessly dry desert. Perhaps the one, anomalously flimsy element of the building—the roof, which must have been replaced many times since the mining company was here—suggests that even as they neared the end of construction, the Americans began to doubt their longevity in this place. Or maybe they just figured the roof would be unimportant, as it might never face rain.

The masonry of the station's foundation extends to form a terrace between the building and the sea. Four rather sparse salt-cedars, or tamarisks, are evenly spaced across the terrace, and they provide a spotty shade. In more habitable parts of the world, such trees might be felled: they are scraggly; they shed long, dry, glaucous needles that infiltrate your hair, book pages, or lemonade; and when the sea exhales a thick humidity into the hot air, the tamarisks sweat an acrid turpentine that can take the paint off a car. In a sense, they should not even be here. Tamarisks are Eurasian trees that were introduced to the desert Southwest, where they are now blanketing riverbanks and displacing native species. But at two in the afternoon, when the unimpeded sun feels like a branding iron against your skin, and the interior of the field station heats to a stuffy ninety-five, you view the salt-cedars not with scorn, but with gratitude. They seem to be a necessary condition for human life.

The five students who several hours ago watched a cucumber explode have pulled their plastic chairs together under one of the salt-cedars, and they sit in a half-circle facing the sea. In a separate island of shade, I am paging through my materials for the afternoon lecture, intermittently listening to them laugh and talk. Anoop is at the center of the half-circle, flanked by Cameron and Allie to his left, Chris and Rafe to his right. Rafe has wads of toilet paper stuffed in his ears, and their loose ends stick out like bleached pigtails. Fortunately, his headlong dive did no serious damage; Rafe's eardrums are intact. Nonetheless, for unspecified medical reasons, he has insisted that the toilet paper remain in place.

The five of them are reviewing their moment of shock, and Anoop is making the others laugh wildly with his rangy reenactment of mad, backpedaling retreat from cucumber guts. I wonder whether Veronica could have foreseen this. Was it this very giddiness—with Anoop seated in the middle—that she had in mind as she poked the cucumber resting in Anoop's hands? Even if liquefaction had gone according to plan—if it had never led to swirling innards—the experience still would have given these five students something to relive here on the terrace. And Anoop, who otherwise might have had a difficult time finding his place in the group, would have been at the center of it all. As it happened, the moment was all the more shocking, and that much more effective in bringing them together.

When the hilarity dies down, there is a lull in their conversation. Glancing up from my pages strewn with pine needles, I see the students staring out at the sapphire bay and the scorched umber islands that populate it. In the distance, on the other side of twenty-five miles of water, a mountain range of variegated pinks and pale grays extends as far north, and south, as one can see. It is not the coast of mainland Mexico, all the way across the Sea of Cortez, but rather a single long island known as Ángel de la Guarda. The channel between Guardian Angel and the east coast of the Baja Peninsula, where we sit, is known as El Canal de las Ballenas, the Channel of the Whales. It is four thousand feet deep.

The wind is picking up, as it does most afternoons, and the bay's water, which was dark and glassy just a few hours ago, is now whipped up into whitecaps that march in ragged rows toward shore. This wind comes to us across the channel's cold water, making the temperature here in the shade almost comfortable.

"Hey, Aaron," Cameron says in my direction, breaking their lull, "why doesn't something just mack that thing?" I look over as if I haven't been listening all along. I have no idea what cue might have told Cameron I was sitting here. Perhaps he heard me pull up a chair for lunch, and then never heard me leave. As I drag my chair now from my own salt-cedar's shade to theirs—to talk with them more easily in the rising wind—I am newly aware of the drumroll the plastic makes as it scrapes across the stone. I sit beside Allie, on the left edge of their half-circle facing the sea.

"You mean why don't fish eat it?" I ask.

"Yeah. I mean, it's slow and defenseless. Why don't they just mack it?"

Cameron's eyes are not closed. They are crystalline blue-green, and offer only subtle suggestions that he cannot see: the gaze is immobile, fixed straight ahead; and at the outside edge of each eye socket, a slight indentation curves the bone. These gentle dips are barely noticeable, and they actually make his brow and cheekbones appear slightly more prominent and, it seems to me, handsome. He has short-cropped sandy blond hair, a muscular jaw and neck, and a body as burly as a marble Hercules. I have seen Cameron lifting weights in the campus gym, but I don't think his extraordinary brawn comes from pumping iron. I think it comes mostly from compensating for the absence of his left leg. A high-performance prosthetic, which looks like a sleek robotic limb, extends from Cameron's left thigh. He is remarkably dexterous with this device, but nevertheless, certain situations seem to require pure strength to compensate for the missing leg. On our journey down the peninsula, we spent the night in a landscape of colossal boulders. Watching Cameron pull himself up steep faces of granite, I understood why he looks so powerful.

"The cucumber's toxic," I tell him. "It makes a poison called holothurin."

"But don't people eat them?" Chris asks.

"No way," Rafe interjects, and then he groans and clutches his stomach, as if the very thought of it might kill him. And in fact, I don't really know why the foodstuff called trepang or bêche-de-mer does not kill people, laced as cucumbers are with truly potent poison. I've read, for instance, that holothurins interrupt signals between nerves and muscles, and at sufficiently high doses, they explode blood cells. Several research groups have even tried to turn this chemical nastiness against aggressive tumors. When I tell the students about this work, Cameron says, "That's excellent. I didn't even know about that."

It strikes me as a strange remark—why *would* he know about such work?—and I steal a glance at the other students, wondering if they, too, find it peculiar. Only Allie seems to have registered it. She's looking at Cameron with her brow furrowed and her lips parted, as if she had been on the verge of saying something when she caught herself and paused. In this oddly suspended moment, I realize that her face reminds me of silent movies, where pretty actresses had to be as expressive as mimes: her eyebrows are thin and arched, making her brown eyes look wide awake; her cheekbones are strong; and her mouth seems to move in an instant from bright glee to frowning gravity to—just now—a portrait of speech cut short. I'm still waiting for her to break her frozen pose and speak when Chris says, "Osprey!" and points down the coast.

A sea eagle, flying a beam reach across the winds from the channel, sails toward us along the shoreline.

"Oh, sweet," says Cameron, sitting up straight, as if he's preparing himself for something. "Those things are amazing."

"It's heading right toward us," Allie tells him. "It's coming along the beach." Cameron turns to face it.

The broad wings are virtually motionless—they make only slight adjustments in cant relative to the wind—and yet, in a second or two the eagle has traveled the length of the beach that stretches north of the station, and it is sailing over the rocky coast directly in front of us. Because the rocks ascend steeply from the water up to the station, and the stone terrace is several steps higher still, we and the osprey are at the same altitude, and as it passes, we look straight into a yellow, war-painted eye. It is fierce.

The eagle vanishes to the south, and we are silent until Anoop, under his breath, says, "Osprey," and Cameron whispers, "That was amazing." Cameron, like the rest of us, is facing south—he is not still looking north, toward the beach—and this makes me wonder whether the broad wings sailing the crosswind might have been audible. If I had listened, would I have heard it, too?

"Could I ask a question?" Allie asks, still a little quieter than we were before.

"That's a funny question," Anoop answers, smiling to himself. For an instant, I expect Allie to go back to the question that arrested her, looking at Cameron, right before the osprey appeared. But instead she looks at me and says, "You know how Veronica said the cucumber's in the same family as sea stars?"

"Same phylum," Chris says, correcting her. "Echinoderms."

"Right—echinoderms. And she told us the things echinoderms share."

"Like mutable tissue," Anoop says, again smiling to himself.

Allie smiles with him, but also continues. "And a body divided into five parts, like a sea star's five legs."

"Oh," Anoop says, "that's right. Where are the five parts?"

"And how did it—how did evolution—take something with five parts, like a sea star, and turn it into one part, like a cucumber?"

I hardly know where to begin, because this question could take our conversation in so many directions—the origin of basic body plans; the nature of deep evolutionary change; how we infer relationships between creatures as different-looking as sea stars and cucumbers. When so many scientific threads come together in a tight weave, it can be difficult to pick a single loose end and follow it. If we begin with the simplest inquiry, I figure, it will soon cross other threads, which we can pick and pursue if we choose. So we start with the most basic question: Where are the five parts?

"They're actually still there," I say. "They just aren't so obvious."

The group seems to lean inward, listening; these are clever and ambitious students, and they sense a puzzle to be solved, or even an opportunity to shine.

"On the animal's underside," I continue, "the tube feet are arranged into three rows, from the animal's mouth to its rear end."

"Oh," Cameron says, stopping me. "And there are two rows of bumps on its back."

Exactly. Those tubercles, I tell the students, have got to be the remnants of tube feet. So with three functional rows along the animal's underside, the two vestigial rows along its back recall the five-part body plan of the ur-echinoderm—the same pentamerous symmetry that's so obvious in the five long arms of a sea star. In the cucumber, this basic body plan is obscured not only by the decay of two rows of tube feet, but also by elongation of the animal's body. The sea star is a flat animal, with its mouth on the underside and its five hydraulic pipelines running outward, along the five arms. The cucumber, on the other hand, is stretched out, with its mouth at one end and its anus at the other. The three rows of working tube feet are pushed close together under the animal's belly, and the tubercles on its back are somewhat scattered, making the cucumber look superficially like a long animal with symmetry that is basically *bilateral*—as in a fish, for instance, or a human. Pentamery would become obvious only if you were to cut the creature in half crosswise, revealing the five hydraulic pipelines running from one end to the other. Or if you were to examine the cucumber in the larval stage when it is known as a *pentacula* and looks more or less like an oval with five stubby tentacles.

Anoop raises his hand, causing me to fall silent and look at him. At first, I'm puzzled, because his gesture is so entirely out of context. And when I realize what he's doing, I'm slightly annoyed. *Anoop*, I want to say, *we are not in a classroom. We're sitting on a stone terrace, facing the sea, and this conversation was not scheduled on our syllabus.* The others, too, seem to find the gesture peculiar, and for a moment we stare at him in silence. Perhaps because I say nothing, Allie finally calls on him.

"Yes, Anoop?"

The other students smile at this, but Anoop simply lowers his hand and asks his question: "Does that mean ontogeny recapitulates phylogeny?"

Rafe guffaws—a laugh that sounds a little cruel, dismissive of Anoop for his classroom comportment. The meanness makes me forget my own annoyance—or perhaps transforms it to a twinge of guilt—and I'm tempted to remind Rafe that he has toilet paper hanging out of his ears. But Anoop and the others appear undistracted, focused on the conversation, and I should follow their lead.

Anoop's question comes straight out of his studies of the history of biology. In fact, the very phrase *ontogeny recapitulates phylogeny* is lifted from one Ernst Haeckel, a nineteenth-century German naturalist and philosopher. Among biologists, Haeckel is probably known best for his lavish engravings of marine invertebrates. And he's probably known second best for being rather obscure, mystical, and wrong—or at least less right than another nineteenth-century German, named Karl von Baer. Given these deep and complicated roots in history, Anoop's question is perhaps not right up my alley, and before I venture an answer, I turn to look around the terrace for the person whose alley it rightfully is—my friend and fellow professor in this course, Graham Burnett.

I met Graham in the first class of my freshman year at Princeton. It was HIS 291—The History of the Scientific Revolution. Graham wore a tie, sat in the front row, and often approached the professor after class to pose a question that I might or might not comprehend. I was trying to escape an identity as an athletic recruit, hoping to become a writer or a scientist, and Graham, who was unabashedly intellectual, seemed to hold the keys to the kingdom. For his part, Graham must have been grateful to find at least one fellow student who did not disparage his seriousness. So we quickly became friends, roommates, and companions in our mission— often pretentious, but always quite earnest—to make something of our minds.

In fulfillment of a plan Graham laid out at the end of our freshman year, he is now the professor who teaches HIS 291 at Princeton. And although this precise realization of a blueprint for a life in academia might lead one to take Graham for something of a career company man, that would be a misreading. It is, admittedly, a misreading Graham readily allows, letting colleagues and students see him as the sweater-vested academic historian. But in fact, Graham has the most wide-ranging and diversely capable mind I've ever encountered. He once, for instance, led an expedition up the Orinoco River in Guyana. He wrote a courtroom thriller about a lurid murder case. And every summer he joins his friends to teach a course at the Vermilion Sea Field Station.

Among our students, Graham is widely considered an eccentric genius—a classification he earns through wondrous eloquence, sartorial oddity, and mysterious scarcity. Instead of pulling up a plastic chair to chat beneath the salt-cedars, Graham often retreats to the staff house, a separate little cottage just uphill from the station. There, he lies on a cot in one corner of the covered porch and reads thick or obscure books. This year it's *The Faerie Queene*. And that must be where he is now, when I turn to look for him on the terrace and realize I'm on my own with a rather difficult audience: Anoop must know quite a bit about the historical debate over theories of recapitulation.

On the other hand, Graham's absence will allow me to impersonate him intellectually, which is something I've improved at steadily over fifteen years of friendship. I would feel mildly depraved for pilfering someone else's intellectual identity—or, as he might put it, his *modes of discourse*—but the pilfering is entirely mutual. When Graham is talking about contemporary biology, he sounds suspiciously like his friend the biologist. In fact, we even have a name for our practice of disciplinary cross-dressing. We call it channeling—as in, "I had two biochemists in my seminar today, so I channeled you." My channeled historian of science is, of course, a ghost of the real thing. But with corporeal Graham supine beneath *The Faerie Queene*, a ghost will have to do.

3

To rephrase Anoop's question in words that are perhaps more accessible, if also less concise, we could ask: Does an individual organism, as it develops from embryo to adult, pass through a sequence of forms that reprise evolutionary history, beginning with the organism's most primitive ancestor and ending with its contemporary form? Or, to retain the brevity, if not the zip, of Haeckel's phrasing: Does an individual's development reiterate its evolutionary descent?

If this all sounds too cosmically orderly to be anything but sheer mysticism, just consider the fact that early in development the human embryo exhibits unmistakable gill slits, while later on it has a tail as nicely formed as a monkey's. So the idea that an individual's development reprises its evolutionary descent was not without some basis in empirical observation.

On the other hand, the theory also epitomized the predominant patterns of thought in the intellectual culture from which it emerged. The nineteenth-century German Idealists—thinkers like Goethe, Hegel, and Schelling—had a deep sense that the universe is *unified* and *striving*: a single set of natural laws governs everything (that's the *unified* part), and these laws conspire to cause a persistent push toward progress (that's the *striving* part). To understand such a universe, in which the same process is unfolding at every scale and in every entity, a natural mode of analysis is the detection of parallels between microcosm and macrocosm. For example, Hegel's works describe parallel trajectories of progress, discovered at vastly different scales: *The Phenomenology of Spirit* describes the rise of an individual from low brutality to enlightened perfection; the *Philosophy of History* recounts precisely the same ascending steps, but at the scale of world civilization. Haeckel's theory of recapitulation might be understood as a direct mapping of this same analytical structure onto the natural world: in Haeckel as in Hegel, a precise parallel is drawn between the individual's development and the much larger historical process in which the individual participates.

For me, one of the most interesting discoveries in my conversations with Graham was that scientific theories are *simultaneously* a deduction from empirical observation and a reflection of predominant cultural views. Newton's view of God led him to expect inviolable and elegant laws of just the sort he found. Darwin's understanding of British economics assisted him enormously in his thinking about ecology and evolution. So while we may be tempted to scissor a theory into one part that is echt scientific and another that is a mere contingency of the time and place it was made, such an effort is confused and futile—no more feasible than it would be to divide, say, Monet's *Water Lilies* into one part that is French Impressionist and another that is a reliable picture of the world. To put the point differently: scientific theories are not mistaken insofar as they are the distinctive products of a certain culture's worldview, and they are not true insofar as they are separable from such a worldview. The ways Ernst Haeckel was wrong were no more distinctively German Idealistic than were the ways he was right.

If we are to consider Haeckel's ideas seriously—without dismissing them as an Idealist's air castle—then one of the first questions we must ask is how the parallel he observed could come about. As both Darwin and

Haeckel realized, there is a straightforward way evolution could engender its recapitulation in embryological development. First, say an organism has a developmental program that passes through an immature morphology we'll call A and ends in an adult morphology we'll call B. This organism, with developmental program A → B, gives rise to two new species:

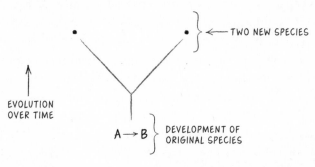

Suppose each new evolutionary change occurs by *adding* a new segment of development *to the end* of the existing developmental process. So, for example, one species adds a new stage that ends with adult morphology C:

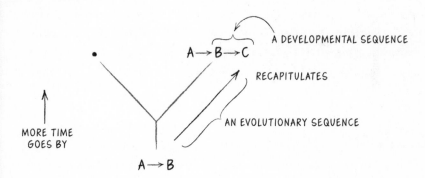

The change from B to C is now the last step in development as well as evolution. Our recapitulation is brief, because we've imagined only one step of evolutionary change. But you can see that if evolution were to carry on adding new segments to the end of the developmental pathway, the recapitulation of ancestral adult morphologies could become extensive, stretching from ancient ancestors to contemporary species.

But would evolution actually work this way? Would it really tend to add new segments at the *end* of the developmental program? Here's a metaphor that illustrates why it might. You have in hand a set of directions to drive from point A to point B in a large city—and it's an old city, like Rome, with many winding, one-way streets. As it happens, you actually need to get somewhere that's one block north and one block east of point B; call it point C. One way you might proceed is to drive to point B first, and then drive one block north and one block east. Alternatively, you could alter your route somewhere in the middle. But of course that's a risky proposition. Driving one block north and one block east before you get to point B might put you on a circuitous route in the wrong direction. You might find yourself merging inescapably onto an *autostrada*, or driving inadvertently into the Pantheon's pedestrian zone. You might never find your way to C at all.

The development of an embryo, like the navigation of Rome, is a complex series of steps in which a small adjustment made early on could cause disruptions that grow as the process unfolds. In Rome, you end up lost; in development, the final morphology could be terribly dysfunctional. On the other hand, adding a short segment to the very end of a working set of directions—travel two short blocks from B to C; transform shape B into a slightly different one, C—could be far more likely to succeed. If it is, then evolution by natural selection probably does show a strong tendency for appending new segments at the end of development, rather than inserting them somewhere in the middle. And if that's the case, then development would indeed offer a quick reprisal of evolutionary history.

4

"Aye—but what's all this got to do with a sea cucumber?"

It's Rafe. And as soon as he speaks, sounding rather impatient, I realize I've gotten carried away with my imitation of Graham and digressed too widely into German Idealism. This Attic setting—stone terrace overlooking the windblown sea; students gathered in a semicircle; filtered sunlight beneath our scrubby proxy for a plane tree—it has all lured me into philosophical mooning. But Rafe has not been similarly beguiled. In fact, he looks irritated. And the toilet paper in his ears suddenly reminds me of the steam that shoots from the head of a vexed cartoon character.

In his application for this course, Rafe registered a definite confidence in the clear distinction between hard science on one side and every other sort of intellectual activity on the other. He was certain, for instance, that physics and chemistry would soon explain everything—including global climate and human consciousness—in terms of its physical constituents and their interactions. Interestingly, this view did not translate into diminished intellectual wonder. To the contrary, he expressed plenty of it, but only in a language of stern scientific reductionism. His application proclaimed awe, for instance, for what he called *the incredible negative entropy of a giant squid*. So whether he was talking about a salt solution or a sea monster—or even, actually, politics—Rafe seemed positive that there is one best way to talk about the world, and it is guided by modern science. Not by humanistic diversions to mystical Germans long since abandoned by the progress of research.

But before I too can abandon the Germans and answer Rafe's question—*what's this got to do with a sea cucumber?*—Chris says, "But I think I've always learned Haeckel was wrong."

And Anoop adds, "Me too, but then I—"

"I've never even *heard* of the guy," Rafe says, and his exasperation seems to demand what business a dusty German Idealist could possibly have in a *science* class. But for Chris and Anoop, Haeckel's relegation to the dustbin seems only to make it all the more intriguing that something actually sounds right in his ideas. And when I glance at Cameron and Allie, I find that Rafe is alone in his frustration: Cameron is smiling slightly, facing directly into the stiff onshore breeze; I have no idea why he's smiling—or even if he's listening—but he certainly doesn't appear impatient with our digressions. And Allie is looking kindly at Anoop, evidently waiting to hear where he was going when Rafe interrupted. Joining her, I turn to Anoop and wait to hear more.

"Oh," Anoop says, endearingly surprised to have our attention. He inches forward in his plastic chair, brushes a few pine needles from his shiny black hair, and then explains that although he, too, had always learned that recapitulation was basically an outdated notion, he couldn't help wondering about it when I told them that larval *I. fuscus* has obvious fivefold symmetry. For it would seem—wouldn't it?—that the developing cucumber first displays the pentamery of the ancestral echinoderm, but then moves on to the more recently evolved bilateral symmetry of cucumbers. So doesn't it appear that, just in this case, ontogeny really does recapitulate phylogeny?

The truth is, I've never really contemplated cucumbers in the context of recapitulation. But now that we're trying, the homely echinoderms seem to serve the theory beautifully. In fact, now that Anoop has outlined one step of recapitulation—the pentamerous immature, growing into the bilateral adult—I can't resist speculating about two more.

After the sponges, the most ancient of animals are the Cnidaria—the jellyfish, anemones, and corals. Their symmetry is neither bilateral nor pentamerous, but radial: they have infinite axes of symmetry, like a circle. To see this, just imagine the lovely, translucent bell of a floating jellyfish. Cutting this form into equivalent parts would yield not five pieces, as in the sea star, nor two pieces, as in the cucumber, but rather a number of pieces that could range from one to infinity, and would depend only on how narrow you decided to make each wedge of the circle. Intriguingly, in a number of other phyla, including the cucumber's Echinodermata and our very own Chordata, the earliest stage of development is a radially symmetric ring of cells. Until about ten years ago, uncertainty in our understanding of the tree of life had obscured what is in fact a direct relationship between the radially symmetric Cnidaria and the early ancestor of echinoderms and chordates. Now that this direct connection in deep evolutionary time appears likely, we must wonder: Is the radial symmetry of our first developmental stage inherited directly from our earliest ancestor?

Shortly after we are radial, we become, as you might expect, bilateral. What's more surprising, however, is that most echinoderms undergo the same transformation. When we imagine the bell of a jellyfish, our emblem of radial symmetry, alongside, let's say, a sand dollar, which is a nicely pentamerous echinoderm, the two shapes do not look so different; indeed, the one would seem to be just a block or two away, developmentally speaking, from the other. But many echinoderms don't take the direct developmental route. Instead, the nicely radial embryo grows into an early juvenile that is as legitimately bilateral as a tadpole. So the pentamery of adulthood might be only a few blocks away from the embryo's radial symmetry, but the developmental directions for walking around the corner lead first to the other side of the city.

The young juvenile echinoderm—bilateral tadpole waiting to be a sea star, say, or a cucumber—floats around, eats a lot, and generally leads a life well adjusted to the open water. When it has finally eaten its fill, it experiences a complete rearrangement of its body. To get a sense of what it takes to transform a bilateral animal into a pentamerous one, just consider a sea star's rocky puberty: The juvenile's mouth, which is situated, like ours, near the top and on the front of its body, migrates down to the animal's middle, and then over onto the left side. Meanwhile, the limbs are absorbed, and the main bilateral body cavities branch out like growing saplings, becoming the hydraulic system of the adult. As transformations go, Ovid's metamorphoses are not more drastic.

But how is this bilateral youth, which comes to such a dramatic close, an instance of recapitulation? We've mentioned that the phylum most closely related to echinoderms is in fact our own, the Chordata. As we know from personal experience, chordates are bilateral. What's more, the other phyla that are close mutual cousins of echinoderms and chordates are also bilateral. This tells us that if we were to inspect the common ancestor of echinoderms and chordates, we would find it to be a bilaterally symmetric animal. In short, before the ur-echinoderm discovered pentamery, its ancestor was bilateral. And before an individual echinoderm becomes pentamerous, it is a bilateral juvenile. Once again, we have a major transformation in individual development that seems to parallel a major transition in evolution.

Following Anoop's lead, we've now put together a four-step series in symmetry, which we can trace through the development, as well as the evolution, of the sea cucumber. It goes: radial, bilateral, pentamerous, bilateral. As we come to the end of this four-step series, it is Chris who ventures to suggest, with some hesitation, "So we're saying Haeckel was actually kinda right?"

Rafe is leaning forward, squinting. He looks from Chris, to me, to Anoop, and then back to Chris.

"I thought you said he was wrong," Rafe says to Chris.

"I said I'd always learned that, but how else do you explain this?"

Rafe sits back abruptly and runs his palms back across his blond hair; then he starts reinstalling his tight ponytail, as if the act itself might help put things back in order. I glance at Anoop, and he seems to be looking at me out of the corner of his eye, though the lenses of his glasses are so small that his line of sight misses them entirely, and I must be very blurry to him. The slender fingers of one hand are covering his goateed chin and his mouth, as if he's about to speak but is not certain he should. He may be thinking there's another explanation, or at least the possibility of one, but he's holding his words—perhaps because he can't imagine why I haven't already spoken up about Karl von Baer. For the moment, however, Haeckel's main rival will have to wait, because Allie breaks our taut silence: "Cameron," she says softly, "what are you smiling about?"

It's true. He's still wearing the same bemused expression, and he seems, I'm afraid, somewhat detached. He did follow the first few twists of our conversation—it was he who recalled the critical clue of vague rows of tube feet—but as we began to unroll Haeckel's theory of recapitulation, he peered into the wind and headed off in his own direction. Presently, he looks toward Allie as if she's just roused him from a daydream, and he says, a bit sheepishly, "I was just thinking about how all this got started."

"What do you mean?" Allie asks. Her tone, sweet and encouraging, counters the note of diffidence in his voice.

Cameron laughs a little and shakes his head. "Do you realize," he asks, "that this whole gnarly discussion started with that knobby thing that puked on Anoop?"

Allie laughs, and Anoop says, "Anything to advance science."

"Oh sure," Allie says. "Why don't you ask the cucumber about that."

5

I did not think much more about Cameron's bemused utterance until Veronica and I returned home, weeks later, and I began to review our days on the bay. Then, as I paged through my journal, reading entries here and there, Cameron's peculiar and seemingly trivial observation—that our entire conversation had begun with a sea cucumber—kept coming back to me. Naturally, I recalled his words when I found us testing out the rival theories—Haeckel versus von Baer—in other organisms: the needlefish; a stray dog; even our own species, *Homo sapiens*. But I also heard a sort of echo, a reminder to notice a conversation's point of origin, in other entries, too—like when a large desert cactus somehow led to the erotic fantasies of conquistadors; or when a vast skein of devil rays twisted itself into a glyph that seemed to demand interpretation; or when a frigatebird, hovering above us, induced Veronica to reveal how one of her heroes lost his life on the bay. In all those varied echoes and modulations of Cameron's initial observation, I began to sense the topography of an underlying idea, and I have come to believe that it needs to be explored.

What lies behind this belief is in fact a suspicion, an awful hunch, that while the places we live and know have been scorched and destroyed, we have been blind to the destruction. And although we now live amid the smoldering wreckage, we hardly manage to notice that something has changed, because we hardly recall what these places once were. That may sound shrill or paranoid, but take it, just for a moment, as a hypothesis to be tested, a claim to be weighed against the evidence. We can begin with the evidence close at hand—a mud-brown loaf, rising to the surface on Veronica's palm—and work our way outward from there.

In the summer of 1995, Veronica was a teaching assistant at the Vermilion Sea Field Station. On the first afternoon of her class, she watched the small local fishing boats, known as pangas, returning across the windswept bay. She stood from her chair on the terrace and walked down to the beach, where the fishermen would slide their boats ashore to unload their catch. The summer before, the waters had been rich with yellowtail, and Veronica was expecting to watch the fishermen heft from the open hulls of their boats the silver fish with scythe-shaped tails of dark yellow. What she saw instead surprised and puzzled her. The hull of each returning boat was filled from floor to rail with the dark brown ovoid forms of *Isostichopus fuscus*, massed like swarms of oversized larval insects.

In retrospect, it is the timing of this story that seems most surprising. Three years earlier, in 1992, fishermen on the Pacific coast of Ecuador had finally depleted what had been an exceedingly productive *I. fuscus* fishery. With their local livelihood vanishing, the fishermen moved their efforts farther offshore, to the Galápagos Islands. But the Galápagos, as a prominent tourist destination, was under closer scrutiny than other areas of Ecuador, and when the cucumber population began to crash there, too, the government responded publicly and dramatically. In 1994, Ecuadorian regulators imposed a moratorium on the harvesting of cucumbers. The government of Mexico must have been watching, because regulators there immediately declared their own ban, and they did so, as best I can tell, without any real indication that Mexican populations of *I. fuscus* were also in jeopardy. The following summer—the first season *I. fuscus* was absolutely off-limits—was when Veronica watched panga after panga full of cucumbers arrive at the beach.

So why was a ban met with a binge of extraction? The answer might be that the new laws in Ecuador and Mexico caused a spike in the price of cucumber on Asian markets, luring more fishermen than ever into the business. And although extraction of a cucumber was just as illegal in Bahía de los Ángeles as it was in the Galápagos, Mexican enforcement was weaker, resulting in a shift of fishing pressure from one place to the other. I could be wrong about the steps in this chain reaction, but in any case, the summer of 1995 wrought a noticeable change on the reefs of the bay. Veronica has told me that before that summer, many reefs supported about one brown cucumber every square meter. In every summer since, they've been so rare that the discovery of one has been cause enough to gather the students.

The claim we set out to test (beginning, humbly enough, with a brown cucumber) was that the places we live and know have been despoiled, and we've hardly taken notice. It's clear that the cucumbers are gone, but is it true that their absence is overlooked? When Veronica first showed me a sea cucumber, she had never told me about the catastrophic summer of 1995. Without that historical memory, I took the discovery of an exotic organism, a creature I had never before encountered, as an indication of the reef's ecological health; certainly it did not occur to me that the number of cucumbers was in fact a sign of devastation. And I have heard many students misjudge the reef's condition in much the same way. Upon emerging from her first dive in the Vermilion Sea, Allie said breathlessly, "I can't believe how much *stuff* there is!" Why would it occur to her that ten years ago there was so much more?

Of course, the students and I are but visitors to Bahía de los Ángeles, and one might suggest that it is merely our recent arrival that makes us so ignorant of what was here before. But there is an entire generation of Bahíans, now teenagers, who also did not see the bay before 1995. They, like us, have no recollection of a reef replete with cucumbers. We once hired the teenage son of a local fisherman to help us with the class. He often sat in the panga and listened to music while the rest of us explored the reef, but we occasionally brought creatures over to the boat to discuss with him. When we brought over a cucumber and asked him if there used to be more, he said, "No, there used to be less." In a way, he was right.

The past few years had brought a slight expansion of the cucumber population, perhaps on account of a sudden spasm of severe enforcement by
the Mexican government. But the state of the population that he took as
his starting point—what he considered the natural condition—was in fact
a state of catastrophic depletion. So the cucumbers are gone, and their
absence from the reef is unremarked—not only by those of us who are
late-arriving visitors, but also by younger generations of locals who, like
us, first encountered the reef without its former inhabitants.

Admittedly, *I. fuscus* is a modest example. It is, after all, an ugly and
inconspicuous animal, and perhaps that is why its decline is so easily overlooked. But what of the Vermilion Sea's more charismatic creatures?
Surely their disappearance would leave more noticeable lacunae. I might
have thought so, until I encountered the writings of Joseph Wood Krutch.

Krutch was a professor of drama, which is perhaps not the sort of professional bio one expects for a man who traveled Baja in the 1950s, a time
when the only other American visitors were a few intrepid botanists. But
he had taken an interest in the desert around his home, in Tucson, and he
had heard that the most extreme of all deserts—drier, more isolated, more
punishing than any other—could be found about three hundred miles
down the Baja Peninsula. So he went, and, being a man of letters, he
wrote about it. On his second visit to Bahía de los Ángeles, a small plane
deposited him at an airstrip that used to run just behind the beach:

> *Ten minutes later the airplane was taxiing across the strip again and
> I, standing alone on the beach, waved it away. Yes, the scene was as
> intriguing and as beautiful as I remembered it. A few hundred yards
> from shore the turtle boat rode at anchor on the blue water. The
> great beach was empty except for two ten-year-olds plodding across
> it and carrying between them a fish (Yellowtail) as tall as they.*

In some ways, the scene today is the same. The water is still very blue.
And the beach remains broad and beautiful, despite the row of spartan
cinder-block cabanas that the Diaz family has built behind a knee-high seawall. But the details—the biological details—are different. Yellowtail are
still sometimes caught by the sport fishermen who spend their weekends in
the Diaz cells, but I have never seen a yellowtail as tall as a ten-year-old. In
fact, I have never seen one as tall as a four-year-old. And, more striking,

what is now a sportsman's prize catch was formerly an overabundant staple: *Go out a few hundred yards*, Krutch wrote, *and in an hour you will take a dozen yellowtails averaging eighteen or twenty pounds each.*

But it is another detail that, for me, provides the more poignant contrast with today's bay. Krutch mentions a turtle boat. A *turtle* boat. When I first read this, it sounded almost fantastical, because I had never encountered a turtle in the Vermilion Sea. To this day, I have seen only eight—and three were ones that we ourselves released at the beach, assisting Antonio Reséndiz, a local man whose mission in life is to bring back the turtles. But according to Krutch, on a typical evening in Bahía, *a twenty-five-foot turtle boat may come in loaded with fifteen or twenty huge turtles, which will be added to the others, collapsed in sad resignation in a covered pen near the water.*

When Veronica first brought me to Bahía de los Ángeles, fifty years after Joseph Wood Krutch had arrived by plane, I found a place that was unimaginably alive. At sunrise of my first day, I watched from the terrace as whale spouts erupted and hovered in the morning haze; before one had evaporated, the next would appear elsewhere in the bay. When the sun had become so strong that the spouts would burn off before they could take shape, I put on my mask and snorkel and swam out over the rocky reef. Twenty feet from the research station, I found beautiful, alien creatures: Christmas tree worms and brittlestars, crystalline shrimp and velvety green anemones. Later, as we ate lunch on the terrace, a dark shadow slid across the station reef, and I looked skyward, thinking it odd that a single cloud could survive such heat. But there were no clouds, and Veronica laughed at me, because the cloud shadow was in fact a whale shark gliding through shallow water.

Over the next few days, Veronica told me about the creatures she had seen when she herself had first arrived, but which, for one reason or another, didn't seem to be around anymore: the cucumbers; the elegant terns; the sharks. When Lane McDonald, Veronica's college mentor, arrived at the research station, I asked him what had vanished before Veronica's time. He mentioned the sun stars, *Heliaster kubiniji*, and he said he thought there had once been more bottlenose dolphin, too. Then, reading Krutch, I learned about the turtles, and from Antonio I heard that it wasn't just one sort of turtle, but four. And suddenly the bay, which at first had seemed only full of life—straightforwardly, unreservedly full of life—

was now growing crowded with the ghosts of its former inhabitants. And the further the stories reached into the past, the longer became the list of the missing until, eventually, the ghosts seemed to outnumber the living.

It is disturbing to learn from a witness of history that a beautiful place, which one took to be pristine, is in fact despoiled. But it is more disturbing to realize that the witness himself confronted the very same lesson long ago. Just as I have come to see that the bay is now but a remnant of the one Krutch encountered, Krutch himself realized, with great dismay, that the bay he knew was but a remnant of its past. And, charting another turn in the spiral, Krutch recorded his realization by citing his own witness of the past, who had learned the same lesson a hundred years before. Krutch quoted Henry David Thoreau:

> *I take infinite pains to know all the phenomena of the spring, for instance, thinking that I have here the entire poem, and then, to my chagrin, I hear that it is but an imperfect copy that I possess and have read, that my ancestors had torn out many of the first leaves and grandest passages, and mutilated it in many places. I should not like to think that some demigod had come before me and picked out some of the best stars. I wish to know an entire heaven and an entire earth.*

Over the past ten years on the bay, I have tried to gather up scraps torn from the poem. I've collected them from Veronica; from her mentor Lane; from the turtle man Antonio; from the wandering drama critic Krutch; and from many others. With my hands full of ragged bits and scraps, I might be inclined to believe that the bay has been more severely rifled—more drastically abridged—than other places I've lived or visited. But the opposite is true. In the bay, many passages were lost so recently that people I knew could recite them, and the poem, pared and slashed though it was, remained sufficiently coherent to betray certain gaps. In other places I've been, so much was lost, and lost so long ago, that there are hardly reminders of the poem's former content. Who on the Eastern Seaboard is struck by the absence of migrating gray whales, or nesting sea turtles, or playful and clever sea mink, or river-frothing runs of salmon and eel? Yet they were all there, and there in great abundance.

We live amid the wreckage, yet we hardly notice that something has changed.

6

Why are we blind to the destruction—so forgetful of what was here?

What have we lost, and what can be done?

These are the questions, broad and unapproachable, that seem to lie just beyond Cameron's modest observation, like a vast landscape accessed through a keyhole.

It seems to me that our blindness to destruction is partly a problem of memory, and partly a problem of perception. We've already touched on the problem of memory: knowing only the natural world we've encountered in the short interval of a life, we fail to notice the substantial transformations wrought by previous generations; and so we overlook the absence of all that was already gone when we ourselves first arrived on the scene. The problem of perception, for its part, is like one of those Greek gods who take the form of a different animal for each of their misdeeds. It will show itself, in several different guises, in the chapters to come, but for now its general shape can be vaguely conjured with a comment once made by a friend. She is a Kantian philosopher, as well as a devoted denizen of cities, and she once described for me her experience of entering the woods. Her first thought, she said, is generally something like, *There's nothing here. I want to go home.* What happens, it seems, is that she faces an unfamiliar and overwhelmingly complicated world, and she doesn't know what to look for or where to focus her attention. As a result, the profusion of detailed complexity blends into a surface of monotony, which presents itself to her as *nothing*.

A mud-brown lump on the rocky bottom is easily overlooked. But *I. fuscus*, resting on Veronica's palm, in the light nearer the surface, became the starting point for innumerable strands of narrative and investigation. There were straight scientific questions, about collagen or evolution or holothurin; there were histories, threading back to nineteenth-century Idealists; and there were Veronica's memories of former fishing seasons—recollections that merged in turn with the paths of causation linking Ecuador and Bahía. This unfurling of *I. fuscus* suggests that the interest of an organism—broadly speaking, its meaning and value—can become clear only when it is seen and approached as the origin of our stories. When we follow the threads of these stories outward, tracing them to other organisms, or cultures, or individuals, we begin inevitably to reconstruct a kind of dense historical ecology, a thick description of a creature's place within webs that are not only biological, but also conceptual and social. And it is in this dense weave that we have our best chance, I believe, of understanding what has been lost and what can be done. In Claude Lévi-Strauss's words, animals are good to think with. In Cameron's, the whole gnarly conversation starts with a sea cucumber.

PART II

Pachycereus pringlei

LA ISLA DE CALIFORNIA

1

While I've been moving box after box of gear from our motel room out to the parking lot, the sounds of the nearby San Diego Freeway have grown, from lonely and fleeting—the occasional predawn commuter—to crowded and ceaseless. But now, as I hesitate in the doorway, holding the very last load, the morning seems gentle in spite of the steady thrum. Leaden clouds hang low, making the early light hazy and vaporous, more like a fading of night than a real break of day. And the only sound other than the highway is the hiss of sprinkler heads that moisten small squares of lawn and stain the sidewalk's edges. Our cars, which are embarrassingly huge, have been backed out, and stand now in a vaguely martial column: Excursion; pickup truck; Expedition. As if to make sure the convoy won't leave without them, our students have gathered with their baggage directly in front of the Excursion. They look chilled and sleepy. Allie is hidden beneath the hood of her sweatshirt. Anoop rubs his bare arms. Rafe has made a pile of several other students' duffel bags, and is sprawled across them in a theatrical portrait of exhaustion. The bandana that was holding back his hair is pulled down over his eyes.

"Okay," I yell, arriving beside them, "wake up, team!" My exuberance is forced, but at this stage, when the students hardly know one another, there is a palpable need to fill up some of the silent spaces among them. Over these next few days, I'll do my best to be extroverted and cheery, and a few of my more obvious foibles—my peerlessly bad sense of direction, for example, and an innate deficiency in the area of logistics—will become fair game for the whole class, our themes for those all-important group jokes. Veronica and I have never really talked about it, but I think this role falls to me because it's simply not in her nature—she's too composed, too instinctively poised to be overly gregarious—and maybe also because those foibles of mine leave her to worry, all on her own, about which dirt road we're on or how much time we have until sunset. Graham, for his part, prefers to maintain his distance. At the moment, he is leaning against the pickup truck, eyes hidden behind dark sunglasses, and with his crew cut and lean face, he looks part sergeant, part hornet. This too will eventually become a caricature open to student amusement, but for now, foolery is exclusively mine.

"Come on, Rafe!" I yell, kicking his toe.

"Easy, mate," he groans, lifting the edge of his bandana to look out with one squinty eye.

I can't resist mimicking his accent, and I shout, "Move it, ya bludger!"

He smiles, enjoying the personal attention. "Good on ya, mate," he says as he rises.

As if they've been awaiting their cue, Chris and Allie hop to the task of loading the truck. He climbs into the bed, and she swings bags of gear up to him. They work with such sanguine vigor that the other students start dragging their bags into a pile at Allie's feet.

Several months ago, Veronica sent the students a list of things they should bring. It was only a page long, so I don't think anyone—except maybe Graham—could have known that it was, in fact, not just a packing list, but rather a highly wrought poem in the language of pedestrian items. This year, for the first time, she elected not to revise her poem—not to change out any pieces of clothing or gear—but to annotate, adding interpretive glosses on key terms. Then she handed me a page printed in ten-point font and asked if I thought she'd forgotten anything.

"No," I said, handing it back.

"You didn't even read it," she complained.

I looked at it again. In truth, the annotations seemed superfluous—none of the items needed explaining—but I didn't say so, because I was secretly sort of charmed by the way each little note seemed to capture a certain side of Veronica's personality. The entry that followed *sleeping bag*, for instance, revealed the woman who knows she really shouldn't cinch up your scarf for you, but just can't seem to help herself: *Do bring something warm*, she pleaded, *because the Cataviña desert, where we sleep on the way down, can be surprisingly cold.*

Farther down the list, following *open-toed shoes*, I found a very different Veronica, the biologist who never quite understood my squeamishness about minor bodily injuries: *You'll get bad blisters,* she warned, *so bring two different kinds of sandals, which will cut into different places on your feet.*

"Isn't that kinda graphic?" I asked.

"Only if you have a foot fetish," she replied.

As I neared the end of the list, I realized that all the annotations—in fact, the very project of writing them—testified to a strong yen for planning. And the final two entries epitomized it: *wetsuit (at least a shorty, a.k.a. springsuit, 3 mm in the torso, 2 in the arms and legs); weightbelt (about 2.5 lbs. of lead for every mm of average thickness in your wetsuit).*

After the list went out, Veronica received e-mail queries: *How much weight do I need? Is my sleeping bag gonna be warm enough? Could I maybe just sleep in my wetsuit?* In my view, the annotations had backfired, creating doubt where there'd been none. But Veronica disagreed. She thought the notes had elicited important questions, and she scheduled a separate meeting to discuss equipment issues. I missed the meeting, but I can see its consequences, I think, in the bags now piling up at Allie's feet. Almost every student has packed snorkeling gear in a bright yellow mesh bag, which precisely matches the ones Veronica and I use. And while I don't doubt that there is a certain rationale for keeping dive gear in yellow mesh—the holes let water drip out, and the color makes it easy to distinguish from everyone else's stuff—it also occurs to me that half the convenience has now been rendered obsolete.

Also, I assume, per Veronica's instructions, most students have packed the rest of their stuff in a large nylon duffel. But even amid the common themes, occasional exceptions seem to reveal deep differences in the students' lives and backgrounds. Anoop, for instance, is just now struggling to lift an enormous hard-vinyl suitcase, which he hauls several quick steps in Allie's direction and then abruptly drops, resting a moment. The vinyl is powder blue and crowded with smudges and scrapes, and beside the handle are little chrome latches—the kind that lock closed with a key. Again Anoop heaves it up, shuffles swiftly forward, clunks it down. I imagine it has taken long train rides across the subcontinent, and at least one journey halfway around the world, but certainly never a camping trip in the desert.

Oddly, Cameron too has his snorkeling gear in Veronica's highly visible yellow—and now that I think of it, I don't actually know whether there would be a dive mask in there. Beside his yellow mesh rests a

rock-climber's backpack, covered with nylon straps and a few spare cara-
biners; in his application, Cameron revealed the astonishing fact that he
is a passionate climber. The pack's two largest straps secure a blue fiberglass leg, which tapers to a slender foot pointing balletically upward. It
must be an alternate for his usual high-performance prosthetic. A sea
leg, perhaps, built for saltwater and sand.

Beside Cameron's muscular heft, Isabel looks especially petite. Her
features are distinctly Latin American: her hair glossy black; her skin like
polished terra-cotta; her cheekbones high and wide-set. In her application, Isabel explained that her parents, who now own a small grocery in
Los Angeles, came to the United States as migrant farmworkers, and their
struggle to care for family on both sides of the border is a memory that
figures in Isabel's determination to become a doctor. I would like to know
more, but she is so shy that we have yet to exchange a word. At her feet,
what appear to be a comforter and pillow, both in the same pink floral
print, are stuffed into a clear plastic bag.

"Hirsh," Graham whispers abruptly, "quit zoning." His black lenses
are close by my face.

"Burnett," I whisper, "you look like a hornet."

"Good," he says. "You can't look vulnerable on your way into Tijuana.
Now wake up and go help your wife."

On her way across the parking lot from the motel office, where she's
been paying our group's bill, Veronica is carrying a large cardboard box.
She also has a dog leash looped around her wrist, and at the leash's
other end, Yukon, my joyful golden retriever, gambols spastically, causing
Veronica to walk a mildly drunken trajectory. Taiga, Veronica's white
German shepherd, follows as closely as she can while remaining just beyond the range of Yukon's wild lunges.

If it weren't for the heavy box, Veronica would appear to be heading
for the beach: her blond hair is held back by sunglasses atop her head; she
is wearing a tank top, nylon swim trunks that locals call board shorts—as
in surfboard—and flip-flops that those same locals refer to as "Reefs." It's
not exactly news to me that my wife fits in here, in Southern California,
but her overnight assimilation still takes me by surprise. At home, where
I see her in khakis and a lab coat more often than swim trunks, her blond
hair and green eyes and lanky frame seem to recall not her college days
in San Diego—three years of surfing every day—but rather her origin in

Central Europe. At present, though, the only clue she's not local might be her pallor: she was terribly determined to gather a certain batch of data before we left campus, so she hasn't seen the sun in weeks.

"Come on," I say, pulling Graham by the elbow—if I take Yukon, Graham can take the box—but suddenly Chris and Allie are striding past us on their way toward Veronica. Allie takes the box, Chris the leash, and Veronica appears surprised to find herself empty-handed. As she looks over at me and Graham, she seems to squint in inquiry, as if to ask why we're standing still, hitched at the elbow. Allie is just passing us again, on her way back to the truck, when Veronica says, "You can just set that down, Allie." And Allie, ever helpful, places the box directly at our feet, as if it were some sort of gift. Graham pulls his elbow free of my hand and slips away. When I look back to Veronica, she points to the box at my feet and nods, as if to say, *Get on with it.*

"Gather 'round," I yell, and I remove the box's top to reveal two stacks of spiral-bound volumes. Veronica takes some satisfaction, I think, in these homemade books, which she assembles each year from scientific articles, textbook excerpts, and—Graham's contribution—intimidatingly long historical documents. This year, the book's cover shows a silver-gelatin image of a towering cactus. You could easily take it for a giant saguaro, that hackneyed icon of Arizona. But in fact this is an even larger species, which is found almost exclusively on the Baja Peninsula. It is the giant cardón, *Pachycereus pringlei.*

What catches my attention now, as I take out a stack of books, is that you can just see a faint tracery of curving lines beneath the image of the cactus. It looks like a road map under vellum, but I know that what the lines actually show are chemical reactions, which Veronica has printed on the back of the cover, since she will talk about them when we stop to camp in the desert. I first saw the diagrams several weeks ago, when I walked into Veronica's study and looked over her shoulder at her computer screen.

"What's that amoeba thing?" I asked.

"It's a plant cell," she said, annoyed at my daft question.

"Oh, right. A cell."

"It shows Cam photosynthesis," she added, pointing at a train of arrows that traced a figure eight inside the amoeba thing. Sprinkled around the figure eight were the names of certain molecules: pep-c; malic acid; rubisco.

"Right," I said. "Cam photosynthesis."

"Do you think the students will get it?"

"Of course."

When Veronica left for the lab, I pulled down one of our botany books. I certainly remembered how photosynthesis works, because it has always struck me as something of a miracle: Basically, a plant draws water from the earth and carbon dioxide from the air, then mixes them with sunlight to make sugar. As magical as that sounds, it's really not an inaccurate description, though one can also zoom in on the process, and Veronica's diagram was several steps closer to the molecular details. A figure in our book's chapter on photosynthesis seemed to match one of the cycles—the left half of the figure eight—that I'd seen on her computer screen.

But what was the right half doing? In its chapter on cacti, our book explained that they have a clever trick for making sugar in the desert. Of the three ingredients for photosynthesis—light, air, and water—deserts are short on one. But the scarcity of water, it turns out, is closely tied to the inhalation of air: to let carbon dioxide in, a plant must open microscopic pores, which inevitably lets water vapor out. The cactus escapes this dilemma by opening its pores only at night, when the desert air is cool and relatively moist. As the carbon dioxide trickles in, a certain protein incorporates it into a liquid and stores it away in cellular cisterns. Come morning, the pores close, protecting precious water, and the cisterns begin to leak their contents, supplying carbon to the familiar sunlit cycle.

We must be there, in that transitional light, right about now, as copies of the black-and-white cardón make their way around the circle of students. And the passing of books feels like a quiet inaugural ritual, until a volume arrives in Rafe's hands, and he exclaims, "Mates! This is huge!"

"If you start reading now," I say, "you might be ready for Graham's lecture tomorrow." The students laugh a little, but several also glance uneasily at Graham, who has returned to his place by the truck. He offers no assurance that my remark was a joke, but rather frowns at his fingernails, which he appears to be trimming with a pocketknife. If my tendency is to ease the students' academic anxieties, Graham's is just the opposite: he wants them to know that this course, despite its proximity to the beach, is not to be taken lightly.

When a volume makes it around to Isabel, she seems to hesitate, and then the next one arrives, so she has two in her hands. "Cameron," she asks quietly, "do you want one?"

"No, thanks," he says. "I've got it all on my iPod." A month ago, Veronica submitted our course-reader to a certain university office, where volunteers read every page into a recording. "It's not as heavy that way," he adds, smiling in Rafe's direction.

"And," I say, "we can play Graham's assignment on the car stereo."

"Shotgun Veronica's car," Rafe says, and the students laugh. Again, Graham is stony, though he does seem to bend his muscular neck ever so slightly to stare at Rafe through black, polarized eyes.

As the students stow their readers in their daypacks, Veronica explains how they should divide themselves for the day's drive: six with her and Taiga in the Excursion; four with me and Yukon in the Expedition; and one with Graham in the pickup truck. At once Anoop raises his hand and declares, "I'll ride with Dr. Burnett." But Veronica seems to hesitate, as if she has misgivings. To me the pairing makes perfect sense—at least from an academic perspective—but the awkward silence endures until Graham, at last, breaks his rigid bearing. "Good," he says. "Let's go, Anoop." He flips closed his knife and climbs into the truck's driver's seat. Hurriedly, Anoop makes for the passenger-side door.

The other students start milling, monitoring one another as they move uncertainly toward one vehicle or another. For Veronica and me, this process feels like drawing lots, because the students' decisions now will choose our company for the entire day. I can already see that my hope of getting Cameron and Isabel is not to be: Isabel feels less shy, I think, with Veronica, and she's heading for the Excursion, with Cameron by her side. To my surprise, Rafe, who just declared he'd ride "shotgun" with Veronica, walks toward my Expedition. Puzzling—until I realize: he's following Haley.

Haley is striking. Six feet tall; a high forehead, like Vermeer's luminous young women; and blond hair pulled back into a pair of French braids, which end in pale blue beads beside the corners of her jaw. Her nose has a scattering of freckles across the bridge, and her mouth is very broad, with just a hint of pout, which one instinctively wants to comfort. Each time she and I have talked, it has been exclusively about basketball. She's a varsity player, and my love for the game seems to make her feel relatively comfortable with me. So it's not surprising that she has chosen the Expedition. Nor is it especially surprising that Rafe is trailing her closely.

Chris, who still has Yukon, also heads for the Expedition. I feel lucky for this one. And he's joined by the quiet but amiable student who goes by the nickname Ace. A tall young man, Ace wears his hair in a spiky tousle and walks with an easygoing shuffle of his flip-flops. What I remember from his application is that he plays in a rock band. Yukon, delighted with his new friends, bucks gleefully between Chris and Ace, bouncing off their sides as they walk.

With the students' dispersal, Veronica and I are left on our own, watching them climb into their chosen vehicles.

"Poor Haley," I say, looking toward the Expedition. She has taken the place behind the driver's seat, and Rafe has positioned himself directly beside her in the otherwise empty vehicle.

"Don't worry," Veronica says. "She can handle him."

Chris and Ace are deferring to each other politely, both declining the front seat. Finally, Ace folds his large frame into the seat beside Rafe, and Chris sends Yukon into the space at the rear passengers' feet. Haley greets Yukon lovingly, letting him nuzzle her nose.

"What about Graham?" Veronica says quietly, without looking over at the truck. "Is he okay with Anoop?"

"Burnett? I thought you were worried about Anoop." As if on cue, Graham starts the truck engine, revs it impatiently, and grins at us out his window.

"Be safe," Veronica says, squeezing my hand. "Don't get lost."

As we pull from the motel parking lot, the students peer quietly out their windows. There's nothing much to see, but everyone seems to feel the same need to observe our moment of departure—our casting-off. Chris turns to me and gives a single silent nod, as if to say, *We are under way*. I return the nod: *Yes, we are under way.*

Chris's complexion is pale, but his hair and eyes are black. His looks are at once handsome and somehow sorrowful, almost gothic. Several months ago, when Veronica first handed me the stack of applications, the first two I read happened to be from students of Mexican descent. Chris and Isabel were both terrifically impressive, and yet I could not help noticing a certain contrast. A number of details in Chris's application seemed to bespeak a life of ample opportunity: he had traveled widely; he was into scuba diving and sailing; and, consistent with a happy absence of financial concerns, his choice of courses showed no hint of preprofessionalism, but rather a passionate pursuit of his interest in the oceans, together with a dabbling of odd, unrelated topics, such as Zen Buddhism and modern architecture.

In this respect, Isabel's application was different. Her courses seemed less freely exploratory, more seriously targeted toward her future as a medical doctor. She had even managed to satisfy her humanities requirements by taking seminars on the politics of health care. At the time, I connected such unwavering professional focus with the little I knew about her family and background. But now, as I watch Veronica's Excursion accelerating ahead of us onto the San Diego Freeway, southbound, it occurs to me that there is, in fact, something off about my tidy account. To start, what's Isabel doing here, with us? How exactly does an excursion to the Sea of Cortez move her any closer to medical school?

In truth, at the moment, my notion of every student is awfully pat. Based on a page-long essay, an academic transcript, and now, I suppose, a few pieces of luggage, I have told myself a story that seems to hold together. And yet I know that such coherent accounts are bound to run into trouble—and only then, at the very moment my stories falter, might I see a bit more.

2

Not long after I'd begun to wonder what it would really mean to wander out along the paths of narrative that originate in an organism, I happened to read, in a field guide to desert plants, that the towering cardón had astonished the first Europeans to set foot in Baja. I liked the idea that certain trails might actually cross the borders between disciplines—even, perhaps, the well-guarded one between the sciences and the humanities—so I decided to abscond, just for a day or two, from my office in the biology department. I wanted to visit the main library, which was on the other side of campus, and find out what the conquistadors had said about *Pachycereus pringlei*.

It was midsummer, and campus was serenely abandoned, the red brick pavilions sunbaked and silent, like a Spanish town at siesta. When

I finally entered the library, descending the stairs into the cool, dimly lit stacks, I felt as if I'd discovered secret catacombs where my colleagues would never find me. I figured I knew where to start for references to the desert around El Mar de Cortés: there were two titles written by Hernán Cortés himself. One was called *Cartas de Relación*; the other, *Letters from Mexico*. So I took a copy of each and found a wonderfully secluded table, tucked into a corner lined with dark leatherbound volumes. It was only after I'd opened my books and placed them, side by side, in the pale circle of light that fell on the table from above, that I realized one was simply a translation of the other: they were both copies of the letters Cortés had sent, between 1519 and 1525, to King Charles V of Spain.

The letters offered a running account of the conquest of Mexico, and though I tried at first to skim them for that flash of what I was after—the allusion to the Baja Peninsula or the Gulf of California—the story was irresistible. I found myself following every skirmish and sally, and soon I was even tracing troop movements on a large multicolor map of sixteenth-century Mexico. It was a foldout I'd found in a history of the conquistadors, and it showed the territories of various peoples, each in a different shade of translucent watercolor.

Cortés landed first in the Yucatán—Maya territory, my map told me—where he managed to liberate a Spaniard who had been held captive there for seven harrowing years. According to Cortés, the rescue was not just a feat of daring, but also a sign of providence, for the bedraggled Spaniard soon became a trusted translator. Through him, Cortés explained to the Maya that they would all, henceforth, be the blessed vassals of King Charles V, Holy Roman Emperor. The Maya, not surprisingly, disagreed, and told the Spaniards to leave. In response, Cortés unleashed a savage assault, which he must have considered quite justified, because he described it graphically and unabashedly for his king.

The Spaniards then sailed for the place they called Veracruz, where Cortés informed the natives that they, too, were blessed vassals. Unlike the Maya, these people heartily agreed, saying they would happily obey anyone other than the tyrant Montezuma, who kept stealing their children for sacrifice in his distant capital. They explained to Cortés that this Montezuma wielded power over a vast empire and exacted a terrible toll from all his subjects. That was when Cortés had his epiphany: *in a flash*—that's the expression he used in his letter—he saw that he had to take his demand for obeisance directly to the infidel ruler. For if the great Montezuma were to bow before him, Cortés would, in one fell swoop, double the size of the Holy Roman Empire. And so, reinforced by the warriors of Veracruz, Cortés and his fellow conquistadors commenced their march toward the city called Tenochtitlan.

I followed them on my map, and it seemed that every town along the way gave Cortés more warriors intent on joining the conquest. The ranks swelled like waves moving shoreward, until finally Montezuma felt compelled to send emissaries. In the city of Cholula, the Spaniards were met by hundreds of regal ambassadors bearing lavish gifts. An Amerindian woman told Cortés's translator, the Spaniard who had been rescued in the Yucatán, that Montezuma's ambassadors were in fact spies plotting a deadly ambush. Cortés invited them into a courtyard for discussions, politely allowed them to enter before him, locked the gates behind them, and set the entire town ablaze. While the buildings burned, Cortés and his men rode warhorses through the streets, cutting down the panicked inhabitants.

3

"Check it out!" Rafe shouts. "That is *ripper!*" Our highway has just dipped toward the harbor, affording a view of several battleships and an aircraft carrier. Rafe is at the center of my rearview mirror, and I watch him now as he lunges across Ace to get a better look out the window. Ace leans back to be out of his way, and Haley looks in the opposite direction, peering out her own window as if she were alone back there.

Suddenly there is a tune—a loud tune—and I recognize it even before I understand what's making it. It's the well-worn theme song from *Top Gun*, a cheesy eighties movie about Navy flyboys, and it's coming from Ace. He's humming it. And I probably shouldn't be laughing delightedly at it, because the song can't but be an ironic comment on Rafe's emphatic *ripper*. Haley turns from her window to look at Ace—and the back of Rafe's head—and exclaims, "That's the best movie ever!"

"Awesome movie," says Ace, smiling. His head is cocked back, and one corner of his mouth rises a bit higher than the other, but his smile is nonetheless bright and engaging: he has big, beautiful teeth and a twinkling eye, and that back-tilt of his head looks more puckish than contemptuous. And now that I notice the mischief in him, the spiky hairdo seems to conspire in it, too; maybe it's just because we're thinking of a corny film from the eighties, but the hair seems to make him look about twenty years out of place.

Beside me, Chris turns in his seat to look back. "*Top Gun?*" he cries, shrill with dismay.

"Okay," Haley concedes, "maybe not the best *ever*. But *so* good." Her voice is very young, and she also seems to have a tendency to use strangely girlish formulations—like *the best movie ever*.

"*Top Gun?*" Chris repeats.

"Definitely an awesome movie," Rafe says, leaning back suddenly between Ace and Haley. Haley leans forward.

"Steve Stevens," says Ace, to no one in particular. "Pretty awesome."

"Who?" Chris exclaims.

"Steve Stevens. You know—" Ace hums the theme song again, and this time I realize that the humming is not ordinary: it's precisely on pitch, nearly operatic. Chris too seems surprised by it, but after a second he recovers his wits and objects: "But that's so—so—smarmy."

"That is genius," Ace says, and now I'm sure of it: there is plenty of irony in there. But it's not a mean sort of irony; it's a delighted and playful sort, as if we were all conspiring with him in his joke. And the joke continues as he lists other songs on the *Top Gun* soundtrack and estimates their smarminess. "'Take My Breath Away,'" he concludes, "now *there's* smarmy."

"I love that song," Haley says.

"It's awesome," Rafe inserts.

"How's it go?" I ask, mostly because I want to hear Ace hum again. But this time, he doesn't just hum. He actually sings. And he doesn't just mumble a phrase or two, as anyone might. He breaks into full-bore ballad, and his face projects the song's wild passions as he sings, loudly and without restraint, verse after verse of potent sap. It all ends dramatically, with long, warbling, eyes-shut modulations of "Take my breath away . . . ," and as the rest of us break into cheers and laughter, Ace finally opens his eyes and grins widely, as if he himself is a little amazed that he's just been inhabited by an overwrought eighties rock star.

For a moment, we're all smiling, shaking our heads and allowing the giddiness to diffuse. Then Haley says, "What about *Pretty Woman?*" and suddenly the conversation becomes a game, in which she names a favorite blockbuster, Ace lets fly with the title track, and Chris shrieks in outrage, defending his filmic standards against the populist mob in the backseat. Ace's renditions are so captivating that we're on the third or fourth theme song before I realize what's actually happening. It's uncanny, but every year, somewhere near the Mexican border, the students make their way to movies.

The first time I experienced the conversation, it was a roll call of "Did you see . . . ?" and I tried to redirect it, only to realize it had unstoppable momentum. The second time felt like a cruel coincidence. But by the third time, the recurrence of the same dialogue, at the same place on the road, cast an almost eerie light on the proceedings, and what had struck me before as prattle seemed suddenly more like ritual—a practice that had to be serving some kind of collective purpose. With various musicians now usurping Ace's voice, perhaps his very being, the game feels different, but I think its function might be the same: Strangers to one another in the parking lot, the students have quickly discovered some ground that is not only common but also—maybe more important, as our Expedition careens now toward a foreign border—soothingly familiar.

When highway signs warn us that we are about to leave the country, the students grow briefly louder—the game's last gasp—and then fall silent. Larger signs warn us again, and then again, as if they mean to communicate to us the gravity of our act. Yet our passage into Tijuana, when it finally occurs, is marked by no more than a stoplight. As we pull forward, we can see to our left, on the other side of the road, thousands of cars awaiting access to the land we've so easily departed. On our right, in a pullout area, Mexican soldiers, boys with smooth faces, lounge against their automatic rifles and watch us roll slowly by. From there the border zone is a rapid succession of vaguely threatening images: a wall of corrugated metal, splattered with furious, vibrant graffiti; beyond it, a swath of dry brown field, strangely vacant but for the white SUVs parked at regular intervals; on the highway median, inexplicably, a person lying prone; above Tijuana, a flag so vast that waves travel through it like slow Pacific rollers, starting in the green, moving through the white, and dissipating in the red.

4

Montezuma was cowed by the massacre at Cholula. He sent word that Cortés and his men would be welcomed to Tenochtitlan. Several days later, the Spaniards entered the capital as royal guests. Cortés described for his king a city more magnificent than any in Spain. It rose from the center of a sparkling lake and could be reached only by long causeways. Without the ingenious dikes and waterworks, Cortés noted, the city would be submerged. The marketplaces, vast and bustling, reminded him of Granada. And the palaces, with their fountains, reflecting pools, and aviaries, were more splendid than those of any sultan. Montezuma seated Cortés on a throne and explained, in a sort of royal soliloquy, that his people had long known that a great ancestor would one day come from the east and reclaim his rightful rule. Cortés, Montezuma understood, was that ancestor.

For a week I read the letters. Each evening I emerged from my enclave just in time to see the last sunlight lifting free of the tiled roofs, and I felt then like a nocturnal animal blinking awake from a long dream. Having lost myself in the story, I would step out onto the warm brick and realize I was missing something: Why would a woman from Cholula warn the Spaniards of an ambush? And how could she warn them? Cortés wrote that she had spoken to another woman, whom he'd brought from the Yucatán, and she in turn had spoken to the rescued Spaniard—but the woman from the Yucatán must have spoken Chontal Maya, whereas the inhabitants of Cholula, according to my map, spoke Nahuatl. And when the story reached the climactic meeting of Cortés and Montezuma, things became only more confounding: If Montezuma had taken Cortés to be his returning ancestor, why had he plotted to ambush him at Cholula?

I couldn't keep on like this; I had to get back to my office. So the very next morning, as I descended to my enclave, I swore I would skim ahead, scanning only for any mention of cacti, or Baja, or even just the sea around it.

But the plot took an irresistible turn. Cortés rushed out of Tenochtitlan, bent on vanquishing a rival conquistador. While he was away, the city's people attacked the Spaniards he had left behind, trapping them in a fortress at the city's center. Cortés returned, and this time, he crossed the causeway not as a guest, but as a conqueror. Tenochtitlan, the lacustrine capital more magnificent than Seville, was reduced to islands of rubble, smoldering amid floodwaters that poured through the dams.

Cortés and his men were settling in the wreckage when I encountered this passage:

I had received news a while before of the Southern Sea and knew that in some two or three places it was but twelve to fourteen days' march from here . . . Once the route to the Southern Sea has been discovered we shall find many islands rich in gold, pearls, precious stones and spices, and many wondrous and unknown things will be disclosed to us. This is confirmed by men of learning and those tutored in the science of cosmography.

At last, there it was: this *Southern Sea* had to be the Pacific, and from there, Cortés would surely make his way northward, into the Gulf of California. I had finally hit upon a trail that would lead to Baja—perhaps even to *Pachycereus pringlei*. But then it dawned on me: I had no idea what Cortés was saying. Which *islands* could he mean? And what sort of *wondrous things*? And what was this *science of cosmography* that had already confirmed it all? It was as if I had pursued Cortés all the way across my map, and when we'd finally reached the sea, where he was supposed to pause, turn around, and say something about Baja, he had whispered only the vaguest riddle.

5

All of a sudden, we are at the end of the earth: The Pacific, open and endless under the morning's cloud layer, feels like deliverance from the border zone. The toll road runs south, atop high cliffs, and the blue and silver expanse on our right is so alluring I almost fear I'll veer for it. We are on the same hovering plane as the distant horizon, where low clouds merge with the ocean, yielding a luminous depth. Up ahead, along the coast, the rocky headlands are layered one behind another, fading from the slick wet black of the closest to the ghostly blue of the one farthest away. The swell breaks into mist across their prows.

The mist rises higher and the cloud layer drops lower until they meet as a vapor around us, and now we can see that it moves with definite direction, from the ocean on our right into the hillside on our left. The ocean breathes out, the plants breathe in, drawing their water from the fog that drifts among them in horizontal cascades. It rarely rains here, but the slope is nonetheless covered with smoke-colored sagebrush, pale green chamise, and coastal agave, which resembles the crown of a giant pine-apple, as if someone had planted the sweet part, leaving only the bluish blades above ground. It's a well-adapted form, this crown, a shape nicely suited to a life imbibing fog, because a dewdrop that condenses on a stiff, grooved leaf will seep reliably to the plant's central root.

In a good year, a wet year, after two or three decades of gathering dewdrops, an agave will make the momentous decision to spend itself. It will draw its water, its sugar, and its many years of accumulated nutrients into its core, which will grow tumescent with nectar. The blue-green blades will shrivel from the edges, and the roots will retract from the tubes they've bored, as the agave gathers its life savings to shoot forth the straight, high trunk of a whimsical sort of tree, a child's drawing of a tree, with branches that extend like the arms of a candelabrum, holding out their flame of pale yellow flowers. At night, the flowers draw long-nosed bats, which hover, plunge their snouts into the tubular blossoms, and emerge with vulpine faces dusted gold. The bats live itinerant, moonlit lives, tracking the mobile mosaic of blooming agave, living on a wandering wave of nec-tar, and shaking grains of golden pollen into each flower they visit.

Then, the agave dies. Having sent its last drops of nectar into the flow-ers of its extravagant tree, it chars in the sun and dissolves in the sand. A life of three, perhaps four decades, prodigalized in a matter of days.

I am tempted to mention the agave and its way of life to my passen-gers, but I fear it's not yet time. Last year, I tried too soon to turn our at-tention to biology, and my efforts backfired badly. The students were not yet ready to interrupt their social rites, and so they retreated to a polite, attentive silence until the lesson was clearly over. That might sound at worst passingly awkward, but in fact the repercussions lasted: I had cre-ated an indelible boundary between science-time, when I talked, and social-time, when they talked. I doubt this could happen with this group: Chris's intellectuality, as well as Ace's manifest indifference to certain social constraints, would surely buck the rules. But even so, now that I do finally venture to say something about those pineapple crowns on the hillside—the way they bloom so gorgeously they die—I try my best to sound offhand, because my comment is in part a tentative probe, an in-quiry about whether it's safe yet to talk about science.

"Wow," Haley says, looking out at several agave in bloom, "they're re-ally goin' for it." Ace sings out something about *throwing it all away*, and

Haley goes on to wonder aloud if this is how pineapples grow, too—small signs of engagement, but just enough to suggest that I might try, ever so gently, to nudge us toward biology. I tell Haley that the description I of-fered—I myself called them pineapple crowns—was a bit misleading: pineapples don't grow with their sweet parts in the ground. In fact, pine-apples and agave aren't even close relatives; they're in different families, far apart on the tree of life. "But even so," I add, "there are some amazing things they share."

Obligingly, Chris asks, "Like what?"

I ask them if they happened to peek under the cover of their course-readers.

"Yeah," Chris says, "it's chemistry." He reaches down into his daypack to pull out his reader, and I realize immediately I've made a mistake: I was supposed to tap our conversation lightly toward biology—some nifty natural history would have been good—but now I've heaved us abruptly into biochemical diagrams. The students in back are already leaning for-ward as Chris considerately holds up the reader for them to see. It's such a didactic arrangement I could hardly fault them for retreating right now. And Rafe, as if on cue, points at the oval full of curving arrows and says, "Oh yeah, I had to memorize that for advanced organic chem."

"You've only seen the left side," I say, almost defensively—and again I'm dangerously off track. How is it going to help my cause to point out that the biochemical reaction is one they haven't seen before?

"Calvin cycle," Rafe says. "Right? A real beauty." Pained as I was by his previous comment, I'm grateful in equal measure for this one.

"It is," I say. "It is a beauty. But Haley's question about agave and pine-apple made me think of the other cycle—this one." Taking a hand off the wheel to point at the page Chris is holding up, I tell them about the bio-chemical trick it shows: the plant opening its pores at night, storing away carbon dioxide, sealing up for the hot day. It's called Crassulacean acid metabolism—Cam for short—because it was discovered in the family Crassulaceae, the jade plants, and the carbon is stored in the form of a liquid acid. "The same cycle is used by the cardón," I say, closing the reader cover so the students can see the photo, "and by those agave over there, and by pineapples. It's an adaptation they all share."

No one responds. Rafe's *real beauty* made me think we were in the clear, but now, with the silence growing longer, I'm not so sure. I might have done it again, doomed us all to weeks of campus banter, punctuated by lectures.

"But not exactly the same," Chris blurts out suddenly.

"What's not?" I ask.

"It's not really the same molecules," he says, and he abruptly opens the cover again, returning us to the biochemical diagram.

"Yes," I say, "all the same."

"But I thought you just said agave and pineapple are really distant." He looks puzzled, frustrated that something—I'm not sure what—simply isn't making sense. Ordinarily I would regret confounding a student like this, but in truth, at the moment I feel only relief, because now it's clear: There will be no indelible boundary between different kinds of conversation.

"That's right," I say, "they're pretty distant. And the cardón is even farther away. But Cam is something they've all got—an adaptation to hot, dry places."

Chris is still holding up the reader, but now, in his state of knotted mental effort, he's waving it a little, as if he were fanning the rearview mirror. At last he exclaims, "But it *can't* be convergence." And suddenly it dawns on me: Chris has made his way to something profound. While I've been worrying about pushing science too soon, he's been contemplating the arcing arrows of Veronica's diagram, and they've led him, like some sort of vortex, right down to one of the deeper questions of evolution.

Cam is rather complicated, and in all those curving arrows and minuscule names of molecules, it's hard not to see some kind of intricate machinery. But here's what's troubling Chris: Evolution by natural selection is supposed to be a chancy form of change. At bottom, the whole process depends on the appearance of new genetic mutations—these are the raw material that selection preserves and builds upon, one after another—and mutations, famously, appear more or less at random. An organism cannot choose to have exactly the mutation—or series of them—that it needs to build a new, terrifically clever adaptation. So how on earth, Chris is wondering, could so many different families of plants make their way to exactly the same elaborate cycle? Doesn't it seem just too damned lucky?

One plausible answer would be that they didn't actually get there separately. Instead, an ancient desert plant long ago discovered Cam, and then passed it on to its various evolutionary offspring. So there was no exact repetition of a chancy process—just one adaptation, bequeathed to a variety of descendants. If this had been the case, however, we would expect to find Cam in a certain collection of relatives—the descendants, that is, of that desert-dwelling ancestor. It would be as if one and only one bough of the evolutionary tree ever bore the fruit called Cam. But I have already made it clear to Chris that Cam appears on branches that are far apart in the tree of life; it is a fruit found here and there throughout the crown, not clustered together at the end of a single laden branch. So there's no avoiding it: Cam evolved more than once. And we've still got to wonder how a game of chance could have been so precisely replayed.

Another answer that comes to mind—the one Chris has just denounced—is convergent evolution: when different organisms adapt to similar environments, their separate evolutionary paths may converge on analogous forms. Just think of the front flippers of penguins and seals. They look alike, but one was derived, over evolutionary time, from a bird's wing, while the other started as a carnivorous mammal's foreleg. And yet this classic example of convergence does not challenge credulity the way Cam does. A fingerless paddle is one thing; an elaborate biochemical cycle seems to be quite another. In other words, I understand why Chris has just ruled out convergence.

But I've misled him. When Chris flipped open his reader and asked if the very same molecules are at work in different families of plants, he was already testing the hypothesis of convergent evolution. If this is a case of convergence, he was thinking, then the underlying details should vary from one family to another. Take our flipper example. The bones inside a penguin flipper match those of a bird wing: the index finger is terrifically elongated, while the other digits are shrunken, almost vestigial. The skeleton of a seal flipper, by contrast, looks a lot more like the five-fingered forelimb of a landlubbing carnivore. And this difference in underlying architecture tells us that the flippers of penguins and seals did not evolve from one and the same ancestral flipper, but rather from forelimbs that were already adapted to different tasks. Chris figured the components underpinning the Cam cycle—in this case, molecules instead of bones— ought to be similarly revealing.

And, in a sense, they are. Say we were to pick a molecule in the Cam cycle—the protein, for instance, that bears the sprightly nickname pep-c and performs the critical job of plucking carbon dioxide from the night air. And say we were then to look quite closely at little pep-c, first in agave, then in cardón. Would the two molecules look *exactly* alike? No. In fact there would be small but detectable differences, confirming that each group—the cacti and the agave—found its own way to Cam.

So I should not have told Chris that exactly the same proteins are present in different families of plants. And yet, cardón pep-c and agave pep-c do serve exactly the same function. What's more, if we were to measure the similarity between the two molecules—using percent identity, as evolutionary biologists sometimes do—it would be above ninety-eight percent over large stretches. In that sense, pep-c is pep-c, and we are still faced with the same fundamental puzzle: How did different families of plants draw essentially—if not precisely—the same lucky hand?

I don't actually know. But I have a hunch, and it arises from this curious observation: Most of the proteins that underlie Cam, including pep-c, can be found in plants that simply don't do Cam—no inhaling at night; no sealing up for the day; none of it. But if the proteins are not there to do Cam, what are they up to? Most of them, it turns out, are working together to perform a certain non-Cam function. Specifically, they're regulating the level of acid inside cells. And they're doing it, evidently, in all the different Camless plants that have them.

Here our forelimb metaphor starts to fail us. The penguin's ancestor flew while the seal's walked on land; but what if those predecessors had never deployed their forelimbs for such different tasks? Would the skeletons of today's penguins and seals look so different? Perhaps we could modify our metaphor to make it more useful: we could, for instance, set aside the seal and replace him in our comparison with a great auk. Sadly, the planet's last breeding pair of great auks were strangled by museum collectors in 1844, but we have enough preserved specimens and recorded observations to know something about the species. It looked a bit like a penguin, and earned its living in much the same way, by swimming after fish. But the ancestral auk that first renounced flight was distinct from the ancestral penguin that took the same bold plunge. So our small comparative study now comprises two different ancestors that used their forelimbs in the same way—to fly—until separate evolutionary transitions sent them both into the sea. And interestingly, when we compare the flippers of today's penguins with those of the last great auks, they look alike not just superficially—like seal and penguin flippers—but in the under-lying architecture, as well. In short, even their skeletons seem to match.

I suspect a similar story may explain how so many separate families of plants hit upon the same molecular machinery. Here's how it would unfold: Before the Cam cycle was ever invented—before agave or cardón took that first nocturnal breath—the cycle's molecular components were already working together to regulate levels of acid, just as they do in to-day's Camless plants. So whenever a Camless ancestor found itself in a hot, dry environment, it could call upon that entire, articulated apparatus of acid regulation, and put the whole system to work in a new job. Much as the bones beneath ancestral auk and penguin wings were already nicely integrated, and therefore did not require complete reconstruction to work well in natatory steering, the elaborate machinery of acid regula-tion was already functioning as a well-assembled whole, and would not have needed to be rebuilt from the ground up to fulfill its new function. In both cases—wing bones and Cam components—the specific muta-tions required to confer new function were not the many that had gone into the original construction of the system, but just the few that would permit its repurposing.

What's important to see here is that this scenario does not suffer from the problem that drove Chris to rule out convergence: we need not ask natural selection to build precisely the same intricate system many times over. Nature, Darwin pointed out, is prodigal in variety but frugal in innovation. She creates in the style of a composer who is assiduously faithful to her theme yet endlessly inventive in its variation. And here we can see why. When a species faces challenging novelty—a bird from the sky moves into the ocean for good; a plant finds itself in a desert so dry it can't breath without desiccating—natural selection cannot come to the rescue with big new ideas. But it can always borrow from its own old notions.

6

Renaissance cosmographers gathered sources omnivorously—the Old Testament; the Greeks and Romans; travelers' reports of pagan lands—and incorporated them into syncretic accounts of the nature of the universe. Graham explained this to me when I called him to ask about Cortés's mysterious views on the Southern Sea. He also gave me the names of some of the cosmographers' favorite sources: Ctesias, Herodotus, and Pliny; Marco Polo, Prester John, and John Mandeville. So now, back in my dusky enclave, I sat before a stack of books and expectantly opened the cover on an old, leatherbound translation of Pliny the Elder. Skimming across the chapters on astronomy and Europe, I made my way to the section on faraway lands.

It was deranged. There were descriptions of people with the heads of dogs, people with their faces below their necks, and people with a single large foot, which they sometimes held aloft for shade in their exceedingly sunny homeland. This, I felt certain, was not the sort of *wondrous thing* Cortés had invoked. But when I set aside the classics and turned to the European travelers, they were no less fanciful. In fact, some of their figments matched their predecessors' so precisely it was clear where they'd found their inspiration.

Still, there was one respect in which the travelogues seemed to presage Cortés's vision of the Southern Sea. They all reported thousands of islands. But even here, on a point of general agreement, it was hard to believe the travelogues could be the source of Cortés's expectations. The conquistador's letters had struck me as rather gritty, down-to-earth documents, whereas these islands were dreamscapes. The travelers placed them in the vicinity of the *Terrestrial Paradise*—the Garden of Eden itself—and populated them variously with Pliny's monsters, lusty sultans, and beautiful Amazons. And these last inhabitants—the warlike women who thrived without men—lured the travelers into truly exuberant flights of fantasy. In *The Realm of Prester John*, for example, I encountered an island called Grand Feminie:

> *In that land there are three queens and many other ladies who are their vassals. And when these three queens wish to wage war, each of them brings a force of one hundred thousand armed women, besides those who drive the carts, horses, and elephants with the equipment and provisions. And know that they fight bravely like men. No male can stay with them more than nine days, during which he can carouse and amuse himself and make them conceive. But he should not overstay, for in such a case he will die.*

This was too much, and I called Graham to tell him so. Cortés, I said, was not caught up in such delirium. But Graham, unfazed, said only that I ought to consult the work of a historian named J. H. Elliott. So the next day, I stood in the basement stacks, before a shelf of Elliott's books, and since my humanistic fervor was dimming by now, I chose the very thinnest one to take to my enclave. As I soon discovered, however, *The Old World and The New* was a little book with a very big thesis. When Europeans first encountered the Americas, Elliott argued, they were so overwhelmed by novelty that they actually failed to perceive what was before them. Instead, as Elliott put it, *they saw what they expected to see.* They arrived at the edge of expansive vistas—new views on geography, natural history, and humanity—only to stagger back and, Elliott wrote, *retreat to the half-light of their traditional mental world.*

I set down the Elliott and picked up the letters, thinking perhaps I could perform an experiment right there in the pale cone of light that fell on my corner table: Would the letters bear out Elliott's claim? Did Hernán Cortés himself see what he expected to see?

Almost immediately, I began to see it. Cortés seemed strangely insistent about comparing the cities he entered to places in Spain: Tlaxcala was just like Granada, Tenochtitlan like Seville; the food market was no different from Salamanca's, the textile bazaar as great as Granada's. And whenever Cortés described Amerindians, he reached for his most familiar version of foreigners: the Muslims of North Africa. Montezuma himself was likened to *the richest sultan or infidel lord.* And the people's clothes, the first letter asserts, *are like large, highly colored yashmaks . . . thin mantles which are decorated in a Moorish fashion.* This struck me as a perfect example of Elliott's claim. Not only did the letter say the clothes looked Moorish; it even stated, quite explicitly, that from the Yucatán clear up to Veracruz, people donned the same Saracenic outfit. And this reminded me that Cortés had also failed to mention a change in the language from one place to another. Could it be that he had failed to notice it? Were the people all just Moors to him?

The more of Cortés I reread, the more compelling Elliott's claim seemed. I was now reading from a heavily annotated translation of the letters, the work of the eminent historian Anthony Pagden, and many of the footnotes on Mexican civilization revealed that, even in the details, Cortés described familiar things that could not possibly have been there.

Montezuma's house, for example, had *a large patio, laid with pretty tiles in the manner of a chessboard*; and the roof of this house was *half-covered with tiles*. But wait—Pagden interjected—roof and floor tiling, so common in Spanish villas and Saracenic palaces, was entirely absent from ancient Mexico. More strikingly, the same could be said of the monarch's throne: there was simply nothing like it in Mexico, and yet Cortés described in great detail how Montezuma had seated him upon one. Could it be that Cortés had actually mistaken—or somehow reimagined—the very thing beneath his bum?

Illuminating as it was, however, Elliott's little book didn't seem to reconcile Cortés's letters with the sources of cosmography: the expectations Cortés stubbornly deployed came from Spain and North Africa, not from the confabulations of Pliny and Prester John. But then, as I followed Cortés just a bit further—beyond his revelation of the Southern Sea—I encountered this:

> *He brought me word from the lords of the province of Ciguatán, who affirm that there is an island inhabited only by women, without a single man, and that at certain times men go over from the mainland and have intercourse with them; the females born to those who conceive are kept, but the males are sent away . . . They also told me that it was very rich in pearls and gold.*

I could hardly believe my eyes. It was the Amazons, just as Prester John had described them. Not only had Cortés seen what he expected to see; his expectations had come partly from the fantastic tales of cosmography. And when I turned to Pagden's commentary on Cortés's island of women, the thread I'd been following—for many days now—finally wound its way to Baja.

Cortés, Pagden explained, had probably read about Amazons in Rodríguez de Montalvo's *Sergas de Esplandián*, a novel that was wildly popular around the time Cortés sailed for Mexico. Like most chivalric novels, this one told of a Christian knight who battled a powerful sultan for a holy city. Adding a little exoticism, Montalvo had worked in some legends of far-off lands. So when the pious hero, Esplandián, was on the verge of victory, the infidels received reinforcements from a certain island:

To the right hand of the Indies, very near to the region of the Ter-
restrial Paradise, was an island called California, which was popu-
lated by black women, with no men among them, and almost like
the Amazons was their style of living . . . Their arms were all of gold,
and also the harnesses of the wild beasts, on which, having tamed
them, they rode . . . And sometimes when they had peace with their
adversaries, they intermixed with all security one with another, and
there were carnal unions from which many of them came out preg-
nant, and if they gave birth to a female they kept her, and if they
gave birth to a male, then he was killed.

Montalvo invented the word *California*—at least, no earlier use of the
term has been found. But he endowed his neologism with a plausibly Saracenic etymology: the island's ruler, *a queen of great body, very beautiful for*
her race, was named Calafia, which seems to me unavoidably reminiscent
of the Arabic Calipha.

Over the next few days in my enclave, Cortés began to appear more
and more obsessed with the idea of California. He built ships, launched
expeditions, and when—at last—one struck Baja, he took the peninsula to
be an island and personally led the expedition to settle it. Baja, however,
was not California as Montalvo had imagined it. It was a barren place,
inhabited only by Paleolithic Amerindians. And yet the name California
would remain, commemorating for posterity a certain exotic dream.

7

Our first sight of Ensenada is a luxury cruise liner, glassy and huge like
a high-rise in repose. Two more cruise ships come into view behind it,
composing a small but modern floating metropolis. At once our toll road
swerves and dumps us into a busy downtown street, where we inch along
beneath bright placards that testify to Ensenada's fame for a certain sort
of tourism: *La Mirage, A Gentlemen's Club*; *WetLips*; *13 Negro* (I'm not
exactly sure what the name means, but the glossy black sign is tellingly
shaped); and finally, the name that's always been my favorite, on account
of its atavistic poesy, *Club Le Uh*. The large building that houses Le Uh
has onion-domed towers and ogive windows that have been filled in with

brick, lest passersby peer curiously inside. It's intended to look exotic, of course, but without the breviloquent sign, one could easily mistake it for a mosque.

Just beyond Le Uh, Ace breaks into laughter, and Chris and I both turn to look back at him. He points at a sign, a head-high post to which someone has fastened weathered pieces of wood. The slats hang at slight angles—there is something beachy and tropical about it—and hand-painted lettering spells out a different flavor of margarita on each slat: *lime, strawberry, piña colada*, and last, *Viagra*.

Rafe exclaims gleefully, "This place is so raunchy."

Farther south, past the towns of Colonet and San Quintín, the road swings east, toward Baja's protruding spine, and the flat plain crinkles into steep and involved hills. The buckwheat and sagebrush give way to barren red earth strewn with sharp rocks. And the jagged scree is populated, sparsely and haphazardly, by botanical specimens that seem to waver—not only in the shimmering heat, but also in a kind of middling zone between the living and the dead, the organic and the lithic, and even, it seems, the earthly and the otherworldly. In phytogeographic terms, we have just passed out of coastal sage scrub and into Vizcaíno Desert. But I think Joseph Wood Krutch, the drama professor, found terms that were more evocative, if less scientific: his essay on this part of Baja is entitled "Plants Queer, Queerer, and Queerest."

Some of the species we saw farther north look newly strange in this setting: each grows in isolation, surrounded by nothing but rock, and many seem to be dying, scorched by the sun to a charred and gnarled remnant. Here, the agave are not wildly blooming; they are waiting for rain. And they may have to wait another year, perhaps two or three, until it finally arrives. Some, it appears, won't last that long, because they are already turning into coal-black rosettes against the dark red rock, as if each plant were being gradually blasted into its own shadow. There are also some stunted candelabra cactus, and a few sour pitaya—tangles of thorned snakes, writhing up around one another—though here some of the snakes are wizened and brown for want of water.

"Are those, like, agave?" Haley asks, sounding vaguely dismayed at their existence here.

A little later, Chris asks, "Is that just a candelabra cactus?"

It's as if the names they've learned are just barely keeping hold of a landscape that has suddenly turned bizarre, whimsical, and grotesque. And only once the merely queer creatures have all been named, confirmed, reclaimed, can certain variations come into view: "Hey," Haley says, "those candelabra have hairy heads." But they're actually *garambullo*—old-man cactus, named for the shaggy gray and gold hair that crowns the top of each green column. They've been there, interspersed among the agave and pitaya, since the moment we turned inland, but the best way to catch sight of one, it seems, is first to grow familiar with candelabra, and then to find a strangely hairy specimen. And a moment later, Chris notices the fuzzy golden cactus called teddy-bear cholla; the name is misleading, I warn them, since the blond fur is in fact a dense coating of finely barbed spines.

Then, somewhere between the queerer and the queerest, the students get a brief respite: the road wraps clockwise around a hilltop, and on the uphill slope stands that familiar symbol of the desert.

"Hey," Chris exclaims, "it's a cardón." And he reaches down to pull out his reader. Again he holds it up for everyone in the back to see, as if they needed to verify, collectively, the resemblance between the black-and-white image and the first real specimen we've encountered.

"Look," Ace says, pointing at the hillside cardón, "it's got Cam. The leaf pores are totally closed."

"They're microscopic," Rafe says dismissively. I'm not sure whether Ace's habitual irony has actually escaped Rafe, or just annoyed him so much that he now refuses to acknowledge it.

"And besides," I add, "there aren't any leaves." And I turn to smile at Ace, just to show him that I know he wasn't serious—that I don't think he said something silly. He smiles back, looking not a bit worried about it.

"Yeah," Chris mutters thoughtfully, peering out the window, "no leaves." He sounds mildly surprised, as though he's just now registering the oddity of a familiar fact. "Is it, like, one really big leaf?"

For me, this hillside sentinel is a sort of mile marker. On my first drive down with students, this very cardón started a remarkably fruitful conversation about adaptation. I had wanted us to start reading the forms of organisms—the fluted columns of this cactus, for instance—as ledgers written in a code of shapes, three-dimensional transcriptions of the environment's most strenuous demands. So I asked everyone, as we passed this point in the road, "What makes that plant fit for living out here?" And what I learned, as answers came forth quickly and easily, is that the harshest desert on the planet actually offers the most fertile ground for good stories of adaptation. Out here, there is one overwhelming environmental demand: withstand heat without water. And this sheer simplicity of physical challenges likewise simplifies our interpretation of biological forms: Whatever features we might perceive, odds are, they have something to do with coveting moisture or enduring the sun.

The leaves, I tell Chris, have been modified into spines, which grow in small rosettes from the ridges between the columns' flutes. And without proper leaves, all the action of Cam must instead unfold in the skin of the columns themselves: they are effectively huge photosynthetic stems. "But why," I ask, "would a plant trade in its leaves?"

"It needs the spines for defense," Rafe asserts immediately, leaning forward. "Right?"

I meant to invite a variety of plausible stories, but it seems I posed my question too much like a teacher seeking the right answer, and now I've sparked Rafe's competitive instincts. And before I can backtrack, Chris says, "Defense against what? There's nothing here."

"Not now," I say, "but there used to be bighorn sheep."

"Exactly," Rafe says, as if he'd been thinking of the sheep all along.

"Where'd they go?" Chris asks.

"They got shot," I say, adding that when we're trying to explain adaptations, it's important to remember that the place we find a plant or animal today may differ from the environment in which it evolved; and often, of course, the difference is due to humans.

"Well it had to be hot," Chris says, "so maybe leaves were just losing too much water." I can feel Rafe lean farther forward, between me and Chris, and without looking, I suspect he's staring at me, awaiting a verdict: Who wins?

"That's possible, too," I say, and Rafe exhales sharply. But even if the uncertainty annoys him, I'm determined to hang on to it, at least for a little while. And so, as we make our way through other features of the cardón, speculating on their functions, I try to keep the tone equivocal: the more convincing a story seems, the more I'm inclined to cavil; and whenever one reading seems spot-on, I try to offer a second. That fluting, for instance: Does it allow the cardón to expand, accordion-like, and store away the rare but diluvial rain? Or do those deep grooves simply offer thin but precious strips of shade? It's hard to say, really. And what about succulence? Surely that plump, fleshy tissue of cacti, agave, and jade plants must be an adaptation for storing water? Probably so, but we mustn't forget that those plants also have Cam, and Cam, as we know, involves storage of liquid acid. So maybe succulence isn't just for hoarding water, but for holding the liquefied carbon that the plant inhales at night. It's hard to say for sure.

"Man!" Rafe exclaims suddenly. "We can't say shite for sure!"

I'm a little stunned by his vehemence. Still, what has provoked him— *such waffling!*—is itself almost the point, one of the more important conclusions to come from our own storytelling: namely, that it is remarkably hard to tell what any single attribute of an organism is really *for*.

Thirty years ago, Stephen J. Gould and Richard Lewontin reminded their colleagues in evolutionary biology of a character from Voltaire's *Candide*. His name was Dr. Pangloss, and he cheerfully maintained that *everything is made for the best purpose*. To prove his rosy doctrine, Pangloss could find a redeeming function for each and every aspect of his world. For instance: *Our noses were made to carry spectacles, so we have spectacles. Legs were clearly intended for breeches, and we wear them.* Even venereal disease, in Pangloss's view, was closely tied to its very own bright side: *For if Columbus, when visiting the West Indies, had not caught this disease . . . we should have neither chocolate nor cochineal.*

It all sounds quite silly, and yet, according to Lewontin and Gould, evolutionary biologists had fallen into a habit of similar fatuity: they would look at an organism, divide it up into attributes—such as spines, grooves, greenness, and succulence—and then assign each of those traits some sort of adaptive function. And while such speculations might well be correct, the problem was that no one ever bothered to test their hypotheses with experiments or even just further observation. So the organism inevitably came to seem a perfectly adapted creature, each piece of which was clearly doing something for the better. Merry as that picture might be, and fun though it is to invent tales of adaptation, the *Panglossian paradigm*—that's what Lewontin and Gould called the doctor's view as it manifested itself in science—actually threatened to undermine the rigor of evolutionary biology.

But what was one to do? What paradigm should one apply in place of the Panglossian? How could one properly test all the plausible stories? In the decades after Dr. Pangloss was presented, admonishingly, to evolutionary biology, the discipline developed several methods meant to vanquish him. And one of them happens to build nicely on what we've said about convergent evolution. When several different organisms follow separate evolutionary paths to the very same structure or behavior, we can search those various paths for common features—certain conditions that appear along all of them—in order to discern what might drive the evolution of the shared trait. Just a moment ago, for example, we were wondering what function succulent tissue might serve. Is it for hoarding water, or for storing the liquid acid involved in Cam? Well, tonight we will sleep among the cirio and elephant trees—the queerest plants of all—and they offer a way to test our alternative stories. With their plump trunks so weirdly reminiscent of pachyderm skin, they appear succulent. And yet, unlike other succulent plants—the cardón, the agave, the jade plant—these desert dwellers have never evolved Cam. They suggest to us, therefore, that the need to store those rare desert rains is sufficient to drive the evolution of succulence.

Still, there are difficulties. Environmental conditions tend to come in interrelated clusters. Places where rain is rare, for instance, are also likely to be quite sunny. So how could we use convergent evolution to find out whether fluting, as we find in cardón, is for accordion-like expansion, or rather for reducing direct sun? Even if we could find several species that separately evolved deeply grooved skin, their separate environments would almost certainly share the very conditions we're trying to disentangle. It can be hard, even with the help of convergent evolution, to confirm what a trait is really for.

"I've got one!" Rafe declares suddenly.

"One what?" I ask, startled from my explanation.

"A case where convergence shows us exactly what something's for." He leans back in his seat, as if a dilemma has just been resolved. It's like he's been intently tracking my words, hoping to get a jump on the answer—any answer—and now that he's got one, he can sit back and relax.

He remains there, in silence, until Chris finally says, "Okay, so, what is it?"

"It's Cam," Rafe says.

"Cam?" Chris asks.

"Right. The Cam cycle itself. It evolved all those times in the desert. It can only be for saving water."

As soon as he says it, I'm hit with a wave of amusement, followed closely by sharp embarrassment. Amusement because, despite his confidence, Rafe is sort of wrong. And embarrassment because, in truth, he's wrong on my account. For the second time today, I've misled a student. Before, with Chris, I spoke too casually when I said the same molecules underpin the Cam cycle in cardón and agave. But this time it's worse. This time I've committed the very error I've caricatured. I've fallen into the Panglossian paradigm.

8

The idea of a scientific paradigm was developed by Thomas Kuhn, a philosopher and historian of science, in his essay entitled *The Structure of Scientific Revolutions*. As Kuhn defined the term, a paradigm was a scientific achievement—an actual work of theory or experimentation—that was so compelling and revelatory as to become both a model of investigation and an irresistible invitation to all sorts of derivative projects: spin-offs, extensions, tying-up of loose ends. Kuhn's exemplars of paradigms were mainly the usual suspects: Newton's *Principia*; Lavoisier's *Chemistry*; Darwin's *Origin*. But what may seem more surprising than Kuhn's picture of revolutionary science is his depiction of all the nonrevolutionary stuff—that vast majority of scientific research that follows from any given paradigm. Kuhn calls this great preponderance of research *normal science*, and his description of it is, ironically, one of the more revolutionary aspects of his work:

> Closely examined, whether historically or in the contemporary laboratory, that enterprise seems an attempt to force nature into the preformed and relatively inflexible box that the paradigm supplies. No part of the aim of normal science is to call forth new sorts of phenomena; indeed those that will not fit the box are often not seen at all . . . The areas investigated by normal science are, of course, minuscule; the enterprise now under discussion has drastically restricted vision. But those restrictions, born from confidence in a paradigm, turn out to be essential to the development of science. By focusing attention upon a small range of relatively esoteric problems,

the paradigm forces scientists to investigate some part of nature in a detail and depth that would otherwise be unimaginable.

Kuhn's depiction of normal science is simultaneously a penetrating assessment of paradigms: *What do they do for us?* he seems to be asking. *How do paradigms affect our powers of scientific perception?* And his answer, surprisingly, is starkly bivalent. He seems to think paradigms affect scientific perception the way a flashlight affects your vision at night: You can see a small area with great clarity, but the rest of your surroundings, which just a moment ago may have been softly illuminated by starlight, are now lost to you, plunged into a deeper darkness. Or, to call on a metaphor close at hand: when the students and I were focused on Veronica's diagram of Cam, we were thinking with the paradigms of molecular biology, and meanwhile, the image of the whole cardón was but a faint silhouette, barely showing through the page.

But is Kuhn right? Must every paradigm that sharpens our scientific vision also delimit it? To be sure, many facets of the cardón are inaccessible to our molecular perspective: the strangely restricted range of the species, for instance; or its preferred pollinator; or even its peculiar flavor. But how would our perception of molecular cycles actually *prevent* us from investigating these other aspects of the cactus? One answer—a rather practically minded one—is that anyone busy with laboratory dissection of photosynthesis simply doesn't have the time or tools to contemplate much else; just ask a doctoral student in molecular biology. But there is, I think, another way in which our paradigm may delimit our scientific vision. It operates not at the level of our schedule, our setting, or our instruments, but actually at the level of our thought and perception. And for this very reason, it's harder to detect. At least, I personally didn't perceive it until I found myself suddenly and embarrassingly costumed in Pangloss's breeches.

When I first discussed Cam with the students, I described it as a beautiful adaptation to dry environments. Such a story came quite naturally: we were venturing into a torrid desert; our attention had been turned to plants that somehow manage to survive there; and Cam plants do in fact preserve precious water by sealing pores during the day. But in telling my tale of splendid adaptation, I omitted an important part of the story.

You'll recall that one side of Veronica's diagram represents the basic mechanism of photosynthesis—the Calvin cycle. This evolved long before

Cam, which was really just a nocturnal addendum to the central sunlight-to-sugar marvel. In fact, the Calvin cycle evolved in bacteria, about three billion years ago, and it is now present in every plant on earth. In view of this ubiquity, it may seem surprising that the Calvin cycle exhibits a certain deficiency. The site of the problem is a protein called rubisco, which does the same job for Calvin by day that our familiar pep-c does for Cam at night: it apprehends carbon dioxide from the atmosphere and passes it down the assembly line that will eventually incorporate carbon into sugar. But in comparison with pep-c, rubisco is a ham-handed worker. All too frequently, it mistakes a molecule of oxygen (that's O_2) for one of carbon dioxide (that's CO_2). And when it tries to shove O_2 where CO_2 belongs, rubisco succeeds only in snapping the incipient sugar in half. As a measure of the cost such clumsiness incurs, consider this: Fully half the carbon that many plants manage to absorb gets junked on account of rubisco's blunders.

For those of us who are even mildly Panglossian—who can't help swooning now and then over the beauty of biological adaptations—this is all sort of horrible. How can we accept that a protein as important as rubisco—perhaps the most abundant single protein on earth—is a maladroit oaf? How could natural selection, whose skillful work we love to admire, fashion such a singularly unskillful worker? The answer, in a way, harks back to those bighorn sheep I mentioned. Their contemporary absence from the cardón's ecology reminded us that an organism's current home may differ from the environment in which it evolved. We were thinking of the Vizcaíno Desert and the cardón's evolution of spines, but the same note of caution could be sounded with regard to the planet as a whole and the evolution of photosynthesis itself.

The air we breathe today is about twenty-one percent oxygen and less than half a percent carbon dioxide. In a sense, then, poor old rubisco has the odds stacked against it: when it reaches out to grab CO_2, it's a lot more likely to hit O_2. But rubisco's air wasn't always so rarefied. You'll remember, perhaps, that when organisms photosynthesize, they "inhale" carbon dioxide and "exhale" oxygen. In fact, it was mainly this process that made oxygen abundant in the earth's atmosphere. So in the earliest stages of the evolution of photosynthesis, oxygen was still quite rare, and therefore a protein that failed to distinguish between O_2 and CO_2 paid no penalty whatsoever. Solving the problem at hand, natural selection settled on rubisco.

But then, of course, the Calvin cycle's first turns minted the very counterfeits that would eventually deceive rubisco. And as the earth greened and the millennia passed, the protein grasped oxygen ever more frequently. By the time the situation became grave, however, rubisco could not be improved. This is not to say that better proteins are inconceivable; if rubisco had only evolved in the presence of oxygen, natural selection surely would have designed a more discerning molecule. Just look at young pep-c: it did evolve amid oxygen, and it's far better than rubisco at picking out CO_2. So the problem was not that a better protein could not possibly be made. The problem was that natural selection could no longer make it, because rubisco had become the central cog in an elaborate molecular cycle. The slightest tinkering with that cog would disrupt photosynthesis, which was by now the organism's sole means of survival. In a sense, then, natural selection was stuck.

And yet, as we know, when natural selection is unable to fiddle with the middle of an intricate system, it might still add something new on the system's fringe. Old rubisco was bad at plucking CO_2 from a sea of oxygen; but young pep-c was better; so why not let pep-c do the plucking, and then pass the CO_2 down the line to rubisco and the Calvin cycle? Well, that's exactly what Cam does. It's a kluge.

But if Cam is really just a post hoc repair for an old and ubiquitous defect, why is it found only in desert dwellers that close their pores to preserve water? Well, it's not. It is also found, tellingly, in a few plants that live at the bottom of lakes and bogs. No shortage of water there. But there is, it turns out, a terrible shortage of CO_2. And similarly, when a cardón seals up in the daytime heat, all the O_2 exhaled by photosynthesis gets trapped inside, and CO_2 gets even harder to find. So whether it's in a deep lake or a sealed-up succulent, old rubisco, who in any case has a hard time picking CO_2 from O_2, doesn't stand a chance. And that's when the Cam kluge is exceptionally helpful.

So if you had to pick which sort of scarcity—CO_2 or H_2O—is the prime mover in the evolution of Cam, you might have to go with CO_2. And yet, when I told the students about Cam, I talked only of scorching deserts and precious water. I'd like to say I made a pedagogical choice to keep things clear and simple—not to trouble us with the rare exception of a Cam plant in benthic mud. But the fact is, I didn't even think of decrepit rubisco or how Cam patches the problem. And the reason for this, I think, has to do with scientific paradigms. The paradigms that are the progenitors for my stories about Cam are from molecular and evolutionary biology. Like all paradigms, they offer their own distinctive ways of making their stories meaningful: call it spin; a favorite mode of storytelling; a source of narrative force. Molecular biology likes to unveil elegant, intricate, or complex machines—nature's most beautiful works of engineering. And in evolutionary biology, as we've already said, there is that tempting Panglossian emphasis on perfect fit—the clever adaptation to the most challenging situation. The truth about rubisco—its poor performance, its adjustment to atmospheres long gone—runs strongly counter to the narrative current of the paradigms I was calling upon. So I followed their narrow and bright beam of light—I focused on the intricacy of the Calvin cycle, the adaptive genius of Cam—and I left jalopy rubisco aside, in the roadside darkness.

Does it matter? I mean, is such an oversight consequential? Rafe has just declared that Cam, which is sometimes found on lake bottoms, is indubitably an adaptation to dry environments. So evidently, my omission was at least a bit misleading. But beyond duping an eager student about a certain adaptation, my depiction of Cam perpetrates a more general misrepresentation: it belies the nature of evolution itself. In pursuit of a compelling and coherent story—and, perhaps, touched by the Panglossian spirit—I portrayed natural selection as a skillful and ingenious designer. But the significance of Cam—what it tells us about evolution—could just as well run clearly to the contrary: Yes, sometimes natural selection performs like a brilliant engineer, but sometimes it fashions a molecule like rubisco. And then it must, rather inelegantly, patch up its own shortsighted design. It wasn't just jalopy rubisco I left in the roadside darkness; there was a whole landscape out there—a view of evolution itself.

9

Veronica shifts beside me and Taiga's head bolts suddenly up, monitoring her, then returns slowly, watchfully, to my shins. Yukon takes no notice. In blithe retriever contentedness, he is sprawled on the dry wash. He exhausted himself on a wild jackrabbit chase at sunset, and now snoozes with his nose half-buried in the dust. But just in case the glee of pursuit should possess him once more, Veronica has looped his leash around her wrist. She is such a hardy sleeper that he'd have to tug her halfway out of her sleeping bag before she would wake.

The calm of her sleep seems especially precious right now, because last night, when we first lay down here, she was still quite upset. Her day of driving had been less happy than mine, and as soon as I saw her climbing down from the Excursion, I knew something was wrong. She didn't perform her usual rites of arrival—her wide-armed stretch toward the cloudless sky; her deep inhalation of air resinous with creosote; her sudden declaration that the students should be quick, climb a boulder, and watch the sunset. Instead, she walked straight to the truck and set to unloading the students' bags, treating them, I noticed, rather roughly. And since she had resolved, evidently, to skip her usual introductory words on the Cataviña desert, I gathered everyone and told them about the boulders of pink granite and how they're sculpted by the wind. Veronica would tell them more, I said, on tomorrow morning's stroll among the desert plants, but for now, they should take their sleeping bags and, in the hour before dinner, scout out a nice place to spend the night.

"You mean just, like, *anywhere*?" The question came from Becca, a young woman who had been riding in the Excursion, and the quick look Veronica flashed me, along with a well-timed thump of a duffel on the sand, suggested that the source of her foul mood was right here. I tried not to look at the luggage pummeling, which continued unabated, and I explained the trade-off between boulders (warm, but hard and gently convex) and sand wash (colder and perhaps buggier, but deliciously soft and flat).

"Oh," Becca said, glancing sideways at the other students, "this is gonna be a great night."

Another duffel hit the sand, this one so hard it drew the attention of a few students, and Graham, who had been leaning against the truck, was roused to help Veronica unload. He has a tendency to think any irritation she might reveal must have something to do with him, and he often attempts to vindicate himself with exaggerated helpfulness.

"And don't unroll your sleeping bags," I added, trying to draw the students' attention back to me. "Wait till you go to bed."

"Why?" Becca asked.

"Ow!" Graham exclaimed, having been struck by a wayward duffel.

"Sorry," Veronica said.

"Because a sleeping bag," I answered, "makes a good home for scorpions."

"You guys sleep in tents?" Becca asked.

"No," Veronica said abruptly, "we sleep under the stars. It's beautiful."

As the students moved off into the desert, Graham asked Veronica if he should start assembling our camp tables to set out dinner.

"Burnett," I said, "she's not pissed at you."

"Did you see that?" Veronica asked bitterly.

"See what?"

"How she made that into something between you and her, with everyone else watching."

I wasn't quite sure I would have described it that way, but when I said as much, Veronica seemed painfully exasperated: How could I overlook such behavior? How could I even stand it?

"She did seem negative," Graham said eagerly, looking at me to solicit agreement. But Veronica seemed to collect herself, saying that for her, too, it had taken a while to perceive what Becca was up to. At first, she said, she'd known only that the conversation in her car was going very badly: people weren't getting to know one another; discussions weren't developing the way they usually did; and somehow, the talk never moved off campus.

As she went on, the modicum of calm that Veronica had managed to recover seemed to be slipping away again, and she was back on the verge of tears when she exclaimed, "We're looking out at the Pacific, and they're talking about fraternities. *Fraternities!*"

"Vica," I said, "you can't take that stuff seriously. They're just getting to know each other. I mean, my car talked all about movies."

"And mine talked all about Feuerbach," Graham inserted, perhaps thinking such subject matter would sound comparatively torturous to anyone other than him and Anoop. Veronica looked at him—just long enough for him to add, "Well, there was a little Derrida . ..."—then she chose to ignore him and go on. South of Ensenada, she explained, she had begun to suspect the problem had something to do with Becca. Whenever a conversation was about to get going, Becca would intervene by asking one of the participants a pointed question. Instantly, the conversation's center of gravity would shift to Becca, and from that moment forward, she would be the hub of discussion, with every exchange passing through her. "It was like she had to be the host of our little talk show," Veronica said. "And the rest of us—we were the guests who sometimes got to come up on stage."

Veronica seemed so distressed—so convinced of Becca's manipulative power—that I was reluctant to confess it was all a little hard to imagine. Becca, to me, was a short, scarecrow-skinny girl with protuberant brown eyes, which seemed to dart skittishly sideways each time she spoke. Not exactly the type to overpower a conversation. Though she did, now that I thought of it, have a rather penetrating voice, and also an accent—New York or New Jersey—that somehow made her seem even louder. And it was also true that both Veronica and I had felt some early reservations about her. In our biology department, she was something of a golden girl: she'd been field assistant for several faculty members, and she'd taken virtually every biology class available to her—all of which would seem to suggest terrific preparedness for a course like this. But there had been something a little troubling in all those qualifications of hers, or at least, in the way she talked about them. In our brief conversations after the informational meeting, Becca had seemed to counter every experience someone else mentioned with an achievement of her own. And at one point she'd inserted that ours was the only biology class she had not yet taken—as if to remind us that, in view of all her qualifications, we really had no choice but to accept her. And yet, strangely, we couldn't quite tell why she wanted to join us. We certainly didn't sense curiosity, or interest, or even a desire to become expert. And it was hard then not to suspect that all those achievements she'd managed to interject must have been motivated mainly by some need to compete and climb the ranks of her community. Why that community happened to be a department of ecology and evolutionary biology—that was sort of mysterious.

But even as I recalled those signs of ambition, that voice, the accent, still I had a hard time envisioning how anxious little Becca would have overwhelmed everyone else in the Excursion. Some extraordinary students had been in there. Granted, Isabel was terribly shy, and Cameron too seemed content sometimes to listen in silence. But what about Allie? She seemed so naturally engaging, and it would have been so easy, I thought, for Veronica to ask her about anthropology, or about her experiences last year—she'd been abroad, I recalled, in Buenos Aires. And then there was Lucy, the beaming, bright-eyed pre-med who—this had been the unforgettable detail of her application—performed regularly with major opera companies. And just the vivid physical contrast between Miles—the strapping, towheaded triathlete—and slight, skittish Becca made it seem strange and improbable that she could possibly silence him. But when I ventured, finally, to ask Veronica about the other students—whether she hadn't been able to talk with them instead of Becca—she seemed to hear my puzzlement as an indictment.

"I *tried*," she pleaded. "You'll see. It's *impossible*. You'll see."

And she was right. I did see. And Graham did, too. When the students returned for dinner, we all settled in a comfortable little arena, bounded on one side by a giant cardón and on the other by an enormous humpbacked boulder. We arranged ourselves naturally into a single large circle, though several more intimate conversations were in play: Rafe was talking with Anoop and Graham; Miles laughed at something with Ace and Haley; and Allie had joined Cameron and Isabel. Voices were low, leaving room not only for one another, but for the desert sounds besides, and as the sliver of moon was just dipping beneath the crumbled stones to the west, the separate discussions slid into whispers.

"So, Graham," Becca said abruptly, breaking off her own dialogue with Lucy, "why is a Princeton professor teaching a Stanford class?"

For the first time all evening, the group was silent, awaiting an answer, which Graham had no choice but to offer. He explained his history with me.

"What's it like being a prof the same place you were an undergrad?" Becca asked, glancing around the circle. I looked at the students who had ridden in the Excursion, and there were telling signs: Miles stared bitterly at Becca; Allie looked at the ground; Lucy wore a wide but anxious smile.

The interview continued until Graham terminated it by rising with his plate. A little pool of silence formed, but before the small, quiet conversations could begin again around its edges, Becca plunged into its center, asking if anyone else's sleeping place was as totally uncomfortable as hers. A few students took the bait and offered details of their own, to which Becca loudly replied, "No way, mine's so much worse—" and as she carried on, Veronica stood, I followed, and we met Graham by the vehicles.

"Okay," he said, "we're going to have to manage this."

Later, when Veronica and I lay atop our sleeping bags, staring up at the

Milky Way, she confessed that she'd felt a strange sense of relief at dinner. At least it hadn't just been her—her own inability to stand up to Becca or guide the group's discussion. Graham is one of the best teachers—and maybe the very best talker—she knows, and Becca had sprung the same traps on him that Veronica herself had fallen for in the Excursion.

"We'll figure it out," I said, feeling suddenly sleepy and wanting to let all of this float up into the clear desert night.

"You know what the worst part is?" she asked.

"What?"

"No one's paying attention. I mean, we have a new moon—amazing stars—and the cirio are in bloom. But no one noticed, because they're all stuck on campus."

When Veronica was a child in Europe, her father would sometimes take her and her mother on long road trips, visiting military bases across the continent. She liked to read books in the car, but sometimes her father forbade it, because he wanted her to see the countryside. I always found that so strange—telling a child not to read. But now, as I drift back to sleep, it's Veronica I see in the driver's seat, enjoining the kids in the back to look—*Divejse! Divejse! You're all stuck on campus!*

10

Scientists, Kuhn wrote, work with *drastically restricted vision*, because a paradigm is a *preformed and relatively inflexible box.* Sixteenth-century Europeans, Elliott wrote, *saw what they expected to see*, and *retreated to the half-light of their traditional mental world.* Admittedly, the parallel lines flank a wide chasm of time and culture. And yet, nonetheless, the parallel seems consonant with both men's views, for each of them indicated that the limitations he detected were not culturally or historically specific, but rather basic to human perception.

When Elliott first documented the European tendency to see the expected, he offered this comment: *This should not really be a cause for surprise or mockery, for it may well be that the human mind has an inherent need to fall back on the familiar object and the standard image, in order to come to terms with the shock of the unfamiliar.* And Kuhn, for his part, frequently allowed his narrow and technically defined term, *paradigm*, to take on a broader significance; indeed, he sometimes adopted the general meaning the word has in *Panglossian paradigm*, where it connotes a certain worldview. I suspect the dual usage was not accidental, for Kuhn often emphasized that works of science are intimately related to the cultural context in which they take shape. And if that's true, then scientific investigation is not cleanly separable from other kinds of perceptual experience. Perhaps for this reason, Kuhn had taken an interest in psychological studies of perception. He was especially intrigued by experiments in which subjects were briefly exposed to an object that defied common categories—a playing card, for instance, in which the hearts were black like clubs. Invariably, the subjects would mistakenly identify the novel object as a canonical one. Kuhn wrote:

> *Surveying the rich experimental literature . . . makes one suspect that something like a paradigm is prerequisite to perception itself. What a man sees depends both upon what he looks at and also upon what his previous visual-conceptual experience has taught him to see. In the absence of such training there can only be, in William James's phrase, "a bloomin' buzzin' confusion."*

Drastic novelty, it seems, poses a grave perceptual dilemma. We have no choice but to call upon the concepts our past has provided, and therefore our paradigms may be as ill-fitted to our world as rubisco is for today's rarefied air. And frighteningly, we might sometimes be driven by our paradigms to remake the world before our paradigms have time to evolve: Cortés charges across Mexico as if it were Saracenic Spain. He takes his seat on a throne of his own imagining, and later sails for La Isla de California, the kingdom of gold and Amazons. More prosaically, a passage through downtown Ensenada—La Mirage, 13 Negro—is haunted by chimeras recognizable from Prester John and his Grand Feminie, Montalvo and his Queen Calafia.

But if our only choice, as Kuhn tells us, is between a paradigm and a *bloomin' buzzin' confusion*, what are we to do? We are caught, it seems, in a sort of paradox: We need a paradigm to see, and yet our paradigm blinds us to so much. And we must ask ourselves: Is this double bind ineluctable? Are we truly fated to dwell in the half-light?

PART III

Pachycereus pringlei

LA ISLA DE CARDÓN

1

Before sunrise, the light is gray and powdery, but from it precipitates—gradually, mysteriously—a landscape of stone: huge gibbous forms of roseate granite, heaped into ridges and cairns. It's as if we have awakened in the crumbled ruins of a city carved from pink rock. Cracked and broken, the ancient walls protrude from flat washes of sand that flow around them, and you can almost infer the map of the city hidden below. Amid the ruin, the only remnants of the giant race that built it are the bizarre artifacts they left behind: the ocotillo, which resembles the cane skeleton of an umbrella that a giant once stuck, top down, into the sand; the elephant tree, whose pale skin peels in sheets of parchment, and whose fat limbs taper abruptly and kink busily, as if it had been the best-loved bonsai of an especially leisurely giant; and the cirio, the strangest of monuments—a tapering greenish column, encased in a crosshatch of thorny twigs and, now that it's in bloom, topped by a tussock of white tentacles.

Veronica, who is just standing from her sleeping bag, points at the highest cairn, where first light has just set the stone ablaze, illuminating a seated figure: it is Chris, who has climbed up to watch the sunrise. As we pack our things in the bluish shade of our swale, the fiery light slides steadily down the crumbled ramparts until, in an instant that is surprising in spite of its slow and evident approach, sunlight strikes the uppermost flutes of the great cardón that stands beside us. Seconds later, a nearby cirio too pierces the ceiling of shade: its high tussock glows like tallow, and the blossoms, warmed by the sun, drip upon us their unusual scent, a savory aroma of meat.

We arrive by the vehicles before anyone else and begin setting up our small camp table in the shadow of the humpbacked boulder. The instant we put out the granola, Becca appears beside us; it's almost as though she's been watching us from somewhere, awaiting the right moment to arrive. She fixes me with brown orbicular eyes and, in a voice that troubles me because it seems knowing, she asks, "How'd you sleep?"

"Like a rock," I say, and I just catch Veronica's eye before she turns and looks down at the table to conceal her smile; she knows my sleep is not rocklike. When she turns back around to offer orange juice to Becca, something about the way she speaks—kindly, but also rather formally—gives me to understand that she has resolved to protect herself, that she won't allow Becca to fracture the sense of calm she has recovered overnight. Still, it's just as well that Becca won't even have time to try: high notes of laughter are coming across the desert like a flight of ground doves, sprung just now by Ace, Haley, and Miles. The three of them are walking toward us, picking their path around thorny plants, vanishing intermittently into the long shadows of boulders. As they step into our own pool of shade, Ace's smile is gleaming, and he starts over with the story that's got them all laughing. Last night, he had carefully selected a large boulder with a top that had looked flattish. But each time he fell asleep, he started rolling, and would wake with a start just before he tipped off the side. Finally, in desperation, he zipped up his sleeping bag and worked himself down into a crevice just wide enough to hold him snugly. He slept there soundly, but at sunrise, he found himself, as he puts it, "in a serious pickle." Fortunately, he was soon discovered by a very strong triathlete, who now falls once more into uncontrollable laughter at the mere thought of his morning find—a young man, wedged into the stone, peering up helplessly.

"Miles just lifted me out," Ace says, imitating the two-handed hoist.

"I just—" Miles begins, but then leans over again in laughter.

"He saved my life," Ace says.

Graham is the last to arrive for breakfast; he slept, as usual, in his own pup tent, and it took him a while to pack up. Ordinarily, Veronica and I would tease him for his tender snuggery, but this year, heedful of Becca's resentment, we remain silent on the subject, though Veronica does eye him knowingly as he tries to arrive unnoticed beside the granola.

"What?" he says defensively.

"You're just in time," she says, and then, speaking more loudly, she directs the group's attention to a slate-green shrub at the foot of the humpbacked boulder.

"Hang on," Graham interrupts. "Is this the plant lecture? 'Cause if this is the plant lecture, they need their notebooks."

The students smile. After only a day, they're beginning to understand that Graham's professorial severity is in part self-parody. But the comment nonetheless serves its purpose: everyone moves to fetch a notebook from somewhere—a daypack, the seat pockets of the SUVs.

"Burnett," I say, "we call it the plant *walk*. Not the plant *lecture*."

"The point is, we're in session."

"Professor Burnett," Veronica says playfully, "would you care to identify this species for us?"

"Certainly," he answers, feigning confidence as he strides forward and kneels to investigate the shrub. "I know it!" he declares with some surprise.

"I thought you might," Veronica says.

"Don't tell me—I definitely know it—it's *jojoba*!"

"Linnaean binomial?"

"Oh, come on—"

"*Simmondsia chinensis*," Veronica says, and she explains that the name *chinensis* is actually misleading, as it comes to us through a peculiar misprint: Johann Link, curator of the Berlin botanic garden in the early nineteenth century, was working through a shipment from the western United States. A certain packet of dark-green nuts, pale ovate leaves, and tight brown flower buds was labeled "*Calif*," which Link interpreted as an indication of oriental origin. Consequently, a species native to Baja was named for China, a place it doesn't grow.

Picking a leaf, Veronica explains that it looks grayish because it is covered with tiny white hairs, which also make it feel soft and talced. She touches Cameron's arm, and he lifts his hand to receive the leaf; he sniffs at it, touches it to his cheek. The hairs, Veronica says, filter the desert sun, and exposure is also reduced by orientation: all the leaves stand perfectly erect, with their edge toward the high point of the sun's arc.

Veronica reaches into the shrub and pulls out a smooth green seed. "And this," she says, "is a magical nut." She looks at the students, waiting to see if anyone knows what she's talking about.

"Isn't jojoba in shampoo?"

"It is, but I don't think that qualifies as magic."

"Does it have Cam photosynthesis?" Rafe asks.

Veronica looks puzzled until I insert, "We talked about it yesterday."

"No," she says, "it doesn't. But that one does." She points at the giant cardón on the edge of our arena and takes a moment to explain Cam for the rest of the group. Rafe appears quite pleased to have prompted the digression. Becca, however, looks intensely dour, and it occurs to me that these two might collide as they stake out their respective academic territories.

"Can you eat it?" Chris asks.

"The Cochimi made pinole from the fruit," Veronica says, scanning the cardón's upper flutes for the spiny pears that sometimes grow there.

"I think he meant the nut," I say.

"Oh, right, my magic nut." She holds it out on an open palm. "You could eat it, but it would go right through you, because it stores its fat as liquid wax, and only one animal on earth—a local pocket mouse—has the ability to digest it."

"*That's* magical?" Becca says.

"Maybe," Veronica replies, unperturbed, "but I think the credit there goes to the mouse. What's magical about the nut—to me, at least—is that it helped save the whales." She looks around the circle, waiting to see if anyone has caught on. Finally she says, "Professor Burnett?"

Graham takes the orb from Veronica's palm and holds it up between thumb and forefinger, like a gemstone under examination. Studying it, he wonders aloud, "What does this seed have in common with the head of a porpoise?"

Everyone is silent, of course, and Graham holds their perplexed attention so long I begin to feel a bit uncomfortable. I glance at Becca, expecting to see terrible impatience, but she actually looks less grim than she did a moment ago, when Veronica was talking. Finally, to goad Graham onward, I say, "Echolocation!"

"Incorrect!" The head of a porpoise, he explains, contains an oil that is the finest of lubricants: it resists infiltration by dust, and it retains viscosity at extreme temperatures, both low and high. As early as the 1820s, the stuff called porpoise jaw oil had become indispensable for protecting the intricate gears of clocks, the joints in firearms, even the pivots in the swinging lamps of lighthouses. Spermaceti oil, which comes from the massive skull of the sperm whale, was known to be less fine than the material from porpoise or dolphin, but it could be separated into two components: a less valuable wax, which was used as a substitute for tallow, and a more precious liquid, which resembled its finer cousin. And with the growth of the whaling industry in the nineteenth century, refined spermaceti oil became readily available just in time to fill the gearboxes of the age of industry.

"Now," Graham says, "jump ahead to the 1930s, when chemists at an American paint company start messing around with this nut." Not much came of jojoba-based paint, but the chemists did discover that the seed contained a liquid ester remarkably similar to the finest lubricants. And when the United States entered the Second World War, the extract of jojoba nuts made its way into machine guns, aircraft, and tanks. Still, spermaceti remained important, both in automobile transmissions and for the military. But by the late sixties, Americans had come to view whales as intelligent, social, even peace-loving, and the idea of filling the war machine's gearboxes with the extract of serene and thoughtful heads would have struck a broad constituency as barbaric. The sperm whale was placed on the endangered species list in 1969, and in the early seventies, plantations of jojoba proliferated across Southern California and Israel.

"And that," Graham concludes, holding up the nut, "is how this seed figured in saving the whales." He tosses the nut to Rafe, who seems to glow with satisfaction at being a part of Graham's finale. "It's a tidy story," Graham adds, "and I wish I could say it were that simple. But to tell you the truth, it's not so easy to show that jojoba had a big impact, because there are other factors to consider: the Soviets carried on whaling as intensely as ever; the Apache Nation embraced jojoba as the crop to turn their fortunes, but then abandoned it; and then an American seed company tried to corner the market and drove every large jojoba plantation out of business."

As Graham concludes, the students look dizzied. Most of these kids are science majors—unpracticed, perhaps, in listening to the sort of story Graham has just told. Anoop looks content, and Rafe is still happy about being singled out. But among the others, squinting eyes, furrowed brows, and worried sideways glances bespeak strained efforts to reconnect all the wildly disparate dots: *Did he say lighthouses? Second World War? Apache Nation?*

2

Graham has taught me, too, to be patient with stories that ramify. When I called him and rushed breathlessly through my findings—Cortés had seen the Mexicans as Moors, the seat beneath him as a royal throne, and,

most astonishing, the Baja Peninsula as an island of Amazons and gold—
Graham did not respond. He was silent, which was a rare and disquieting
situation. At last, he suggested that I might want to consult the work of
his colleague Anthony Grafton. And so, the next morning, I sat on the
stone steps in front of the library and opened *New World, Ancient Texts*;
among Grafton's many works, this title had seemed most promising,
reminiscent as it was of Elliott's *The Old World and The New*.

I was braced for a rebuttal of Elliott's argument. Why else would
Graham have prescribed Grafton as a corrective—that's what it had
seemed—to my own zeal for Elliott's hypothesis? To my surprise, though,
Grafton called Elliott's studies *pioneering*, and he concurred that, in the
face of new continents, the European worldview had remained remarkably intact. And yet, Grafton did seem to prefer a different sort of explanation for such conceptual resilience: Perhaps the Europeans preserved
their worldview not because they failed to see novelty, but because they
were prepared to accommodate it. After all, the Europe that sent forth
Cortés was already grappling with cultural contradictions, which had
arisen not from contact with the New World but from deeper readings of
the Old: Renaissance humanists had recently produced their own transla-
tions of Greek, Roman, and other ancient texts, and in doing so, had disclosed societies that were shockingly different from Christian Europe.
The New World, then, was not the first new world that early modern
Europeans had faced. They had been practicing with their own past.

What Grafton was suggesting, then, was that the same classical
sources that struck me as ludicrous had in fact provided Europeans with
vital concepts and precedents. The dog-headed men and Amazons dem-
onstrated, at the very least, that the foreigner could be many different
things. And what's more, the various foreigners the classical sources described were not purely evil—or even, for that matter, purely foreign.
Ctesias testified that the dog-headed men, if strange and horrifying, were
distinctly human in their intelligence, and more fair-minded, in fact,
than the Greeks themselves. For Herodotus, the Egyptians could be
both the diametric inversion of the Greeks—women stood to urinate,
while men squatted—and also their direct forebears in most matters of
civilization.

Cortés, I recognized, probably never read Herodotus on Egypt, much
less Ctesias on the wise canids. But just as he'd received the legends of

Amazons and exotic eastern isles—transmitted indirectly through explorers and novelists—so too would he have encountered recycled forms of early, nuanced depictions of foreign lands. His *traditional mental world*, therefore, must have been packed with images and concepts of strange lands and peoples, and in Grafton's words, *the images were rich and pliable, and the concepts were often shot through with fruitful contradictions.* In short, Cortés was hardly fated to see nothing but Moors; he had many other models to call upon.

3

As Veronica leads us from one desert plant to another, she walks backward, like a museum tour guide. This not only allows her to continue talking with the group as we move, but also adds an element of dramatic tension to the tutorial, because the Cataviña desert, unlike an art museum, is mined with menacing plants. Yet somehow Veronica guesses her way through. She's backing straight for a chain-link cholla, a snarl of two-inch barbed spines, and several students are just about to speak up when she pirouettes around the hazard and continues on her way.

She has now talked about three species—jojoba, cardón, and ocotillo—and it's already apparent to me that she has a secret agenda. In my efforts last night to convince her that the first day with the students is always hard, I confessed that I had inadvertently played Pangloss and made photosynthesis seem perfectly adapted. In truth I would have been happy just to let the students forget my gaffe, but Veronica has decided instead to give me a chance to backtrack and clear up any confusion. When she stopped in front of the ocotillo, she talked briefly about the plant's ability to sprout leaves within hours of a rain, only to resorb them a few days later, and then she pointed to another head-high bouquet of bare canes twenty feet away and asked, "And that one—what species is that?"

"Um, ocotillo?" said Chris, playing along, since it was clear that the answer could not have been so obvious.

"It's actually a sibling species, palo Adán, and we can use this pair to think about neutral evolution." Then Veronica turned to me and asked that I explain the idea of evolution without adaptation. It was a slow pitch to let me hit Pangloss out of the park.

Ocotillo and palo Adán, I explained, appear to possess the same adaptations, and even to occupy the same place in the desert's ecology. Only a few subtle differences allow us to tell them apart: the red tubular flowers of palo Adán are a touch paler; and where the canes of ocotillo are always straight, those of palo Adán might—or might not—bifurcate. We could be missing something, but it certainly seems that these slight distinctions are adaptively unimportant. And if that's true, then they probably came about by neutral evolution.

The first step in neutral evolution is the appearance of an utterly unimportant mutation—say it slightly modulates an individual's color, but in no way affects her chances of reproducing. Then, for reasons unrelated to color, that same individual has many offspring, to whom she passes on the new mutation. Those offspring—again, no thanks to their color—have many descendants of their own. And so it happens to go until, eventually, every individual in the population is born with the new, slightly different color. The population has evolved, but the evolution has not been driven by natural selection.

With one strange-colored individual after another just happening to get lucky, such a scenario may seem quite unlikely. And indeed it is. If you pick any new, useless mutation, its odds of becoming the future norm are slim. But here's the thing: there are lots of new, useless mutations—they're happening all the time, in every generation—so the number of them that do finally make it into posterity is actually not so small.

"Isn't this sort of anti-Darwinian?" Rafe asked.

"Actually, Darwin himself wrote about exactly this kind of nonadaptive evolution. So I wouldn't say it's anti-Darwinian—just anti-Panglossian. If Pangloss's view is 'everything for the best possible purpose,' neutral evolution implies that some traits come about for no purpose at all."

As she walked backward from the ocotillo toward a nearby cirio, Veronica explained, for everyone who wasn't in my Expedition, what Dr. Pangloss signifies in evolutionary biology. And now, as she stops a few inches in front of the towering taper of thorns, she says, "We can think about him here, too."

Light from the rising sun has now slid three-quarters of the way down the taper, and its boundary is just above Veronica's head. Soon we will all be out of the cool protective shadows of nearby cairns and ridges.

"The cirio's tall, thin shape," Veronica says, "is probably adaptive. But how did it evolve? The first thing you think is probably that this shape was favored by selection because it reduces exposure to overhead sun. But we could think of other stories. See those fantastic flowers?"

She turns now and, arching her back, peers up at the spray of waxy blossoms. It occurs to me that this whole anti-Panglossian reading of the cirio might be mainly a pretext to point out the flowers she could not bear to leave unnoticed.

"Do you smell that?" Veronica asks. "The gamey scent?"

"I thought that was us," Ace says.

"Nope. It's those beautiful flowers, and it's a clue that their pollinators aren't birds or bees, but flies. Rot-eating flies. And maybe, in the competition for those flies' services, taller trees can win by sending their stink farther and wider. But if natural selection pulls a tree taller, it will also—just as a side effect of the change in patterns of growth—make the tree thinner. It's a plausible story—no more—but it does make the point: a trait, like thinness, can evolve as a side effect, without ever being favored by natural selection."

"Darwin wrote about this kind of thing, too," Graham adds. "Under the rubric 'correlated variation' he collected all sorts of examples of evolution in one part of an organism pulling along other parts, as if they were all interconnected by invisible filaments. Elongation of one limb usually involves elongation of the others. Or cats bred for white fur often turn out to be deaf."

"Do you all see," I ask, "how this is relevant to the Panglossian paradigm?"

Ace says, "Deaf white cats are not for the best possible purpose."

"Yes," Veronica says, squinting in the sun, "that's basically the point."

4

Thomas Kuhn emphasized that a scientist is able to embrace only one paradigm at a time: the progress of his research, and in fact, the very coherence of his worldview, depend critically on his commitment to peer through a single investigative lens. In physics, for instance, a quantity as basic as mass has profoundly different meanings under Einsteinian and Newtonian paradigms; one simply can't think of the world in terms of both definitions at once. To evoke such cognitive exclusivity, Kuhn turned to illustrations of visual gestalt: the simple line drawing of a cube, for instance, which can be perceived in either of two orientations; or the inkblot that can be seen as two different objects—duck or rabbit, young beauty or old crone. Much as you register only one version of such drawings at a time, the scientist can see but one paradigm's image of the world. But the scientist's cognitive commitment runs deeper. Whereas you can flip visual interpretations at will, the scientist's mind cannot move so lightly between paradigms. Kuhn wrote: *Scientists do not see something* as *something else; instead, they simply see it . . . the scientist does not preserve the gestalt subject's freedom to switch back and forth between ways of seeing.*

But I wonder if the study of evolution really works this way. I mean, which of the discipline's theories would pass Kuhn's test of exclusivity? What's the paradigm that excludes all others? The candidate that comes to mind is evolution by natural selection. But that theory, illuminating as it is, has the same problem as Goldilocks: it's comfortable only in the middle range. If the change is too small, as in neutral evolution, the theory doesn't fit: natural selection plays no role at all. And if the change is too big, there's a different problem. Like Goldilocks in the big bed, the theory leaves too much room uncovered.

As an example of really big change, take the evolution of photosynthesis itself. It originated in bacteria, and we have a fairly detailed reconstruction of early adaptive steps—those successive molecular changes that gradually assembled and improved the team of proteins responsible for corralling sunlight. In this arena, the theory of evolution by natural selection seems echt Kuhnian: We can trace history without ever getting our head out of the theory, or the theory out of our head.

But then we encounter a remarkable turn of events. A photosynthetic bacterium is swallowed by a large predatory cell, and instead of getting digested, the bacterium somehow takes up residence. The unexpectedly happy result is a new collaboration, in which the bacterium makes sugar, while the predator shelters his new pet from other marauders. Eventually, the predator renounces his old ways and the bacterium shares all she makes, at which point, two have become one, and the one could be called the first alga on earth. In the course of one very small meal, an entire biological kingdom has been born.

Evolutionary adaptation is sometimes depicted as a process of hill-climbing: each time a beneficial mutation arises and spreads, over the course of generations, into the whole population, a small step up a great hill is taken. The evolution of photosynthesis in bacteria fits this image nicely: each molecular change incrementally augments the performance of the primitive photosynthetic system. But if those early steps are like a gradual march up a mountain in the Alps, then that momentous unicellular meal is more like a sudden flying leap from the high slopes of Mont Blanc to a base station somewhere in the Himalayas. Suddenly, a whole new territory, with previously inconceivable heights, is opened up. What this new landscape represents, of course, is the vast evolutionary potential of green plants.

Such a leap does not contradict the theory of evolution by natural selection, since there must have been some kind of advantage gained in the initial cooperation between predatory cell and green bacterium. And yet the theory does not, in itself, cast a great deal of light on such a flying leap. Right now, as I write this, I am following the rules of English composition. But to say I am composing English does not tell you much about what I'm writing. There's more to the story. Similarly, that momentous meal takes place within the rules of evolution by natural selection, but the newfound collaboration seems to call for a story of its own: a story within a story, one could say.

And indeed, there is just such an account. It is called the theory of endosymbiosis, and it explains a lot—like why photosynthesis, to this day, takes place in an organelle, a kind of cell within a cell; or why that organelle has its very own genome; or why that genome bears a close resemblance to those of certain photosynthetic bacteria.

The unavoidable fact is that evolution is not a one-size-fits-all phenomenon. Some changes are adaptive, some are not, and some, though they may be adaptive, are illuminated but weakly by the theory of natural selection. No single paradigm, therefore, can be both comprehensive and informative. Darwin himself seems to have known as much, for he wrote with interest and enthusiasm about many different mechanisms of change—not only those we mentioned in the plant walk, but others, too, including some we now know to be basically false. And so, as Darwin's own stance suggests, and in spite of what Kuhn says about paradigms, our view of evolution must forgo theoretical exclusivity in favor of a far messier variety.

5

My crew of four—Ace, Haley, Chris, and Rafe—have known one another barely twenty-four hours, but our return to the familiar quarters of our Expedition feels nonetheless like a giddy reunion. Outside, on the planet of pink granite and alien plants, the air has just exceeded ninety-five degrees—so my digital dashboard informs me—while here, in our capsule of tinted glass, we have a comfortable and strangely scent-free seventy-two. Ace has taken control of my iPod, which streams music to the stereo, and the intersection of his expert musical tastes with my limited musical library—heavy on classical, light on Ace's dear eighties rock—has so far yielded a series of Beatles tunes. The familiar, sunny melodies have the crew humming, tapping, even wiggling a little as we fly down the highway.

We've been through something now: we've spent a restless night with our backs against the ground, our eyes trained on a wildly transparent sky, our ears attuned to a riot of nocturnal rustling—and though it's but one night in a relatively unthreatening place, it feels nonetheless like a small feat we've shared in accomplishing. No one was bitten by a rattle-snake; everyone saw the same shooting stars; and today, we shall reach the sea.

But there is also, I'm afraid, a thread of our joy that is not quite so innocent: we are all relieved to be free of Becca's overweening presence. Before we departed our desert camp, I tried to hand Veronica my keys. But she declined, saying the others in the Excursion would feel abandoned. And I suppose she was right. If I sense a collective sigh of relief from my own passengers, then surely, among hers, there must be some silent acknowledgment of the challenge they are enduring together. At least, I hope there is—which is unkind of me, I know; but just a few knowing glances or rolled eyes would gird the group and contain the damage.

Somewhere toward the end of the *Sgt. Pepper's* album, we approach the abandoned settlement of Punta Prieta. Like most other attempts at life in this desert, it appears to be blanching, crumpling, finally dissolving beneath the relentless sunlight. It now consists of two cars without wheels, one semitrailer stranded on the sand, and a shuttered truck stop that has been blowing away, piece by piece, since I first drove past it eight years ago. Veronica swears the place used to serve carefully prepared hue-vos rancheros, with handmade corn tortillas, chilies grown out back, and eggs from their own flock of chickens. She says Alejandrina, the quietly strong and loving woman who now cooks for our class at the research station, used to run the place with characteristic attentiveness. But today I can only wonder at the bravery—or recklessness—of whoever it was that discovered those huevos rancheros.

Just beyond the windblown flotsam that trails from Punta Prieta, there rises a surprisingly shiny and modern road sign. In bright white letters on a marine blue background, the sign reads ESCALERA NAUTICA, which means something like "Nautical Stairway." It's a strange phrase to find on a barren salt flat, beside the dusty remains of a truck stop, and the words of glimmering blue ascent in this setting of arid decay strike me as a suspicious clue—a cause for skeptical narrowing of the eyes, not unlike certain bureaucratic euphemisms, like *collateral damage* or *downsizing*. But as our caravan swings left, eastward, onto the side road that will lead us to the sea, none of my passengers seems to notice the gleaming sign. So maybe I'm wrong; maybe it looks foreboding only on account of what I know. In any case, I'd just as soon leave the topic alone until we've spent at least a day or two in Bahía.

On the final climb before we descend to the sea, we often encounter a checkpoint watched by young soldiers holding automatic weapons. I had always found it improbable that the route for smuggling northbound drugs and southbound guns should follow the paved road in and out of Bahía, rather than one of the many dirt tracks through desert, and therefore I thought the soldiers daft for setting up their checkpoint, day after day, on that same hill. But when I read that the smugglers often carry hand grenades and rocket launchers, I realized that the soldiers may in fact have good reason for making their location entirely predictable and their checkpoint easily avoidable.

Now, however, as that last hill comes into view, I see no soldiers. And as we approach, the only traces of the usual checkpoint are black marks left on the pavement by the soldiers' smudge pots—tin buckets of oil that burn dirty, billowing black smoke and, it occurs to me now, politely announcing the soldiers' presence to anyone watching from a distance. At the hill's crest, Veronica veers slowly across the road and leads us bumping off the left shoulder. And with that, all at once, we are looking at the sea.

We are looking, that is, at the Bay of Angels, beyond it the Channel of the Whales, and across the water, the long desert escarpment of

Guardian Angel Island. And though each of us has, I am sure, imagined our destination many times over as we've passed through these many miles of desert, I am also sure that none of us has imagined what we now see, because none of us—not even Veronica, who knows this lookout so well—could possibly hold in memory or create in imagination such teeming, multitudinous *detail*.

We might recall that the water is sapphire beneath the sun, but the way the winds write arabesques of glare across the surface—of that we can remember the fact, but not the exact unfurling curves, much less the million little flares that compose them. Or again, we might remember well that Guardian Angel Island, the distant boundary of the channel, appears to be made of sand in many different colors—pinks and grays, blacks and beiges—layered one upon another. But we could never recall the actual pattern of sedimentary bands, or the places a layer below erupts upward, or one above spills down to the sea. And then there is the way those layers run crosswise with the ripples and folds that carve the side of the island—deposition making the weft of color, erosion the warp of texture, and the two of them together elaborating a tapestry far too involved for memory. Landscape exceeds the mind—it is bristling with wild precision—and I realize, as my vehicle rolls to a stop beside Veronica's, and my gaze comes briefly off the scene before us to glance at the dark tinted windows of the Excursion, that since the very first moment I met Cameron, he has neither asked nor permitted me to feel a pang of regret on his behalf. But now, confronted by this brilliant and infinite view, I can't help feeling it—a sort of surrogate grief, a tinge of bitterness—though somehow I know I should not.

As soon as we pop our doors, the heat rushes in—an unnatural, unbelievable kind of heat, as if you'd stepped accidentally into the path of some sort of industrial exhaust. The extremity of it is strangely exciting, and the sudden stimulus, along with the expansive view of our destination, makes the students a little wild. They scramble loudly out of our SUVs and make straight for the seaward edge of this small flattened hilltop, gasping and exclaiming as they go. Seconds later their awestruck excitement gives way abruptly to busy and hurried handing-off of cameras, a frenzied photo shoot. This too, I suppose, is a sign of enthusiasm, but it also means that many of them are standing with their backs to the bay, striking hammy poses, while others appreciate the scene on their two-inch digital viewfinders, intermittently hollering, "Wait, wait," as they stare down at their gizmos and finger some recalcitrant function. I glance at Veronica, thinking the flurry of distracted activity might annoy her, but she looks happy enough standing beside Graham and pointing at various islands.

A number of students start striking their poses in pairs, and a moment later, the pairs draw together for a team photo: with big grins, arms around shoulders, thumbs up, they look like they've just summited a peak or won a softball tournament. But for all their gestures of camaraderie and victory, the team can't quite vanquish a faint strain of awkwardness, because in truth they are not yet so close as their spirited smiles suggest, and therefore their poses can't but feel a bit forced. Still, it's that same awkwardness—the sense that certain motions must be pushed through a viscous social medium—that makes their gestures seem so touchingly kind: Ace grasps Anoop by the arm and pulls him into the photo; Allie, who last night stared grimly at the sand as Becca commandeered our conversation, now puts an arm around her shoulder. There is optimism in these poses, for they seem to insist that the shared memorabilia now being fashioned will eventually seem honest.

Veronica volunteers to snap a group photo, then takes advantage of the opportunity to turn the group around and map out the landscape for them. Instinctively, they all face straight out to sea—to the east—but she starts instead with the land, the mountain that rises just to the south of us. It's called Mike's Mountain, she explains, after an eccentric American anchorite who retreated to a cave up there sometime in the late fifties. Some people in town say he lived there with rats. At the foot of the mountain, on the sandy skirt between rocky slope and blue water, the town huddles modestly, dwarfed twice over—by the mountain hulking behind, and by the sea extending before it.

From town, Veronica directs our eyes farther southward along the coast, to Red Mountain, a ferrous quoin of rock that presents a sheer face to the east, where the sea strikes it, and a gentler slope to the west, where the mountain falls gradually into the desert. Sometimes, when the onshore wind blows all the surface water against that dark red cliff, milky clouds of plankton build up like thunderheads against a mountain range, and that's when we can find the whale sharks there, swimming slow circles, filtering billows of drifting food.

At its southern edge, Red Mountain drops into a broad alluvial fan, a geological flow of sand that spreads from the mountains down to the sea. We've now followed the coast to the bay's southern extremity, where an arcing yellow beach marks the boundary between pale shallow water and a thin layer of green, which I know to be a mixture of pickle weed and stubby mangroves. When the tide rises, the water floods southward, through a tideway in the beach, filling the flats with warm, clear pools where young puffer fish like to hover. Those tidal flats are called Estero La Mona, though in fact it is an estero—an estuary—only in an everted sense of the word: saltwater pushes inland, but no sweet water enters the sea. La Mona, in Spanish, could be a female monkey, and since many local landmarks are named for animals—from the island called Piojo, Louse, to the Channel of the Whales—it wouldn't be surprising for the estuary, too, to bear such a name. But in this case, the noun is proper, not common, and refers to a mysterious American woman who once inhabited that part of the bay. Evidently, if a place here isn't named after an animal, then it's known for its resident American eccentric.

Beyond the estuary, in the southern distance, barren jagged peaks enclose the bay. The mottling of rust, pink, and gray is deceiving: when your eye picks out a patch of darker shade, it's hard to decide whether you're seeing rock tinted by iron or a valley shaded by adjacent ridges. But once you find the mountain's high creases—the ones that are real—you can make out the wellsprings of alluvial fans: from each valley, a cascade of sand originates; it spreads as it descends, merges with the flows from other valleys, and sweeps like a gown from the mountains to the estuary below. Across those sandy slopes spreads a great cardonal, a forest of towering *Pachycereus pringlei*. From here, they are visible only as a faint verdigris on the hillsides of bronze.

Following the coastline across La Mona and then northward, along the other side of the bay, you come to a narrow crease of white. It is a small crescent of pearly beach backed by limestone boulders, and it has two names. Usually, the place is El Rincón, The Nook. But sometimes it is La Cala del Coyote, Coyote Cove, because on certain nights, cackling and howls carry across the water from that remote corner of the bay.

Moving northward from El Rincón, the coast is composed of dark rusty granite, much like Red Mountain. For no reason I understand, brown cucumbers remain abundant in the small coves over there. The cucumber coast stretches north to a narrow headland, Punta la Herradura, Horseshoe Point, which marks the eastern edge of the bay. That headland shields the bay's southern reaches—La Mona and El Rincón—from wind and waves.

A chain of sixteen desert islands, the angels of Bahía de los Ángeles, extends northward from Punta la Herradura. It might at first seem strange that those barren hunks of rock, populated by a scant sprinkling of cacti, cirio, and ocotillo, and visited mainly by seabirds that splatter the rock with guano, should be called, of all things, angels. But soon enough, you discover that they are your protectors, those isles, because they shelter coves that remain safe through the most violent weather. And from that moment on, even when the sea is placid and the islands linger over perfect reflections of themselves, you look on them as a beneficent presence.

The first islands north of Horseshoe Point are Los Gemelitos, the little twins. To give the students a sense of proportion, Veronica tells them it would take about thirty minutes to swim around one of the twins. Then she pauses, staring out at the pair of islands, and I know she is thinking now about the accident. On that day, the wind was blowing powerfully out of the west, as it sometimes does. It swept five people right past those last hunks of rock, out into the channel. From here, the brief stretch of blue between Los Gemelitos and the next island to the north, Cabeza de Caballo, looks so very calm and narrow that it's hard to imagine you could be shot through—right between protective angels—without a hope of catching hold.

Cabeza de Caballo bears its name because it slopes down to the north in such a way that, when you look at it from town, it vaguely resembles the head of a horse in profile. From Cabeza, Veronica's tour hopscotches northward from one island to the next, and with each one's name she offers a little promise, a detail that will serve to hold the place in memory until we visit: Isla La Ventana, the window, after a flying arch of granite at the island's eastern edge; atop that arch a pair of magnificent frigatebirds build their nest each summer. Cerraja and Llave, Lock and Key, for the way the two small islands fit together, which happens to create a protected bight of turquoise water. Flecha, the arrow; on its steep western cliff, each slim ledge is whitewashed in guano by pelicans and cormorants that stand there in dinosaur stillness. Jorobado, Hunchback, also called El Borrego, the Sheep; there too the seabirds perch against the cliffs, but El Borrego seems to be favored by blue-footed boobies. Their feet, it is true, are preposterous: splayed wide, rubbery, and cyan. But the rest of a booby is in fact quite elegant—her head feathers so smooth she seems fashioned of porcelain, her shape sleek and fusiform—and there is something poignant, I think, about such a lovely figure, clad in her formal finest, with her clown shoes protruding ludicrously below.

East of Jorobado are Pata and Bota—Foot and Boot; the deep channel between that pair points to the open sea, which is perhaps why huge gorgonians—sea fans—are anchored there, swaying inland on the floodtide, seaward on the ebb, catching plankton in their reticulate vanes. A good three miles east, into the channel, lies the Louse, Piojo. Sheer around its edges but flat across the top, it resembles a tableland on the water. Bryde's whales sometimes tarry in the plateau's shadow.

A mile northeast of Pata rises Calavera, the Skull. The dome-shaped island is guano-white everywhere except for three shallow corries, which remain dark because the birds cannot perch there, and which happen to lie in just the right positions to resemble gaping eye sockets and the gash of a nasal passage. Even more unsettling is the way the island hoots, moans, and growls, and wind from Calavera's direction carries not only riotous sounds but also a deathly mephitis. And while all this is perfectly explicable—on the island's northern rocky shore, a colony of at least a hundred sea lions loll, stink, and complain endlessly—the Skull nonetheless occupies a mysterious place in the lore of local fishermen. When disturbing events unfold on the water, the Skull is often somewhere in the background, looking on.

Beyond Calavera rises the large island that seems to lead the northward procession of the others. Isla Coronado is a dormant volcano, and I have often imagined, though never actually witnessed, a foreboding wisp of smoke rising from a high fumarole. The channel between Coronado and the peninsula's shore is two miles wide and very deep, but relatively protected. In the morning, the waters are glassy and black, shaded by the dark cinder cone, and we sometimes find great fin whales there, lunge-feeding on rust-colored clouds of krill.

A few small islands huddle in the shadows of Coronado, but Veronica skips over them. Instead she draws our attention back along the peninsula's shore, until we come to a thin point that reaches out into the bay and shelters, on its southern flank, a pale beach known as La Gringa. An

English translation of *gringa* would be something like "white woman from elsewhere." For years I assumed the name alluded to the color of the shingle, but as it turns out, La Gringa, like La Mona, refers to a particular woman, an American, who was often seen walking that stretch of beach. And in fact, the stories of the two women are so resonant that they may well have blended and traded elements on their way down to the separate moments I heard them.

It was about five years ago that a certain old-timer from town told me about La Gringa. "She was a very attractive woman," he said, and then he looked around. Leaning in, he continued. "She had blond hair, which came down to her waist, and concealed her breasts whenever she strolled her beach." He raised his eyebrows, waiting for me to indicate that I had grasped the significance of his statement. "She lived alone," he said. "All alone."

"Oh, Hirsh," Graham yells, as though he were calling me in for supper. He's standing by the door of the pickup truck, and most of the students are piling back into the SUVs. Veronica and Anoop are crouching to inspect a minute succulent, and only Allie, Isabel, and Cameron are still looking out at the sea. As I walk back to the Expedition, I pass by them, close enough to hear Allie saying, "No, it's really dark out there. It's like you can see how deep it is." I can't resist turning around again, because I realize now that, though I looked closely at the islands and the bay, I didn't really see the water in the middle of the channel. And she's right. It is a color that lets you understand Homer's strange phrase, *the wine-dark sea*, or maybe even that second line of King James: *and darkness was upon the face of the deep.* And at once I know why my pangs of regret on Cameron's behalf were misplaced. What do I know of what he perceives? Why would I presume that his view is but a sampling of my own, when my own, after all, is itself infinitesimal? My particular stupidity, in that moment of uninvited pathos, was failing to remember just how much bigger than our perception the world actually is, and therefore, what limitless room there is for perceptions other than our own.

6

If Hernán Cortés really did have myriad visions of foreigners available to him—including strange but noble Egyptians, Innocents from Hesiod's

Golden Age, and countless other creatures that swirled in the head of a semi-learned man in the age of humanists—why did he, in the event, perceive such a narrow slice of possibility? Why so many Moors?

When I next returned to the library, nighttime rain had rinsed the summer heat from the red brick and terra-cotta tile of campus, and it was partly the morning chill that made one feel like the light had tipped, overnight, into the orange obliquity of autumn. Sitting on a newspaper on the library's stone steps, I turned to a passage in the letters that had mystified me until I'd encountered Elliott's idea that Europeans had seen what they expected to see. It was the place Cortés said an Amerindian woman had warned him of an ambush in Cholula—a claim that had seemed impossible, since Cortés's translators spoke Chontal Maya, whereas the inhabitants of Cholula spoke Nahuatl. In my enthusiasm for Elliott's thesis, I'd decided Cortés had simply misunderstood: seeing every Mexican as a Moor, he had failed to recognize a language barrier.

But what to make of this if Cortés's view of foreigners had in fact been more flexible than I'd thought? Now, on the steps, a lead came in the form of an annotation by Anthony Pagden, my learned translator and guide: interpretation from Nahuatl to Chontal, Pagden explained, could have been performed by Malinche, a bilingual Amerindian slave who proved dear to Cortés—not only as his interpreter, but also as his mistress, and mother of his favored son, Martín.

It was then, as I read what Cortés had not revealed, that I understood something so basic it's almost embarrassing: These letters were, first and foremost, political documents. Cortés was writing *to his king*. And to reap the rewards of his own deeds, he had to demonstrate that he had brought the crown wealth, glory, and many loyal subjects. So perhaps it was in the service of this agenda—and not because he failed to see—that Cortés papered over real divisions of language, culture, and allegiance. For if Mexico had truly been the single unified empire that Cortés made it out to be, then Montezuma's investiture of Cortés would have passed the entirety of that empire to the Spanish crown. And yet, despite Cortés's best authorial efforts, Mexico's seams show stubbornly through his story: every town he entered furnished new warriors for his campaign; inhabitants of one town feared walking into the next. As Pagden helped me to see, the "Aztec Empire," a phrase and notion still current today, was in all likelihood the politically expedient fabrication of Hernán Cortés.

In other places, too, attunement to political exigencies makes the inkblot of the letters look suddenly like a different animal. Did Cortés really perceive the cities of Mexico as matches to their counterparts in Spain? Tlaxcala as Granada, and Tenochtitlan as Seville? Or was Cortés strategically presenting his king with images that would be both familiar and appealing? Such an agenda might even explain why he adorned Amerindian adobe with an entirely fictitious skin of tile.

And of all Spanish cities, Granada and Seville were shrewd choices, politically speaking: both had been centers of Muslim empire and, later, sites of the triumphant Christian Reconquest. In fact, Muslim Granada had surrendered to King Charles's own grandparents only two decades before Cortés conquered Mexico. So perhaps Cortés's canniest move of all was to cast his own campaign in the image of the Reconquest—which would explain why he rendered the Mexicans so very Moorish: a Muslim identity for his adversaries made Cortés himself into a captain of reconquest, not unlike the king's forebears or the hero of Montalvo's popular novel about California.

At the end of Montalvo's novel, Queen Calafia not only surrenders but falls in love with the Christian hero and donates her entire domain. Cortés, taking after that hero, offered his king not just victory, but the final capitulation that the Reconquest had never achieved. In the legend of a long-awaited ancestor returning from the east, Cortés found—or invented—a plausible reason for Montezuma to donate all of Mexico, supposedly a unified empire, to Spanish power. The conquest thus culminates with an image that is perfectly suited to Cortés's political needs: the conquistador sits on an imaginary throne; the Aztec emperor bows voluntarily before him.

And finally, the image that had struck me as the most extraordinary case of Cortés seeing the expected: his report of an island rich with gold and pearls, and inhabited only by women. Could that too shift its shape, from monstrous misperception to politically astute fib? Perhaps so, for King Charles would have known the legend of Amazons from other sources—Pliny, Marco Polo, and also probably Montalvo—and therefore, Cortés's report would have come to him precorroborated; the Amazons would have not only roused his interest but also, ironically, earned his trust.

But La Isla de California still presented a puzzle: If Cortés did not truly expect to find the island of Amazons, gold, and pearls, then why did he pursue the place so madly? I mean, no sooner had he caught sight of the Southern Sea than he was building ships. And when the first two expeditions ended disastrously, he launched a third, which ended in mutiny. As it happened, the mutineers struck Baja, and thus became the first Europeans to set foot on Californian soil, as well as the first to be bludgeoned to death by Californian natives. When Cortés got word, from a few lucky fugitives, of the other mutineers' demise, he immediately declared that he himself would captain a return to the same shore.

But why? Why such dogged pursuit of a hostile desert headland, when he'd already conquered the greatest city of the New World? It was as if he really thought he'd discovered Montalvo's California. But I'd grown skeptical by now of explanations that turned exclusively on delusion. And indeed, as I spent a few more late-autumn evenings on the library steps, reading about those star-crossed western voyages, a different account did, eventually, come to light.

After Cortés had conquered Tenochtitlan, he never quite secured the political authority he felt he deserved. Rather, he was engaged in a ceaseless power struggle with the crown's bureaucracy. And when Cortés finally committed a brazen act of military autonomy, the king officially suspended his governorship: Cortés was no longer to rule the empire he had conquered—or, more accurately, the one he had concocted. Tellingly, it was then, shortly after his official title had been revoked, that Cortés launched the first of his ill-fated westward fleets. And he himself, meanwhile, traveled to Spain, where he secured a new royal contract. What's remarkable about this document is that Cortés seems to have crafted it expressly to protect his future conquests against the same bureaucratic appropriation that had just cost him New Spain. In essence, having lost his first empire, he obtained a royal order safeguarding his second: Now it was just a matter of finding it. If this was, in fact, his reasoning, then what drove Cortés westward was not fantasies of Amazons, but a clear-eyed assessment of his own political reality.

The crew Cortés gathered for his voyage to Baja looks consistent with this interpretation. He assembled a small settlement in ships—a seafaring village that included, according to one observer, *thirty married men, accompanied by their wives*, as well as *a few priests, surgeons, physicians, and an apothecary*. Not the right team for visiting warlike women who consort with sailors; more like the requisite roster for a typical Spanish town. And when Cortés arrived on the coast of Baja, he enacted the same traditional rituals of possession he had performed a decade and a half before, upon his arrival on the coast of New Spain—pacing the beach, tossing sand in the four directions—only this time, he also exhibited the contract he had made with his king, and he commanded all present to acknowledge him as governor.

Some of those who witnessed this display would later write reports of their own. I found an annotated compendium of these sources in the stacks and took it out to my place on the front steps. As I read, a certain detail struck me: the voyagers did not always call the place they'd found La Isla de California. Yes, sometimes they used that exotic name, reminiscent as it was of Saracens, Amazons, and chivalrous adventure. But sometimes, instead, they used a term that was more down to earth and observant. Noticing the abundance of prickly plants, they called the place the island of thistles—La Isla de Cardón.

La Isla de Cardón . . . Looking up from the volume in my lap, I had the strange sensation of stumbling across a distant and mildly admonishing memory. Here I was, the light fading early now, the students ensconced already in their dining halls and dorm rooms, and at last I was closing the cover on the story—only it wasn't the one I'd set out to read, about *Pachycereus pringlei*, but rather the one I'd fallen into, about Hernán Cortés and the limits of perception. And I saw now that my own interpretations of the story, as much as the story itself, showed how expectations can bewitch perception. Eager to apply Elliott's thesis, I had construed every misleading statement Cortés made as a case of authentic misperception. And perhaps I had taken Pagden too far, as well, turning every possible misperception into a canny political move. For it seemed clear now that Cortés, possessed by his political ambitions, actually had misperceived California: Even he—even the man who had marched across Mexico and built ships on the Southern Sea—could never wring life from such a relentless desert. His new township, cornerstone to a second empire, mostly starved. The few remaining survivors, Hernán Cortés among them, were rescued by a search party sent by the conquistador's wife.

7

So precipitously this place changes. From the lookout high above, the bay is resplendent, the islands starkly beautiful, the seascape a wide-open wilderness, and then, descending into town, you are rudely reminded: you are not, in fact, in the wilderness, and the village of Bahía de los Ángeles is anything but picturesque. It feels rather like a construction zone. The earth is barren, half the walls are fallen, and here and there, stray dogs root through small piles of beer cans, Doritos bags, and fish heads, the middens left behind by weekend visitors.

On our way down the final hill, I say nothing to warn my crew. The Excursion, up ahead, is an admonition to hold my tongue, for Veronica would be angry with me for disparaging the town. In truth, though, I think she confuses the charisma and virtue of the local people with the character of their settlement. Undeniably, the inhabitants are resourceful, authentic, straightforwardly honest. But that doesn't change the fact that the town is dilapidated, littered, and dusty. Nor does it change the cruel fanaticism of the climate, which seems to have but three states of being: grossly hot and humid, infernally hot and dry, or, at another time of year, cold and ceaselessly windy. The tolerable hours are moments of transition between different sorts of discomfort. So I certainly wouldn't say the inhabitants are to blame for the place being hardly habitable; I don't even think it's their fault the walls are falling in.

In fact, the main reason I want to speak up about the disrepair is to exculpate the locals. On the near edge of town, one of the first structures we pass is Alejandrina's new restaurant—or rather, the bare foundation and pillars sprouting rusty rebar that were going to support her restaurant, but now lie exposed and dormant. It's not her fault, I want to say. Her mother became ill; she had to pay the medical bills; and then a stranger showed up saying the property belonged to him—a chronic problem here in Mexico, where land title is badly recorded and weakly enforced. I'd like to explain all this, but can't—at least not without crossing that boundary and acknowledging all the ugliness. And besides, I'm not sure Alejandrina, in her quiet dignity, would want the students to hear such things, anyway.

Beside Alejandrina's, the narrow road changes abruptly. It has been widened and repaved since the last time Veronica and I were here, and it now feeds into a new and oversized traffic circle, at the center of which rises a grandiose steel sculpture of a sail catching wind. This is undoubt-edly the latest perplexing manifestation of Escalera Nautica, or whatever it's called these days. But I decided just a few hours ago, at Punta Prieta, to postpone talking about the proposed megadevelopment, and that still feels like the right plan, because if we were to start thinking of high-rises, condos, and golf courses right now, when they're about to confront the dusty town for the first time, megadevelopment might sound rather appealing.

Just past the traffic circle, the road is the same as always—a raised bed of asphalt crumbling at the edge, where there's an unnerving, undercarriage-grating drop-off to dusty yards. On the hardpan, old cars and trucks are parked at haphazard angles. Behind them, the buildings look like bunkers: windowless concrete walls, some painted, some not. In the students' expressionless gazes, I sense their dismay, though I keep try-ing to tell myself what Veronica would say—that it's new to them and they might see it differently. Partly to fill the silence, I start pointing out places they'll need to know about.

"See that dirt track?" I ask, pointing to our left.

"Does it go to the beach?" Chris asks with determined optimism.

"It does," I say. "But you can't take it, because it goes by a little naval base, and they don't like visitors."

Next on our left, just behind a crumbling cinder-block wall, a basket-ball court with netless rims and steel backboards—kids play soccer there, but they'd make room for Haley, I say, if she wants to come shoot.

A yellowish dog, coated in dust, skulks into the road in front of us. Its hip bones protrude from its haunches, and its ribs are like slats in its side. As I steer slowly around it, Rafe says, "Hey, Captain? Was it like this the last time you were here?"

"What do you mean?"

"Kinda shithouse, ay?"

"I think it's really cute," Haley says quickly, as if she were afraid my feelings might be hurt. But in truth Rafe's bluntness comes as a tremen-dous relief, because now I can admit that, yes, I see it, too, and no, I don't find it picturesque.

"It's just impossible to fight these conditions," I say. "But we're not re-ally here to see the town." And somehow, with that out in the open, I find it easier to focus on some of the charming things. On our right, a blue-and-white colonnade, the facade of one of the only pretty buildings in town; it used to be Las Hamacas, Alejandrina's family's restaurant, but is now the centro artistico, where you can buy local handicrafts, like mother-of-pearl jewelry or expressive little animals sculpted from rusted nails. Next on our right, the lime green building where soldiers are always lined up, waiting to use the telephone. One can actually call there from the States, and the proprietress will answer the phone with a ceremonious, "Bahía de los Ángeles," as if you could ask to speak with anyone in town. And I suppose you could. She'd just send her knobby-kneed son running in flip-flops to find whomever you asked for.

As we pass the bright billboard overhanging the driveway for Guillermo's beachfront restaurant and bar, I let it go without mention. And where the road bumps straight into Casa Diaz, turning right then to head up the hill, I tell them about the Diaz grocery, where they can buy cold drinks.

When we turn left, south again, onto the rocky dirty road that meanders around the back of Casa Diaz, even Haley seems to lose her cheerful enthusiasm. The scenes are hostile and desolate: a twisted barbed-wire fence, adorned with dusty and shredded plastic bags; a small cluster of headstones, difficult to notice, since rocks jut everywhere from the hard sandy ground, and the yellowish dust makes everything—even the plastic flowers set there by mourners—the same dull color; and behind the cemetery, a few cardón, atop which vultures perch, their pink, smallish heads poking from the bony black shoulders of their wings.

But then the road turns back down toward the sea and takes us past the small adobe houses of the Ocaña families—Roberto, fisherman; Rubén, auto mechanic—and their homes, as always, are freshly painted in pretty colors. Rubén's front door even has a bougainvillea arching over it. I watch the students faces in the rearview mirror as we pull around Rubén's house and they catch sight of blue water, just behind the field station: it looks so close the waves would seem to be lapping the stone foundation. And no one seems to notice Rubén's little cluster of junked cars, because they're all looking first at the sea, and now at the stone archway, the entrance to the kitchen, where Alejandrina has just appeared, smiling exuberantly. And as she raises both hands in greeting and the students smile back, they look relieved: Maybe it's not such a desolate place, after all.

8

Interestingly, Elliott's beautiful little book and Kuhn's important essay were published only six years apart. It seems there was something of paradigms in the air, a new level of doubt about the essential openness of our minds to experience. The Gestalt psychologists had just begun trying to measure the strength of perceptual biases, and the linguist Benjamin Lee Whorf had recently argued that the structures of grammar shape our sense of space and time. And the way all of this happens in a wave, with many different thinkers suddenly seeing the mind stranded in its preformed world, might suggest that this vision, too, is the product of certain expectations—that it, too, is there when you're looking for it. So perhaps Thomas Kuhn himself was blinkered by a paradigm, but for him, it was the paradigm of paradigms: He found them everywhere, saw them exclusively.

Still, seeing Kuhn's ideas in a broader context does not simply dissolve the paradox he revealed. After all, it remains true that perception calls on concepts—that you've got to have some idea what to look for, or you'll be looking at a *bloomin' buzzin' confusion*. Cortés may have had many concepts to choose among and clear political motives for choosing the ones he did, but still, he did rely on those concepts and, at times, they blinded him to the obvious. Perhaps then we must entertain the possibility that he sometimes saw the Mexica as Moors—particularly when it suited him to do so—and sometimes saw them as something else, even as something rather new. And we could go a step further and suggest that he may have seen them both ways simultaneously—that perception, with its reliance on multiple precedent concepts, is in fact an activity that admits of contradiction.

In those psychological experiments that intrigued Kuhn, where people were asked to identify noncanonical playing cards like a black four of hearts or a red ace of spades, there was often a period in which the experimental subject experienced evident discomfort, strain, or anxiety. Perhaps that was the period in which they were actually seeing more than one thing at a time. And if the example of Hernán Cortés can be generalized, then such a state of unsettled vibration between ways of seeing is not just a rare situation concocted in a psychological experiment, but rather a basic and common condition of human existence—something we all live with, at least a little bit, all the time.

This is a sense of perception that seems consistent with Grafton's view of history—its emphasis on the plurality of concepts available to historical figures—and also with Darwin's scientific sensibility. For the father of evolutionary biology was an intellectual pluralist, through and through. So call the resolution Graftonian, or call it Darwinian, or call it *all of the above*, but in any case, I think the way out of the paradox of paradigms may lie in a suspended state of perceptual tension, an admission of potentially contradictory views. If we can adopt such a pluralistic cast of mind, we might not be fated to the half-light, after all.

PART IV

Leptalpheus mexicanus

LISTENING TO OTHERS

1

One way to make sense of what Cortés did or did not see is to understand seeing as an act performed not by an individual, but rather by a group. What he saw depended critically on what others had prepared him to see, and on what he, in turn, wanted others to see. In this sense, Cortés's view of the New World is the collaborative creation of a network so far-flung and dendritic in form as to comprise the novelist Montalvo, the concubine slave Malinche, and the monarch Charles V. This raises a troubling question of moral agency. The vision of the New World that prevailed in Cortés's letters was one that justified, and perhaps even propelled, horrific violence. But whose vision was it?

There were other Spaniards—Bartolomé de las Casas, Bernardino de Sahagún, Juan de Betanzos—who called upon the variety of cultural resources available to them to fashion visions of the Amerindians that not only contrasted starkly with the one Cortés had offered his king, but also implied that the violent destruction Cortés had perpetrated was criminal and shameful. To suggest that such men perceived the Amerindians clearly and accurately while Cortés saw only a preconceived delusion would be simplistic; the Amerindians' defenders, too, had to rely on the conceptual categories available to them. Instead of seeing the Amerindians as Moors, for instance, they saw them as Innocents before the Fall, or as the closest cousins to another people they thought of as naked and hairless—the ancient Greeks. Those images, too, were misrepresentations of a kind. And yet, Las Casas, Sahagún, and Betanzos did actually take a greater interest in observing the Amerindians. Sahagún and Betanzos even managed to learn Amerindian languages. In his *History of the Incas*, Betanzos wrote that he had been disturbed to understand *how differently the conquistadors speak about things, and how far removed they are from Indian practice. And this I believe to be due to the fact that at that time they were not so much concerned with finding things out as with subjecting and acquiring the land.*

It does seem that those who were *concerned with finding things out*, especially those who managed to extend the boundaries of the collective that influenced their vision, were also those who condemned the depredations of the conquistadors. It may be that the act of careful perception—the hesitation it entails, the social ties it requires, the plurality of paradigms it inevitably cultivates—in itself forestalls violence and fosters restraint. Just bothering to *look* is an act both social and moral.

Having seen this, in its iconic and tragic form, in the Cortés story, I now think of collective vision in many other contexts, as well. It was early 2001 when Veronica sent me an e-mail message with two words— *oh no . . .* —and an attachment entitled *EscaleraNauticaPR*. Whoever had translated the press release from its original Spanish had stumbled into certain pithy truths, creating a kind of inadvertent poem: *President Vicente Fox considers the tourist industry like a passport toward modernity—the Escalera Nautica—crucial and strategic megaproject for the country—to make possible decades of proposals and plans—to detonate the tourist development of the north region.*

The plan for this great *detonation* had been drafted by the federal tourism development fund, which, in another dark poetic twist, goes by the acronym FONATUR. What FONATUR envisioned was a riviera of twenty-two gleaming yacht marinas, ten of which would rise in towns that were presently small fishing communities. There would be seventeen thousand new hotel rooms, ten modernized airports, a dozen large condo developments, and thirty-four golf courses. And finally, to save American yachts the trouble of sailing all the way around the Baja Peninsula to gain access to the Sea of Cortez, a fleet of cargo cranes, a new four-lane highway, and dozens of big rigs would compose a transpeninsular conveyor belt for vessels up to sixty feet in length. The "Land Bridge," as FONATUR had christened it, would run from the fishing village of Santa Rosalíita, on the Pacific, to Bahía de los Ángeles, on the gulf.

That evening, Veronica and I talked. Bahía, we agreed, would be remade in the image of Cancún and Acapulco, the destinations that FONATUR considered its greatest achievements. But such resorts, incandescent stage sets for sunburned Americans, left no room at all for the kind of life that still thrived in the Bay of Angels. What, then, was the best we could hope for?

We tried to tell each other that Baja had always been remarkably resistant to development. Most missions had lasted hardly a generation. The mines not even that long—our own research station, abandoned headquarters of a mining company, could testify to that. And then there were the failings of FONATUR itself. On our way down Route 1, we regularly pass enigmatic parking lots, in which each space is flanked by a short, knee-high wall of concrete. From afar, these places look vaguely like military cemeteries. But when we stopped at one, just out of curiosity, we discovered that each anonymous headstone bore a power outlet and a water tap—and yet we were certain that the nearest electrical or water line was many miles away. We drove off in puzzlement, but would later learn that those sprawling monuments to thirst and powerlessness are in fact the remnants of a grand FONATUR plan: a network of scenically situated trailer parks for droves of rich American RV owners.

When we remembered them, the RV parks seemed heartening parallel to Escalera Nautica: in one plan, the rich Americans were to fill a circuit by land; in the other, they were to come by sea. Perhaps, then, the marine stairway would also amount to nothing more than a mysterious ruin.

But there was something different about Escalera Nautica, and despite our hopes, we could not ignore it: FONATUR's new plan represented fantasy on a far greater scale. The federal government had already pledged 200 million dollars to the project, and was enthusiastically declaring its intention to bring an additional 1.7 billion in private investment. Vicente Fox, the dashing new president, had traveled to Baja to announce the plan.

The next time Veronica and I drove down the peninsula, we encountered fresh blue signs every ten kilometers: ESCALERA NAUTICA 306 . . . ESCALERA NAUTICA 296 . . . and so on. We soon realized that the incrementally shrinking number was the distance to our own turn-off—formerly the narrow road to Bahía de los Ángeles, but now, we presumed, the vaunted Land Bridge. As we drove on, the signs made for a cruel sense of inevitability—the countdown to detonation.

But when the countdown hit zero, at Punta Prieta, there was nothing but the sound of wind, the familiar shack, the fragments of tin siding scattered on the sand. Then we noticed one rather surreal addition: a blue sign indicating we had finally reached our destination, Escalera Nautica. Along the road to Bahía, some potholes had been repaired, but there was no four-lane highway, no procession of big rigs towing white yachts.

Mystified, we rolled into Bahía. Never had I felt so pleased to see those dusty yards, the crumbling walls, the familiar concrete buildings. The townspeople we talked with seemed to be waiting, squinting toward the future in the same way I'd seen them look at the horizon to predict the winds. A real estate company had recently sent agents to visit property owners in town, and it was rumored that a certain magnate, someone with connections to FONATUR, had a stake in that very company. This led to the conjecture that Escalera Nautica was nothing but an elaborate scheme to drum up land speculation. And when all the big blue signs had gone up along the highway before any construction began in town, the rumors had gained momentum.

But the story didn't quite hold together. The land around Bahía was owned by los ejiditarios: In the 1930s, the federal government distributed much of the nation's land among newly formed *ejidos*—essentially, co-operatives of local peasants, los ejiditarios. The ejidos remained democratic communes until the early nineties, when the Mexican government, moved by the spirit of NAFTA, allowed ejiditarios to divide their common land into privately owned plots. By the time Escalera Nautica was announced, the ejido around Bahía had fragmented into several hundred parcels, and each ejiditario had been allotted a small plot in town and a larger one out in the desert. If some plutocrat's plan had really been to buy cheap, excite speculation, and quickly sell, then surely he would have bought before conniving the announcement of a two-billion-dollar development plan. But when those real estate agents first started snooping around Bahía, everyone was already anticipating high-rises and yachts, and yet the land was still in the hands of ejiditarios.

The land-speculation theory was also cast into doubt by reports of real work elsewhere on the peninsula. In Santa Rosalíita, the fishing village FONATUR had designated the Pacific terminus of the Land Bridge, a crew had packed a load of dynamite into the side of a small mountain just behind town. Whether the explosion had been much larger than anyone intended was a matter of local debate, but in any case, half the mountain was now rubble. Huge cranes arrived, and they were presently building a rocky peninsula off Santa Rosalíita's main beach.

But still Bahía was mysteriously quiet. Real estate signs sprouted along the desert coast, advertising the chance to get in early on big things to come. Alejandrina described a businessman, looking "muy importante" in city clothes, sitting quietly at the most recent meeting of ejiditarios. Antonio, the turtle man, had been inquiring with officials at FONATUR and other agencies, but he was given only vague replies about tremendous benefits arriving soon.

Back home in the States, news came from a small but vocal network of Baja bloggers: a Mexican poet; a few journalists from Ensenada or Southern California; a ragtag crew of gruff American expats, who seemed to have spent the better part of their lives wandering the peninsula and the gulf. With the exception of the journalists, these authors made little effort to contain their bile. But their arguments, for all their rhetorical roughness, seemed well-considered. To start, there was no freshwater for hotels and golf courses. One blogger also pointed out that occupancy rates in Acapulco had declined for several years running, so the prospects of filling an additional seventeen thousand hotel rooms could not be good. Another compared the capacity of Escalera Nautica's marinas with the number of large yachts registered on the West Coast: if every last one sailed for the Sea of Cortez on the same day, the marinas would still have room to spare.

Clearly, Escalera Nautica was ill-conceived. And yet FONATUR lumbered on, like a Gargantua afflicted with deafness, or obsession, or both. Every time the agency's director gave a new interview or its websites announced the involvement of another resort corporation, the luminous vision of the peninsula and gulf remained astonishingly undisturbed: the place was a yachter's paradise, a chain of sandy beaches missing only high-rise hotels, a rocky scaffolding for grassy golf courses. Like a familiar film, the images rolled again and again.

Prompted, perhaps, by the profane voices hollering market analyses from far-flung bays and rancheros, several small environmental organizations took up the cause. Defensa Ambiental del Noroeste, a group of lawyers in Ensenada, filed suits against FONATUR for failing to conduct the requisite environmental impact studies. Algalita, an independent marine research organization, took a coastal engineer to Santa Rosalíita to inspect the recent work. He reported that the main jetty had been positioned improperly in relation to local patterns of sand circulation;

consequently, the town would lose its beach, and the marina would require constant dredging to remain in operation. Wildcoast and Proestero used this gaffe in construction to catch the attention of several Mexican and American newspapers. And in their canniest move, the environmental organizations persuaded the Packard Foundation, a giant of American philanthropy, to commission an independent assessment of FONATUR's plan.

The report, published a few months later, reached the same conclusions—and in fact, cited many of the same numbers—as the motley bloggers. This time, though, the warnings were issued not by unshaven sailors, but by economists, civil engineers, and the like. Almost immediately, the story was taken up by publications of greater reach and renown. Even *The Wall Street Journal* covered it.

Remarkably, FONATUR plodded blindly on, reciting its beloved tale. But maybe something had already changed. Maybe the bloggers, or the resourceful environmental organizations, or the journalists who had picked up their story—or all of them, together—had finally gotten through to someone at the agency. Because several months later, and without an obvious cause, FONATUR announced that Escalera Nautica would henceforth be known as "El Proyecto Mar de Cortés," reflecting *an important change in direction*. In the new plan, the grandest reverie of Escalera Nautica had given way to more modest goals. Visions of glassy hotels and golf courses were replaced by proposals for basic infrastructure: power, sewers, seawalls.

Many people in town say the original plan will proceed. That the renaming was merely a ruse to quiet the growing resistance. A paradoxical piece of evidence in favor of this theory is that the gleaming azure road signs were instantly removed—a show of effort and efficiency that is suspiciously atypical of Mexican governance. And then they left that last placard, at Punta Prieta, to continue making its shiny promise. Lately now there is also the new sculpture of a billowing sail, the oversized traffic circle, and some more repaired potholes—though I'm not sure whether those portend megadevelopment, or infrastructure improvement, or something else altogether. One Mexican lawyer who fought valiantly against Escalera Nautica said to me that, under the new plan, Bahía will die by many small wounds rather than one sudden drop of the guillotine. The new infrastructure, he predicted, will facilitate a rush of small-time development, turning the town not into Cancún, but rather into the sprawl of Southern California—a change that could, in the end, leave even less room for the local ecology.

He is not truly so fatalistic; if he were, he would not already be hard at work on ways to prevent that death by countless little cuts. But his warning reflects the same mood of somber uncertainty that makes me hesitant now to offer an unambiguous end to the story of Escalera Nautica. It is fitting, in a way, that just one of those thirty blue placards remains, as if to remind us that we can't be certain what road we're actually on.

2

Waking on the bay is like bursting from the silence underwater into the raucous air above: the nasal laughter of Heermann's gulls, happily deriding one another; the high, quick calls of yellow-footed gulls—the proper accompaniment to boats and salty air; a pair of guttural and prehistoric croaks from a great blue heron, annoyed by the squabbling of gulls at his ankles. And also, beneath all the voices, a strain of silence: Guillermo must have doused his roaring generator in the small hours, when the heat and humidity finally broke.

How I cursed that machine last night. Five hundred feet from where we sleep—or don't—the engine boomed and rattled. Guillermo wants to please the guests in his bar and motel—a few intrepid sport fishermen, probably—and they prefer to drink their margaritas and sleep them off in the dry chill of air-conditioning. Which I understand, because we, meanwhile, lay poaching in a hot fog. Every surface, living or inanimate, was slick with the bay's perspiration. And I knew that if it weren't for the noise down the beach, I would have heard, from that same direction, the sighs of our students as they searched for sleep in the murk. But what really enraged me was the thought of what *they* would have heard. When the air is that still, you can often hear whales breathing on the bay. You might at first think it's a wave sliding up, then down, the rocky shore. But the water, you remember, is still, and then the sound comes again: a deep, hollow exhalation; a pause; a sonorous inhalation. And the sound instantly redeems the wet, still air, because drifting off to sleep with the breath of whales is, in some sense, as close as you'll ever come to them.

But instead we listened to an engine, and I'm certain that even the well-rested sport fishermen are poorer for the bargain.

Sitting up now on my cot, I face the placid waters of the bay. In about five minutes, the sun will flare above the rugged horizon line of distant, reliable Ángel de la Guarda. The sky there is already a furious shade of orange, and the sea before it is vermilion through the middle, cobalt to either side. The islands closer to us—Los Gemelos, Cabeza de Caballo—are backlit silhouettes, resting stolidly on the glassy water. Closer still, the boundary between the sea and our cots is but a narrow strip of intertidal rocks, dark with algae, giving way to a seafront slope of sandy ground that is sparsely populated by prickly saltbush and barrel cactus. The heron croaks twice more now in protest, spreads his huge wings, and lifts himself from the rocks; the way his neck folds as he rises, it's as if his head were reeling up his body and spindly legs behind. I feel compensated already for the petty discomforts of the night—and mildly reproved, too, for my impudent little fit.

Three minutes now, I'd guess, and the flames will crest La Guarda. The bay's middle third, presently an oblong pool of lava, will gather itself suddenly into a narrow, intense line, which will streak across the surface and deliver the first astonishing blast of heat. It will wake Veronica, who sleeps beside me, and Graham, who has set his cot about fifteen feet away from us, courteously leaving the couple their modicum of privacy. I watched him last night as he stretched nylon cord from the posts of Melissa's porch, behind his cot, to the branches of a tamarisk in front of it. He was, I suspected, preparing to hang a mosquito net, and I tried to reason with him:

"Burnett, there are no mosquitoes here, because there's no freshwater."

"But there are bugs, Hirsh. Bugs of other varieties."

"What varieties?"

"Scorpions. Black widows. Scorpions. That sort of thing."

"You're neurotic, Burnett."

"Do you want to borrow some DEET? Stuff really works."

"Because there are no bugs."

"Especially if you're covered in DEET."

The hem of his mosquito net, I see now, is meticulously tucked under the edge of the Therm-a-Rest he has placed atop his cot. He has found a way to cocoon himself. Veronica, on the other hand, is sprawled—two arms overhead, one leg hanging off the side of her cot. Her sleep is head-long, like that of a child, and suddenly I wish I could push back those flames—just by a minute or two. For when she wakes, we will have to move hurriedly, getting ourselves ready before the students. One thing we've learned is that our own levels of energy and rigor determine theirs: if we are five minutes late, they will be ten, and before long, the easygoing disposition will ripple out into other parts of the course—their reading will grow cursory; discussions will become a bit lazy; and when exhaustion sets in, as it surely will at some point, they will suc-cumb to it. Graham once said to me, "It's extraordinary what they can do if they don't know they have a choice." A bit severe, perhaps, but also true.

It will be with some haste, therefore, that we gather our gear—dive bags, marine radios, a hydrophone for listening underwater. In the station kitchen, Alejandrina will commiserate with me about last night's heat and humidity, smiling at the very extremity of it as if it were God's own joke; as we talk, she'll pour coffee into my mug, place tortillas and eggs on my plate. On the station terrace, where we eat, Graham and I will position our plastic chairs precisely behind the trunks of the tamarisks, so that we may center our faces in their narrow stripes of shade. But Veronica will choose instead an unobstructed view of the bay, as she is somehow able to consume an elaborate breakfast while peering constantly through binoculars. These early morning hours are the best for spotting activity on the water: spouts linger in the still air, and whatever might disturb the glassy surface—the caudal fin of a whale shark, the snout of a sea lion, the dorsal fins of dolphin—will leave a diagnostic pattern of ripples, glint-ing in the sun.

Just as the students begin pulling up chairs and sitting down to eat, Veronica and I will depart. We will pick our way down the rocky slope from the terrace to the beach, which arcs gently northward from our station to the stone jetty and boat ramp owned by the Diaz family. As we walk swiftly along the firm strip of sand close to the water, the gulls will waddle before us. They are oddly committed, those birds, to plodding along in their rub-bery yellow galoshes instead of unfurling elegant charcoal-gray wings and leaving us behind. And so they will keep a begrudging five feet between them and us until Veronica and I must finally turn left and cross the soft dry sand higher up the beach.

We will step over a low, crumbling seawall, and then pick our route through the row of cinder-block cabañas rented out by the Diaz family. This part of the walk often seems an intrusion into unseemly privacies: the plastic tables are heaped with beer cans, bottles of yellowish margarita mix, plastic cups, and half-consumed shrink-wrap packages of deli meat. Metal barrels overflow with the same items, already discarded. And half-inflated pool toys lie about like drunks passed out after the party. As we duck between a pair of cabañas, it is difficult not to glimpse, through a small rectangular window, bodies sprawled on the bare mattresses inside.

Behind the cabañas, we cross a hardpan lot and step over a steel cable to enter the Diaz boatyard. There, five or six pangas—twenty-foot fiberglass skiffs with outboard motors—rest on their ramshackle trailers. And five or six men, representing perhaps three generations of the Diaz family, sit in a semicircle of collapsible chairs, surveying the segments of bay visible between the cabañas. At sixty-five, Samy Diaz is still handsome—a dark, leathery, and rather short James Dean. As we approach, he will acknowledge Veronica by name; relative to her, I remain the newcomer. He will then give a slight nod to one of the younger men, perhaps his son Octavio, who is more friendly with us than his brothers are. With sudden alacrity, Octavio will climb into a rusty '71 Ford pickup. The truck will wheeze and cough as it is coaxed into roaring life; then it will peel out angrily on the sandy soil, careen around the pangas, skid to a stop, and reverse rapidly until the rear hitch ball is positioned precisely beneath one panga's trailer hitch. Throughout this procedure, I will nervously endeavor not to forget one of the necessary steps—remove rocks from beneath the trailer wheels; push rubber plug into back of panga so it won't fill with water when deposited in the sea; open trailer hitch and, as hitch ball appears suddenly beneath it, lower trailer quickly; close hitch. If I do forget a step, Octavio will climb from the truck, remedy my oversight, and return to the wheel—all without making eye contact with me.

For Veronica, the boatyard spectators, including even Samy himself, seem to hold a measure of bemused respect: She is the blond gringa who does men's work—runs a panga, fiddles with its outboard—and yet also converses with them, the local men, in a way that is polite, deferential, and suitably feminine. But I'm afraid this makes me the gringo whose gringa does men's work—in general, more competently than he does it himself. And besides, I'm never quite sure how I, for my part, ought to converse with them, since it seems that my own politeness and deference must likewise be perceived as, well, suitably feminine.

Seconds now: the ridge of Ángel de la Guarda wavers, serrated with light, just before the fire—there—scorches the hem of the sky. You can almost hear it, the sound of a torch, as it drives the blue out of everything. The line of light parts the bay, strikes us here on the shore.

"Vica," I say, now that she's stirring, "do you think the Diaz think I'm kind of—I don't know—unmanly?"

She squints at me, not responding.

From the nearby cocoon of netting, gauzy now in the sun, come the words "Of course they do, Hirsh."

3

As we step over the cable and enter the boatyard, it is clear that something is different this year. There was no sign of it yesterday afternoon, when Octavio brought his truck to get our pangas from the field station garage. And the usual spectators are here, in their folding chairs. But now it is Samy who stands—Samy Diaz himself. He smiles and says, "I was wondering when you'd get here. You're a little late." Then he starts walking toward the pickup truck. With some alarm, I glance at Veronica, but she's watching Samy. Then I glance at the spectators, and they're watching Samy, too. Is this as stunning to all of them as it is to me? Is Veronica at least a little worried about it? Making Octavio climb out of the truck to fix my oversight, that's one thing, but Samy? I just need to focus on my tasks: don't forget the plug; don't forget to close the hitch; don't forget the safety catch on the engine tilt.

All goes smoothly, but at the top of the boat ramp, something unexpected happens. When Octavio was at the wheel, I would climb, at just this moment, from the truck bed onto the frame of the trailer; he would then back slowly down the ramp until the trailer, and my feet, were in the water; at that point, I would unhitch the bow line and push the panga backward, vaulting from the trailer into the bow just as she floated free. But now, Samy leans out his window, looks at me in the truck bed, and tells me to unhitch the bow line. At least I think that's what he says, but the truck's engine is extremely loud.

"Undo that?" I yell, pointing to the bow line. "Now?" Since I've never touched it before the trailer was in the water, I don't have an idea what releasing it now will do. For all I know, the boat could slide backward off the trailer and crash onto the boat ramp.

"Sí," he yells, "undo it."

I undo it. The boat stays put.

"Get in the panga," Samy yells.

"In the panga?" I yell. "But I need to—"

"Sí. Get in the panga."

As gingerly as I am able, I climb over the rails, into the boat. It remains on the trailer.

Samy smiles, nods, and then reverses the truck so suddenly I'm thrown to my knees in the bow. Just as I'm standing up again, he slams on the brakes and the truck and trailer halt suddenly, but the panga and I slide smoothly backward, off the trailer and into the sea. Before I even understand what's happened—Samy has launched the panga, with me in it and my feet still dry—the truck and trailer are heading back up the ramp. Samy holds a hand out the window: thumbs up.

The *Cortez Angel*'s outboard ignites without complaint; she is an old but reliable boat. I motor a short distance through the glassy water, then cut the engine and linger, because I want to watch Samy launch Veronica in the *Sea Eagle*. Sitting on the rail, with my back to the rising sun and my toes in the water, I can see a driving rain of silvery minnows as they pass through the shadow of my boat. A brown pelican swoops in and lands like a full-bellied seaplane beside me. With white head feathers, a scaly patch around his eyes, and long, heavy bill folded down against his neck, he looks like a stern old man who must tilt his face downward to peer over his reading glasses. He stares at me this way, expecting a baitfish.

"Not fishing," I explain.

He stares, unconvinced. "Watch the boat ramp," I add, pointing to shore. "It'll be good."

In a few minutes, the pickup and the *Sea Eagle* appear. At the top of the ramp, Samy leans out his window.

"He's telling her to undo the bow line," I tell my pelican. But Veronica appears to refuse. She walks up to the truck's window to talk with Samy. After a moment, she steps back as he opens his door.

"Good for her," I say. "Nothing risky with the *Sea Eagle*."

My pelican paddles closer, sensing that we've become friends.

Samy walks back to the trailer, releases the line himself. Then he points at the panga. But again, Veronica refuses. Samy tries to reason with her. He's explaining, gesturing with his hands.

"Go tell her," I say. "Tell her it's Señor Samy—he knows what he's doing."

Finally, Veronica agrees. She climbs cautiously into the boat. It appears she's understood the procedure, because she's standing at the boat's center console, grasping the steering wheel for stability, when Samy reverses, slams on the brakes, and sends her and the *Eagle* sliding into the sea. As Veronica motors out to where I float, she smiles and shakes her head.

Under the pretext of testing out the pangas before we pick up the students, we head for the sun, making gentle, banking turns that trace sine waves of sea foam across the glassy surface. Then we arc round and head back toward shore, where the students, strung out in a line led by Graham, are just now traversing the brief but steep pitch of sandy rocks between the terrace and the beach. It's a hazardous morning commute, that rocky slope, and as I draw closer to shore, cutting the engine and gliding silently in, I find myself watching Cameron. He's wearing his blue fiberglass leg. In one hand, he holds two dive bags—his own and, I presume, Isabel's, for his other hand rests on her shoulder, directly ahead of him. Each step with the prosthetic appears to include a split second of investigation—an instantaneous test to determine whether the foot will hold. And the other foot appears unusually busy, performing quick lateral explorations to discover what ground lies ahead, which position will be most promising. Every step, I think, must require courage.

Yesterday, when we arrived at the station, Veronica led the students on a brief orientation while Graham and I unloaded the truck: our own bags of gear into the staff house; crates of food into the kitchen; boxes of our books into the room we call the library, though the term is vainglorious, since the small resident collection consists mainly of miscellaneous titles left behind by visitors. I happened to be placing our field guides and textbooks on the empty shelves when the tour entered the room. Passing behind me, Veronica led the train of students through the seminar room and into the dim passageway known as the museum room—another flourish of vainglory, since the little hall is really no more than a cabinet of curiosities, some of which still wear bits of rotting tissue. Amid Rafe's theatrical groans about the stench, I heard Veronica say, "Watch your head," and then Allie, saying, "Duck right here, Cameron." And I realized then that the greatest challenges Cameron will confront in the next few weeks will not arise, as I'd always assumed, in our group's more adventurous activities—diving, handling sea turtles, climbing to the top of Isla Ventana—but rather in our daily life here at the station. For the old headquarters of Las Flores Mining Company is a maze of rooms, some of which have slightly different floor elevations than those adjacent to them. In several passageways, a timber lintel hangs at the altitude of foreheads. More hazardous still, in the sandy yard between the main building and its outhouses, los baños, Rubén Ocaña fixes automobiles—and also stores the rusting remnants of those he was unable to fix. And to flush those baños, one must fill a bucket with seawater from a large cistern that stands behind them, and then pour the water, all at once, into the toilet. Of course, we discussed these daily obstacles with Cameron while we were still on campus, and he assured us they would not present a problem. Not until yesterday, however, when I heard Allie's gentle voice— "Duck right here, Cameron"—did I grasp just how astonishing that assurance of his had been.

Over the course of the day, as I went about my chores around the station, I happened to see Cameron several more times. I was sitting at one of the two large tables in the seminar room, with various parts of the disassembled projector spread out before me, when Allie and Cameron entered. They stood in front of the whiteboard, where Veronica had drawn a simple map of the station, and Allie moved Cameron's hand from one room to the next. As they made their way through the floor plan, Allie dutifully recited not only the labels Veronica had written— lecture room, kitchen, library—but also the annotations that Graham must have added later, after Veronica had completed her map and left the room. When Allie said, "Kitchen, Terra Alejandrinae," she looked my way inquisitively.

"I think he means it's Alejandrina's turf."

"So the staff house is his turf?" Allie asked. Behind her, above Veronica's STAFF HOUSE, I could see *Terra Burnetti*.

"He likes to think so," I said. "He reads his tomes up there."

VERONICA'S MAP OF THE VERMILION SEA FIELD STATION

(annotated by Burnett)

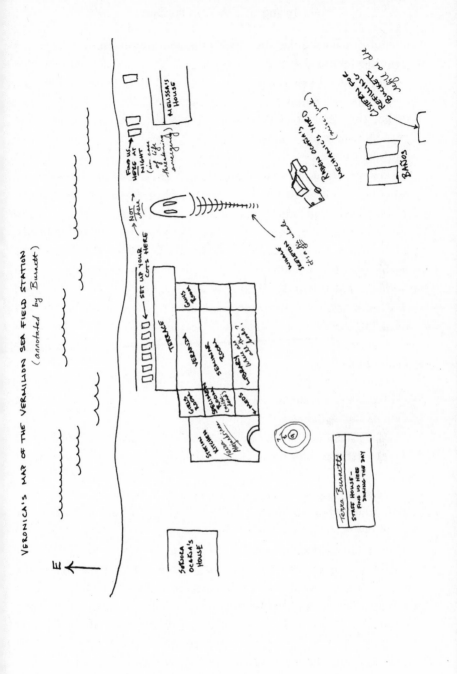

"Who's Melissa?" Allie asked, referring to the label on the house next door to the field station.

"Don't know," I confessed. "I think she was a student here, like you guys, and she ended up leasing that place from the Ocañas."

"That's excellent," Cameron said. "I'd love to have a place here."

"Kinda hot," I said, "don't you think?"

"I think it's prime."

In the afternoon, I was standing on top of the cistern, struggling to open a certain salt-encrusted valve. The heat was alarming, and I felt like a lonely soul at the center of the circles of hell Graham had added to Veronica's map; no one without baños to repair would choose to be out in such sun. I was surprised, then, when I saw Cameron appear at the station's back door, holding his white cane. I watched as he positioned himself carefully in the door frame, touching both sides. He then turned slightly to his left, and paced, very deliberately, up toward the baños. His cane caught a rusting axle, which he easily navigated around, but the maneuver slightly altered his trajectory, and he ended up missing the baños by about ten feet.

"A little to your right," I said from my perch.

"What are you doing up there?"

"I'm on top of the cistern."

"I know. But what are you doing on top of a cistern?"

"Trying to figure out how to open the valve, to pump in seawater."

"Do you need any help?"

"I think I've got it now. Thanks, though."

He turned around and paced his way back to the door. Then he positioned himself in the door frame again, and repeated the trip.

"Twenty-two steps," he said, smiling at me. "Did you get it open?"

In the evening, when the sun had hidden itself mercifully behind Mike's Mountain and the light had eased into shades of blue and gray, I was sitting on the staff house porch, working on a lecture. Again I saw Cameron appear at the station's back door, and I figured he was heading for the baños. But instead he made a sharp turn to his left, walked to the far corner of the station, turned left again, and disappeared behind the south end of the building. Puzzled, I stood and walked a short distance out into Rubén's car yard, where I would be able to see between the station and Melissa's house, all the way down to the shore. What stands in that passageway is the one specimen at the station that truly does merit the word *museum*: It is the complete skeleton of an adult fin whale, fully articulated on a metal scaffolding and forever gazing out at the bay it came from. Its bones are beautifully white, its skull like the overturned hull of a panga, its rib cage so spacious I can stand upright inside it, stretch my arms out to the side, and touch nothing. I have stood there before, in that posture, at sunrise, for no better reason than to enact the combination of Vitruvian Man and biblical Jonah.

But now, in the shade of early evening, it was Cameron who was with the whale. He stood beside the great skeleton, and with his outstretched hands he grasped a pair of ribs as if they were the bars on a window he wanted to look through. Then his hands moved slowly up, feeling the bones arcing away from him, toward the whale's spine. After a moment, he moved each hand one rib toward the animal's head. Five more ribs, and he was at the skull, which he ran his palms across until he reached the eye socket, where his fingers dipped in and grabbed hold. That was when it occurred to me that, when we spot whales on the bay, Cameron will know what we are seeing. Like the moment in the lecture room, like his pacing of Rubén Ocaña's yard, this too was an exercise of orientation, a way of finding his bearings in his new place.

4

Our first dive must be an easy one—good visibility, no current or waves, no tempting depths—so we have decided to anchor our pangas in the cove between the islands of Cerraja and Llave, Lock and Key. A rough-edged quarter-moon of pellucid aquamarine occupies the narrow space where the rusty and jagged key has been tectonically pulled from its equally corroded lock. And though the islands' rocky verge passes at a relatively steep pitch into the clear water, it descends only ten to fifteen feet before encountering a flat bottom of sand. The crystalline whiteness of this sand is what gives the water here such brilliance and clarity. And as we motored, just a moment ago, into the quarter-moon from one of its narrow ends, the students gasped and celebrated our discovery of such a splendid and secluded cove.

Rafe said, "What this place needs is a tiki bar," and everyone laughed—including, to my own mild surprise, me. I'm softening toward him, and his little jokes don't annoy me quite the way they did at first. To be sure, he remains something of a caricature, a send-up of plumped Australian manhood. But his remarks, though sometimes boorish in accord with his image, are also sometimes unexpectedly observant. And I've been surprised, too, by how gracefully he has accepted his small defeats. In our discussions in the car, he was often wrong where Chris was right, and yet, just when I might have expected a flash of petulance, Rafe moved on rather effortlessly. And in his swift loss of Haley's attentions to Ace, he showed only a moment or two of impatience with his rival's unrelenting irony. Of course, such graciousness could stem largely from a self-love so robust as to be impervious to setbacks. But whatever its cause, I'd still call it gracious.

Besides Rafe, the other students who climbed into my panga from the shallows of the station beach are Cameron, Isabel, Ace, Haley, and Miles. In summers past, our students have perpetuated, for a few days at least, the arbitrary teams created when they picked which SUV they would ride in down the peninsula. But this group is a bit different. Cameron is here, in my boat instead of Veronica's, simply because I am to be his guide in the water today. (Veronica will have the job tomorrow, and then, for each day thereafter, a student will be designated to escort him through our dive.) With Cameron came Isabel, since she, along with Allie, has become his most reliable guide. And Miles jumped into my panga presumably to join Ace and Haley, for the three of them seem endlessly entertained by one another. I would say they are like old friends, but that's not quite right; they are, rather, like old friends who have magically forgotten all the charms, jokes, and stories that made them friends in the first place, and therefore have the opportunity to relive all those joys on a greatly compressed schedule.

Becca, surprisingly, has remained with Veronica. At the beach, she seemed to be hovering inexplicably at the water's edge until everyone had settled onto the panga's bench seats, which are directly in front of and behind the driver's center console. Then, as Becca finally came to the *Sea Eagle*'s rail, she asked Veronica if she could sit up front, in the bow. Veronica should have simply said *No, weight there makes the panga ride low and heavy*. But for some reason, she was unable in that moment to produce the simple and reasonable answer; I think maybe she felt such a surge of anger at what she perceived as Becca's selfish and tactical ways that she did not quite trust herself—did not realize that her own desire to refuse came from reasonable considerations, and not personal animosity.

In any case, Veronica managed only a wordless shrug of feigned indifference. Then she started the engine. And Becca assumed her desired position in the bow.

She is still there now, perched on the rail beside the bow cleat, as if she were a coxswain crying out to her crew. And because the air is calm and our pangas are anchored only a boat length apart, I and everyone else in my boat can hear her every word, as if she were our coxswain, too. Strangely, she is once again tarrying a few seconds behind the others: as they all wriggle into their wetsuits and pull on their fins, she carries on with her story—something about a certain South Pacific atoll she visited with her family, a place so remote and idyllic she is having trouble finding enough tropical adjectives.

"Seriously," she says now, "it puts this place to shame."

No one responds. No one, it seems, feels like taking her up on the challenge to name another place more beautiful than this.

"Anyway," she says, changing tack, "Veronica, what species are we supposed to look for here?"

Veronica is seated on her panga's side rail, her fins dangling in the water, and for a second, as she pulls her mask from her forehead down over her eyes, I have a terrible feeling she's about to ignore Becca's question and slip herself into the safe privacy of the water. With her hands gripping the rail, her mask peering down at the water's surface, she appears caught in a moment of indecision—or maybe it's a small struggle for self-control—until at last she reaches again for her mask and lifts it back up to her forehead. She answers Becca's question, but does so without ever looking in Becca's direction. Instead she speaks as if she has just remembered something she'd been wanting to tell everyone: try to recognize some of the fish whose pictures appear in the course-reader; look for sergeant major, leopard grouper, Cortez damselfish; and there's a fairly rare species we sometimes get to see here—the orange-throated pike blenny, which will resemble a small eel with its tail wedged into the rocks. Pulling her mask back down, she adds that anyone who wants to should join her. And with that, she slips from the rail, and the water closes quietly over her head.

Without returning to the surface, she folds over and descends, dolphin-kicking through the bright, clear medium. At the white bottom, she pauses to wedge her anchor deeper into the sand, then moves off horizontally, a few feet above the seafloor. As her shadow slips across the ripples of sand, its leading edge triggers mute explosions of silt—juvenile sting-rays shooting off like round projectiles. A number of students splash into the water behind Veronica, kicking up a froth at the surface as they pursue her. They look like paddleboats madly chasing a submarine, and Rafe is last in the whitewater flotilla. I don't see Miles there, but he too is missing from my panga; he must have slipped in quietly while I was still watching Veronica. Hearing all the commotion, Cameron turns in my direction and says, "You ready yet?" He's all suited up and seated on the rail, and a few things I've wondered about are now answered: Would he wear a fin on his prosthetic? (He does not.) Would he wear a dive mask? (He does; presumably he must protect his eyes, since they are always open.)

"Two seconds," I say. "Sorry to be slow."

"No problem," he says. "I'm just totally psyched for this."

"If anyone wants to join us," I say, struggling into my wetsuit, "Cameron and I are phylum hunting."

"Awesome," says Ace. "Let's bag some phylums."

"You mean, like, spearfishing?" Haley asks. "You'd better get Rafe, 'cause he's been talking a lot about spearguns."

"No spearguns," I say, and I explain that I've been thinking about the best way to use my day with Cameron—my one chance to be his under-water guide—and what I've decided, in the end, is that we'll try to find as many different phyla of animals as we possibly can.

"That sounds good," says Cameron. "I don't think you should be giving me a speargun."

"Less dangerous than Rafe," says Haley.

As I sit myself on the rail beside Cameron and start pulling on my booties and fins, I remind this new team of mine—Cameron, Ace, Haley, and Isabel—what phyla are, and why you'd care to find them. Simply put, they are the deepest evolutionary divisions of a kingdom. But if you want to get a feeling for what that really means, I think it helps to call up that prototypical evolutionary image, the Tree of Life. So just envision, for a moment, all the earth's extant animal species—there are probably about ten million of them—arrayed as the terminal twigs of a great, spreading canopy. And now imagine that you pick a single species—let's say it happens to be the leopard grouper—and from that small twig, you start climbing downward, following the lineage back in time, offspring to parent, twig to branch, branch to bough, down you go. You pass the species that spawned the entire family of bass; you pass the ancient *Agnatha*, which one could properly call the very first fish. And still you continue downward, until at last you find yourself on a bough so deep within the tree that, when you pause to look around, you can count but a few dozen other boughs at your own altitude. Above you, of course, is that billowing profusion of thousands of families and millions of species; below you, the heavy limbs converge on a trunk—the earliest multicellular animal. But at your own level, or roughly so, are those few dozen heavy boughs, each of which represents a single, profoundly promising species, the ancestor of an entire phylum.

Arriving from your place in the canopy of the tree, you have the benefit of truly distant foresight. And you can therefore see, for each of these ordained ancestors-to-be, which properties—which anatomical or physiological attributes—shall remain unchanged forever, and shall thus become the definitive properties of a phylum. You would recognize, for instance, the fundamental architecture of each organism's body: How many axes of symmetry are there? Are the appendages jointed? Is the torso divided into segments? Evidently, these are questions that evolution is allowed to answer only once in the history of a lineage; the body of Echinodermata will remain fivefold forever after, even if subsequent changes cloak the underlying pattern in a superficial disguise. Similarly, each of the ancestors has—or will soon acquire, as it evolves along that single heavy bough you're looking at—a set of key innovations: a special device for scraping food; a form of collagen that can quickly alter its rigidity; a certain kind of suction cup for attaching to the seafloor. And these attributes, too, will remain in place through the eons of change to come—though their persistence may not have the same causes as that of the basic body plans.

Everyone is suited up now, and black neoprene bodies become awfully uncomfortable when they're stranded on a white panga beneath a bright sun. Cameron is palpably longing to plunge: he has inched his rear almost off the edge of the panga's rail and is poised there with all the taut readiness of a swimmer on the blocks, awaiting only the pop of the starter's gun. But I can't fire it—not just yet. Because now that I've invited everyone into this little thought experiment, I'm noticing that the Tree of Life, this famous metaphor, is not quite as straightforward as I'd anticipated;

and I fear that if we ascend once more to the present without taking just a moment more to think through our metaphor, it could actually deceive us. From our vantage deep in the tree, we view but a few dozen boughs, and yet, at this same point in the history of life, there were in fact far more than a few dozen species roaming the oceans. So where are they? Why do we not see them in our tree? The canopy, I said, contains all ten million extant species, and the tree below consists of all the branches that led to such a canopy. What this definition excludes, then, is all the species whose descendants do not reach the present. We have, in other words, conceptually pruned at the stem every lineage doomed to extinction. Had we not done so, we would have found ourselves in a thick bramble of twigs at every altitude along our journey through the tree.

And what of the very lineage we followed? We imagined it not as a series of twigs, but rather as a branch that thickened as we descended. But why? I suppose because that is how a tree looks, but here our metaphor misleads us. If a twig represents a species, then in fact our tree must consist entirely of twigs, because every stage in every lineage is but a single species; it is never something broader than that. The genus, the family, the phylum—these are not real creatures that participate in descent; they are groupings of species that share a common ancestor. Of course, if we are so inclined, we can posit certain meanings for the thickness of a branch—it corresponds to the number of future species an ancestor shall spawn; it is a measure of evolutionary promise—but such metrics are nothing more than efficient use of a potent metaphor.

What we're about to hunt down now, I hope, are points of contact between that metaphor and our immediate experience. When we take in our hands the members of two different phyla, we can actually find the deep, phylum-defining differences between them, and we know that these are differences that must have arisen in that era of a few dozen boughs. And we know, too, from many different methods of estimating the passage of time, that those auspicious ancestors first arose about 540 million years ago—which means, of course, that between the lineages we're holding, we're looking at more than a billion years of descent.

"A billion years!" cries Ace. "In the palms of our hands!" He has come to his feet in the bow of the panga, and with his arms raised up and his hands wide open, he looks like some kind of televangelist in a moment of exaltation.

"You got it," I say. "The origin of phyla, right there in front of us."

"Excellent," says Cameron, "let's do it. Cause I am *so* hot in this wetsuit."

"Hang on—just one more thing to think about underwater."

Cameron emits a groan; it sounds involuntary, like a small child's expression of agonizing anticipation.

"Ten seconds, and you're in. In the tree we imagine, all the big, supportive boughs are low down, near the trunk. But is there really such a place in the tree of life? I mean, why do all the phyla—all the deeply different and promising ways of making an animal—have to originate in the same period of evolutionary history? Why not a few phyla six hundred million years ago, a few more at five hundred, a few more at four hundred? Why do they *all* appear between five hundred forty and five hundred twenty million?"

"Cool question," says Ace, whose arms are still raised overhead. Then he leans back, falling off the bow and into the sea.

5

Cameron's strong hand grasps my elbow as we swim slowly over the sand, toward the rocks, and it is his presence, I think, that causes me to notice sounds so soon in the dive; ordinarily, they would not push through all the other sensations until much later. Loud and close by is the splashing of everyone's fins—I must remember to tell them to keep their feet below the surface. But even through the frothing tumult, I can hear the familiar crackling, like pebbles being sprinkled into an enamel platter. I pause and we lift our heads from the water.

"Do you hear the crackling?" I ask.

"Sorta like speaker static," says Ace, and Cameron, snorkel still in mouth, nods his agreement. Haley and Isabel both put their faces back in to listen.

"Oh, yeah," Haley says through her snorkel—then something like "I ear et. I odally ear et."

"Is it the rocks?" Ace asks.

"It's pistol shrimp," I say.

"Shrimp?"

"Anywhere you go diving, you'll hear that same crackling, because pistol shrimp are everywhere."

"You're telling me that static popping"—he leans his head back to dip his ears, then says loudly to the sky—"is *shrimp*?"

"There are something like four hundred different species worldwide—and those are just the ones that have been named. There could be ten times more."

"Are they just, like, snapping their claws?" Ace holds a hand up to his own ear and claps thumb against fingers, imitating a claw; then he does the same with the other hand, and makes a quick little rhythm out of the wet clapping sound he's just discovered.

He's got it about right—that is in fact what the shrimp are doing, though what we hear is not actually their claws smacking shut. Every pistol shrimp has one claw that's enormous—about half the size of its body. One of this claw's two fingers has been modified into a square little hammer, and the other is molded into a socket that the hammer fits neatly into. The shrimp cocks its claw by contracting two powerful and opposing muscles, so the hammer pulls back slowly but under great tension. And then, when the shrimp is ready, one muscle releases and—*wham*—the hammer strikes. But the sound doesn't come until a split second later, and that's what tipped off some researchers—what showed them it's not the smack of the hammer that makes the noise.

"So what is it?" Ace asks. He looks vaguely disappointed that we're not hearing the clap of shrimp claws.

"As the hammer slams into the socket," I say, "it squeezes out a little jet of water. And this jet moves so fast—over a quarter of a million miles per hour—"

"*What?*" Cameron says, spitting out his snorkel just to express his dismay.

"It's true," I say, "it's not a number I'd forget." In fact, when I first read that number, I was so incredulous that I phoned a friend who is an engineer and teaches fluid mechanics. He found the velocity astonishing but not supernatural. And he also explained that if a jet were to shoot through water that fast, it would inevitably open up a little chasm, a rupture called a cavitation bubble, which would stretch out behind the shooting water. And then, a short distance in front of the claw, the bubble would start to collapse under the pressure of the liquid around it. But by now the bubble would have a little water vapor in it, so as it collapsed, the pressure inside would rise, higher and higher, and the temperature too would climb, hotter and hotter, until at last the gas would be shoved—in one final violent instant—back into the liquid. And that moment, my friend explained, that split second of resorption, would produce a great shock wave, which, yes, you would certainly hear from a long way away.

"Bubbles?" Ace says.

"Not just bubbles—cavitation bubbles."

"*Shrimp* bubbles?"

"*Cavitation* bubbles. They're different."

"Why are they doing it?" Haley asks.

"I'd do it all the time," Ace says, and he smacks his finger cymbals, reawakened now, in a rapid little rhythm.

"I don't really know," I confess. I've read that pistol shrimp use their pops to communicate, and also that some species live in big social colonies, like ants. But other pistol shrimp use cavitation bubbles as weapons. That fiery collapse is destructive—so dangerous, in fact, that when two pistol shrimp communicate, they have to stand a certain distance apart just to avoid killing each other.

"That's awesome," Ace says. "Their words are weapons."

"When they're not love songs," I say, because I bet one form of communication that popping serves is the calling and identification of mates.

"Really?" says Haley.

"Just a guess."

"I'm gonna write a song about this," Ace says.

"Phylum?" I ask, returning us to our dive's official mission.

"They deserve their own," Ace says, "just for being so awesome."

"Arthropoda," I say, "same as insects." We spend basically our whole lives around but two animal phyla—Arthropoda and Chordata—while in the next twenty minutes or so, we'll see at least another dozen. Strange as that seems, it might have a simple explanation: the phyla were, after all, born in the sea, and so, if we want to see them, to the sea we must return.

6

What about that last question on the boat? Did all the animal phyla really arrive on the scene at the same geological moment? In 1836, the Reverend William Buckland, author of the first paleontological account of a fossil dinosaur, published his most thorough report of an abrupt transition at a certain depth in the geological strata, those discrete layers of dirt and rock that preserve, piecemeal and scattered about, a chronological history of the earth. At the very base of a dark shale layer called the Cambrian—it was named for Cambria, the Latinate term for Wales, where the layer happened to be exposed sporadically at the earth's surface—the rock seemed suddenly to change: an inch below this critical plane, there was not a sign of life; an inch above it, the ground veritably teemed with fossilized shells, skeletons, miscellaneous bits and pieces of a wide variety of animals. And though these creatures were not identical to any Buckland knew, they were also not entirely foreign or deeply primitive in relation to the modern animal phyla. To the contrary, they were easily assigned to known groups: there were molluscs; there were arthropods; there were annelid worms. There was nothing mysteriously between or outside familiar phyletic boundaries.

For the good Reverend, the clear transition at the base of the Cambrian shale made perfect sense: it simply revealed God's creation of sea life. The deeper, barren rock preserved the time before the Divine Act; Cambrian rock, the time just after. But for Charles Darwin, that same indication of sudden biological change presented a haunting problem, one that would trouble him for his entire professional life. He first met Buckland in 1837—less than a year since young Darwin's return on the *Beagle*, and also since Buckland's report on the Cambrian transition. Three and a half decades later, in the sixth and final edition of *The Origin of Species*, Darwin wrote, humbly, *The case at present must remain inexplicable . . .*

The problem, of course, was that evolution was supposed to happen gradually, with one form leading to another, but in the Cambrian shale, the modern phyla seemed to explode onto the scene fully formed, wildly diverse, and without warning or predecessor. Where were the remains of their ancestors? From what parental forms had the Cambrian phyla diverged? The best Darwin could do—and he was by no means satisfied with this—was to suggest that animals had in fact swarmed the oceans long before the Cambrian, but their fossils had somehow been destroyed, or perhaps deposited in places we could not look. Maybe, he ventured, with palpable halfheartedness, they were hidden beneath the ocean floor.

He was partly right. There were indeed older fossils yet to be found. In the 1960s, paleontologists realized that a certain category of fossils, which had been turning up here and there around the globe for the past century, resided in rock that had in fact formed well before the start of the Cambrian. (Our best geological dating techniques now place the base of the Cambrian at 542 million years ago; those fossils were turning up in rocks that date to various times in the interval between 578 and 542 million years ago.) But there was a rub: whereas Cambrian fossils looked, in a sense, too much like modern phyla, these older fossils looked too different. Darwin and every evolutionary biologist after him would have anticipated a fossil that was primitive or transitional—something, for instance, kind of like an arthropod but not quite there; or, better yet, kind of like an arthropod *and* kind of like a mollusc, an intermediate form that could have been the ancestor of both modern phyla. But what the pre-Cambrian fossils offered instead was something entirely *other*, something neither molluscan nor arthropodish, but rather more like life from another planet.

Such ancient aliens had shown up in England, Newfoundland, and Namibia, but they were named for the Ediacara Hills of South Australia, where they make an unmistakable and diverse showing. If you're not frustrated by the Ediacarans' resistance to classification, they are actually rather lovely creatures. They resemble roundish quilts, in which long, tubular pillows converge toward the midline. At first glance, they might look most like Cnidaria—the jellyfish, anemones, and corals. But where Cnidaria are truly radial, most Ediacarans exhibit threefold symmetry: they are elongated like ovals, so their front and rear, while identical to each other, can be distinguished from their sides. This arrangement is fundamentally different from that of the chordates (we're bilateral), the echinoderms (they're pentamerous), and in fact, every other living phylum. So weird are the Ediacarans that paleontologists have earnestly debated which kingdom ought to admit them. Are they really even animals? Or plants, perhaps? Or do they deserve a kingdom all their own?

The Ediacarans were not exactly rare; otherwise they would not have been found in so many locations around the world. But it also does not appear that the oceans *teemed* or *swarmed* in quite the same way as they would later on, in the Cambrian period. The fossils are often isolated little blooms, preserved all alone on the rock. And even when they are close together, they do not seem to be equipped to interact in any way at all. There is no anatomical part extending in search of another creature; no evident mechanism for rising from the bottom and going to look around; no sign, even, of sensory perception. Strangely, insofar as there are traces of motion, it is exclusively two-dimensional: the traces never turn down to bore into the earth; and there's not much evidence, either, that any of them turned upward, into the water column. If I were inclined to mapping the stages of Genesis onto geological time, I would take the early and middle Ediacaran, from 578 to 550 million years ago, to be the days in the Garden: creatures were quiet, peaceful, innocent of contact, but also sort of lonely and boring. It was, as one paleontologist has put it, the Garden of Ediacara.

So how did we—we animals, that is—make our way out of the Garden? What turned a bed of retiring wallflowers into such a crazed tumult of stalking, fleeing, swimming, and digging? Just in the past decade, that clean horizon beneath the Cambrian dawn, the unbroken plane Darwin found so troubling, seems to have been punctured: a number of small but momentous fossils have lately revealed thin lineages that leak, like wellsprings, from the dark quiet of the Ediacaran below into the bright hubbub of the Cambrian above. If you were to see, on the rocky shore of Russia's White Sea, the delicate impression of the creature called Kimberella, you would not think much of it at first. Compared to middle-Cambrian fossils, which are almost like silver-gelatin prints in their meticulous clarity and flash-frozen activity, Kimberella looks rather bland, a bit short on detail and dynamism—in a word, Ediacaran. But when you look closely, guided by your understanding of the tree of life, you do begin to see a few things.

To start, the overall shape is not a perfect oval. Instead, the animal's head is slightly broader than its rear, which sweeps gently back in the suggestion of a tail. And that subtle difference matters enormously, because it means this creature is actually *bilateral*, unlike the earlier, ovoid Ediacarans and more like the Cambrian phyla. Then you might notice that the impression has a softly scalloped halo, which looks very much, actually, like the outer edge of the suction cup on certain molluscs, like sea snails and cowries. That resemblance might prompt you to look just behind the animal's body, because all molluscs feed with two close rows of scraping teeth—and sure enough, there, in the stone, you would see the telltale marks: repeated etchings of the roman numeral two. And by now you could believe it: this animal, Kimberella, was in fact a primitive mollusc, though the rock on the edge of the White Sea dates with certainty to the interval between 558 and 555 million years ago—still firmly in the Ediacaran, and a good 15 million years before the dawn of the Cambrian. And that means the Mollusca, phylum of snails, cowries, and clams, did not appear abruptly and full of diversity at the base of the Cambrian. They got a quiet start, at least 15 million years deep in the Ediacaran.

So too, it seems, did the Arthropoda, phylum of crustaceans and, of course, our terrestrial insects. For on that same shore of the White Sea, you could find Parvancorina. Its head looks a bit like a horseshoe crab's; behind the head rises a prominent central ridge; and from this ridge extend about ten appendages, which appear to split at their ends into a pair of fingers, just as the legs of shrimp and lobster do. Your first phyletic guess—Arthropoda—would be correct, and the affinity is even more obvious when you compare Parvancorina not with contemporary shrimp, but rather with Skania, an arthropod whose form was found in the Burgess Shale, a famous bed of fossils from the middle of the Cambrian.

Kimberella, Parvancorina, a few other small fossils from Russia and Australia—these appear to have been the evolutionary predecessors of modern phyla. And if Darwin had known about them, he would have worried a lot less about diverse legions of animals splashing fully formed into the Cambrian sea. Still, the discovery of manifestly Molluscan or Arthropodish Ediacarans does not explain away every aspect of the abrupt transition that has troubled evolutionary biologists for so long. What looked to be a great and sudden flood of animal evolution still looks to be one, only now it appears to have rushed from a deep Ediacaran spring.

Just to review the facts as they now stand: First, in the lowest Cambrian stratum, which dates to 542 million years ago, trace fossils—paths left in the mud by ancient organisms—become suddenly more numerous. Oddly, they also change trajectory: whereas before they were exclusively horizontal—the unaspiring and sparse tracery of an Ediacaran Garden—now, all of a sudden, they bore down into the mud, making it as porous as worm-eaten wood. Second, in that same transitional layer of rock, body fossils—the mineralized forms of animals—become suddenly more abundant, bigger, and more diverse. They also appear to be newly equipped for a busy life in the crowded ocean: they are endowed with armor, appendages, and a variety of sensory systems. Third, moving up from that basal stratum, into the early and middle Cambrian period, from 542 to 510 million years ago, the diversity of animals continues to increase rapidly: the number of genera climbs from about 50 up to about 1,200; and we have also the first unambiguous fossils of the phyla Chordata, Echinodermata, Brachiopoda, as well as several others. In the history of life, such sudden expansion has never been repeated. The number of genera may have climbed nearly as fast during the brief periods of rebound from the greatest mass extinctions, but the higher taxonomic categories—the orders, classes, and, especially, phyla—have never again proliferated as they did during the first half of the Cambrian. So, yes, something strange and spectacular took place. And the next question has to be: *Why?*

7

Our shadows, stretching out ahead of us across the white seafloor—for the sun is not yet high in the sky—appear to be magnetically drawn by the dark horizon farther on: the mottling of coal black and iron red that is the island's verge. As we close in on it, the mottled band resolves into a scree of angular and sharp rocks. Closer still, and the rocks present their living skin: white barnacles speckle the black faces; the gray coils of snails are like squat temples in a jagged landscape; dark green anemones cluster in irregular patches. Three different phyla, but I resist the temptation to pause, and we continue on, moving more slowly now that we are over the reef, edging in toward shore when the water grows deeper, keeping the rocks always about a fathom away.

But that little jewel there—small but precious, unassuming but exquisite—is irresistible, so I pause and lift my head from the water. Cameron feels my movement and does the same, and the others, who have been trailing behind, watching the reef go by, pop up beside us one by one.

Spitting out his snorkel, Ace says, "There are *so* many of those—what are they?—surgeon generals?"

"Sergeant majors," I say, laughing. And then, mainly as a way of offering a description to Cameron, I add that the peculiar name might derive from the chevrons that indicate an officer's rank: the fish are yellow on their backs, navy blue on their bellies, and striped with vertical bars of black. And while the resemblance to a shoulder patch may be vague, it's definitely closer than that to any surgeon general.

"What are the tiny blue ones?" Haley asks.

"Cortez damsels. Those are the juveniles." Amid the rocks, they are like sparks of static electricity in a rumpled dark blanket. Only in youth, however, do they shine so brightly. As they age, the iridescence fades, giving way to a drab chocolate brown. But their temper follows an opposite trajectory, burning hotter as they age. There's probably one defending every good-sized rock beneath us, and at some point, when we dive down, a student's mask will be attacked by a furious damsel.

"The babies of those other ones," Haley says, "the surgeon generals—they're super cute."

"Sergeant majors," I plead, "sergeant majors." Fish aren't often called cute, but I do know what Haley means—there is something very dear about perfectly miniaturized versions of bigger fish. Sergeant majors are unusual in this respect. In fact, we could easily spend the rest of the dive just figuring out which trios of flamboyantly and diversely colored fish compose nuclear families—male, female, and juvenile—and the rest of the afternoon discussing how and why different sexes and ages would appear to be entirely separate species. But right now we ought to peer past the fish, for they are the reef's gaudy and loud welcoming party, the crowd in the foyer you must sidle through if you are to access the dimly lit salons farther back, where the more retiring yet intriguing characters conduct their quiet conversations. After all, we are out to find phyla—a deep form of diversity that Cameron can actually put his fingers on—so we shouldn't linger among the flighty Chordata.

I start describing the look of the rocks for Cameron, and the others push willingly through the reef's busy entryway and begin talking about what's beyond.

"It looks like a big rock slide," Ace says at one point, "like the island is crumbling into the water."

"And the rocks are kinda scary-looking," Isabel says. "There's lots of stuff on them. Little fuzzy plants and snails."

I tell them that what appear to be plants—the fronds of ferns, feathery plumes, small tufts of moss—are in fact animals. The fronds and plumes are hydrozoans, and though their membership in the phylum Cnidaria might seem a little confusing, since they don't appear to be radially symmetric, in fact each little plume or leaf is composed of hundreds of minute polyps, which are just as nicely radial as one of those green anemones fastened to the rocks nearby. The mossy things, for their part, belong to yet another phylum, Bryozoa, which is one of the few that did not make its first appearance during the Cambrian. Growing side by side, Hydrozoa and Bryozoa—both vaguely botanical in form, but evolutionarily as far apart from each other as two animals can possibly be—are a reminder that appearances can be deeply misleading when it comes to detecting common descent. And the same point is illustrated by the tiny white cones besprinkling the black granite: barnacles look rather like limpets or chitons—molluscs that adhere tightly to rock—but in fact they are arthropods, the same phylum as shrimp and insects. They've merely cloaked themselves, both physically and phyletically, in a hard protective casing. Again and again, the evolution of form to suit certain functions—filtration of passing seawater, protection against predators, and so on—makes even the most deeply diverged animals look like kindred species. Only when you open them up and look inside do you see how different they really are.

Isabel has been listening to me with her mask in the water and her head tilted to one side, so that the ear turned toward me remains in the air. I don't know whether she's poised like this just because she wants to keep looking, or because she is assessing the evidence for my claims even as I make them. Either way, it strikes me as a wonderful way to listen, and I ask her now what else she sees.

"There's a lot of starfish," she says.

"Yellow-spotted stars," I say, "and tan sea stars. Echinodermata."

"Tan?" says Ace. "Those aren't tan." His face splashes into the water for a second, then reemerges. "They're more like purplish blue, with rusty stripes down the arms."

Cameron says, "Don't be short-changin' me, Aaron."

"It's just the common name," I say hurriedly, "just what everybody calls them." He must hear the alarm in my voice—the dismay that he might think me a lackadaisical guide—because he immediately smiles and, knowing just what to do to let me feel pardoned, asks if we can pick one up.

I dive down and pick up a tan sea star, then pull myself a short distance along the rocks and grasp, between thumb and forefinger, the precious little specimen that first caught my attention a moment ago and caused us to pause here. As I float back up, I can feel its small but strong suction cup pulling on the insides of my fingers.

I place the sea star on Cameron's open palm. As it attaches hundreds of gentle tube feet, he smiles and says, "It's totally got me. It thinks it's gonna eat me." I take it off, turn it over, and let the sea star's top side rest against his forearm. At once he yells, "It's yanking out my arm hair!" He sounds astonished, and I think he is feeling a certain contrast that has struck me too, albeit in a different sensory register. These sessile creatures look so immobile, so stony, that when they actually *do* something it can be quite startling, as if you've just witnessed the animating breath of life itself—a tiny reprisal of that mysterious Cambrian moment. But what has always been for me a purely visual experience—watching a stolid invertebrate suddenly rouse itself—is for Cameron a feeling in his hands. The attachment of tube feet is gentle and slow, a primeval sort of process, whereas this tugging of hairs feels spirited, almost mischievous, and so Cameron has just felt the animal become more clearly animal. In more prosaic terms—which I now offer the students, since I won't venture a connection between the plucking of arm hairs and the Cambrian explosion of animal diversity— what Cameron feels are pedicelleria, tiny little pincers that cover the surface of the sea star's body. Their job is to keep the star well cleaned, and Cameron's arm hair is to them an irksome contaminant.

In my other hand, too, life is stirring. My prize specimen has just begun muscling its way around my closed fist, plotting its escape with slow but formidable stubbornness. So I hurry through definitive attributes of Echinodermata, hand over the star to Ace—he immediately pushes its surface against his cheek, saying he forgot to shave this morning—and, at last, lift my other hand from the water to reveal a glossy, gorgeously patterned cowrie.

Cameron touches the shell. It is as smooth as polished marble—but not quite like marble, because there is something distinctly not-dead about it. Some cue—a thin membrane, perhaps, or microscopic cilia— lets you know: this little dome is not a stone; it is alive. I start to describe its appearance, but the shell is so intricately colorful—there's so much to notice and say about it—that as soon as I pause, others jump in with more details, and we end up offering observations almost all at once:

"It's pale pink around the edges," says Isabel.

"With black Dalmatian spots on the pink," adds Haley.

"I think it's got, like, two little eyes poking out from under its shell," Isabel says.

"The top is like a web of copper," Ace says, "on a bluish pinkish background."

"Bluish pinkish?" Haley asks.

"Yeah," Ace says, "it's kinda both. Can it be both?"

"I think it looks like a giraffe," says Haley.

"A giraffe?" says Cameron, reaching out again to touch the dome. "What are you looking at?" He's smiling at the non sequitur.

"Not the whole thing," Haley says. "Just its back. It has the same pattern." I would never have thought of such a comparison, but it's perfect: with its copper tracery, this little hemisphere might well be the seed for a reticulated giraffe.

"That's a very interesting observation," Ace says, smiling impishly, "because the giraffe is actually its closest evolutionary relative."

"That's correct," I say.

"Really?" Ace says, and this delights me, because it means the bryozoans and barnacles convinced him—maybe even a bit too thoroughly—that real evolutionary relationships can be hard to detect.

"No," I say. "Not really." Though the cowrie does wear a different sort of taxonomic cloak. Within the phylum Mollusca—the great bough that begins with Kimberella and ramifies into branches as far apart as scallops, chitons, and snails—the cowrie doesn't sit where you might at first suspect. It is a gastropod, which is to say, basically, a snail. And if you want to see how such a smooth dome descends from a spiral, just imagine, as the starting point, one of the tight conical snails populating the rocky landscape beneath us. And now imagine the opening at its base—the snail's round aperture—growing in one great, widening whorl, and thus wrapping all the turns that came before it. In development and in evolution, this final sweep of the cape is essentially how a cowrie comes about.

I pull the creature off my own fingers and give it to Ace, who is still holding the sea star in his other hand.

"So there it is, Ace. A billion years of descent in the palms of your hands."

"Yes!" he says, his smile widening. "I can feel the electricity running through my arms."

"Actually, with you in the middle, that's another five hundred forty million years—at least."

"There's that number again," he says playfully, as if he knows he's offering me an opening.

"Nuh-uh," says Cameron, grasping my elbow. "We gotta get some more phyla first."

8

Rubisco, you'll recall, is that hapless but vital molecule who must pluck CO_2 from the atmosphere and pass it along to the sun-wielding workshop that manufactures sugar. And you may remember, too, that there was not much oxygen around when rubisco first evolved, because it was rubisco's own work, the process of photosynthesis itself, that would eventually pump oxygen into the atmosphere. That process got going a very long time before the Cambrian—a few billion years, actually. But there were a great many places for freshly minted oxygen to go: some of it bubbled up into the atmosphere; some dissolved into the vast oceans; and much reacted with all the chemical compounds that had been lying about in an oxygen-free world for millions of years. Hypothesizing that it might have taken billions of years for primitive photosynthetic organisms to fill such capacious earthly urns with oxygen, some paleontologists have argued that what animal life was waiting for—what it needed before it could proliferate and diversify on a Cambrian scale—was simply enough oxygen to breathe.

The amount of oxygen an animal's body requires depends on three important factors: size, activity, and types of tissue. Size is straightforward: the bigger an animal is, the more energy its body uses and, therefore, the more oxygen it needs to burn food. Activity too is easily explained: digging, swimming, thinking—all the verbs of post-Ediacaran life—increase energy use and thus suck up more combustive gas to feed the fire. The third factor, tissue type, is a bit more subtle. As it happens, the construction of certain biological tissues requires inordinate amounts of oxygen. In fact, one fairly well-developed theory holds that the critical oxygen-intensive substance was collagen—the connective fiber that binds together all modern phyla, from sea cucumbers to human beings. Without it, the theory goes, complex and dynamic animals could not be assembled, and yet its oxygen budget is so high that it could not be synthesized in substantial amounts until sometime very late in the Ediacaran.

What's elegant about the idea that rising oxygen levels unleashed the diversification of animals is that it seems to accommodate the features of fauna on both sides of the Cambrian transition. Beneath the horizon, the Ediacarans were both sedentary and sparse, which you would expect if oxygen were in limited supply. They were also very flat—virtually two-dimensional, in fact—and one function such a geometry might have served is the maximization of surface area for absorbing precious oxygen. Rising into the Cambrian, all three determinants of the animal oxygen budget change at once: animals grow bigger; they become suddenly mobile and busy; and they are now built of oxygen-intensive tissues. They are also, of course, far more abundant, and that too could follow from an increase in available oxygen.

All very tidy. As of late, however, the evidence seems to be interfering with our elegant story. One of those capacious earthly urns that took up fresh oxygen also happened to store away a very detailed and durable ledger, which geochemists have recently learned to read. Imagine a pebble of iron sitting on the seafloor beside an Ediacaran animal. If there is oxygen in the seawater, the pebble rusts; if there is not, it doesn't; and in fact, the process is nicely proportional—the more oxygen there is, the more the pebble rusts. The animal dies; time passes; and much as the animal's body might be blanketed by the sediment that could preserve a fossil, the pebble too can be covered over. So when a biogeochemist, working about 600 million years later, draws a bit of iron from a stratum of rock, he will be able to measure, with surprising precision, how much oxygen was in the water around that Ediacaran animal.

What such measurements have shown is that deep ocean waters filled with oxygen not at the start of the Cambrian, but actually at the start of the Ediacaran—which is intriguing, to be sure, but not consistent with the tidy evolutionary tale we were telling. Worse still for our story, about halfway through the Ediacaran, deep-water oxygen levels again declined, and they remained very low throughout the first 25 million years of the Cambrian. Partly in response to this observation, some paleontologists have suggested that the really creative Cambrian evolution was happening not in the deep water but rather on shallow shelves. But there we have a different sort of discordance between measured oxygen levels and the story of animal evolution: as far back as 800 million years ago, shallow waters were already relatively replete with oxygen, and they appear to have remained that way throughout the Ediacaran and the Cambrian. And so we find ourselves presented all over again with the very question we had thought oxygen levels would resolve: What were animals waiting for?

Failing to find a plausible answer in the animals' environment, we might look for one in the animals themselves: Was there perhaps some key evolutionary innovation, some new capability, that released such a sudden flood of animal life? Kimberella and Parvancorina show us that, as of 570 million years ago, the basic bilateral body plan was already in place, and at least two of the phyla that would go on to become highly successful—the diverse and abundant Arthropoda and Mollusca—were already evolving as separate lineages. This raises two important difficulties for the idea that a key innovation released the flood. First, if the bilateral body plan was already up and running, what could the key innovation have been? And second, since the Cambrian flood clearly issued from more than one deep well into the Ediacaran—more than one phyletic lineage—the only way a key innovation could have uncorked the waters would have been for it to occur independently yet simultaneously in more than one species. And that seems highly unlikely.

The same points are echoed, albeit in a different biological language, when we compare the genomes of Mollusca, Arthropoda, and other bilateral phyla. All of these animals share the same basic tool kit of genes that direct embryonic development. If a particular gene is critical in the development of, say, a chordate like you or me, we can be nearly sure that a very similar gene serves a very similar function in the embryo of a mollusc. For example, the gene that is activated at the tips of human fingers as they sprout from the nubs of embryonic limbs is quite similar to the gene activated at the tips of the growing legs of a shrimp (an arthropod), the tube feet of a sea star (an echinoderm), and even the eye stalks of a cowrie (a mollusc). This strongly suggests that a version of this gene was already present, and probably even deployed in the very same developmental occupation, when the various bilateral phyla diverged; otherwise it would not be there, doing the same job, all over the Tree of Life. And we could find hundreds of other such examples—so many, in fact, that it is tempting to think we could reconstruct much of the genome of the earliest bilateral animals based only on the observation of genes held in common across diverse phyla.

What's more, we have an opportunity here—in fact, many opportunities—to estimate when the phyla diverged from one another. If we take, for instance, the gene that is activated at the tip of a sprouting finger, and we compare it with the very similar gene that's at work at the tip of a growing shrimp leg, we can count up the number of small differences between the two genes; and we can then use that count to estimate how long the Chordata and the Arthropoda have been evolving as separate lineages. In practice, this is a complicated business, fraught with small technical challenges and sources of error. But since we have so many chances to make comparisons—all those hundreds of sets of genes that have a member in every phylum—we can at least try to compensate, with a large number of measurements, for the uncertainty inherent in any one of them. When we do just that, making as many comparisons as we can across as many different phyla as we can, the estimates that emerge place the divergence of some bilateral phyla deep in the Ediacaran. Thus we encounter, in the coded language of genomes, the same conclusions suggested by the fossil record: the bilateral body plan was already well established, and at least some phyla had already diverged, long before life flooded forth in great abundance and diversity across the lowest plain of the Cambrian. A key evolutionary innovation could not have uncorked the flood.

If we do not find an answer in the animals' environment, and we do not find one in the animals themselves, where else could we possibly look? It may sound vexingly like a Zen koan, but I suspect the answer may in fact lie neither in the animals nor in their environment, but rather in both. Ediacaran animals do not appear to have been present to one another in any consequential way. The environment was a physical place, a world of seafloor, seawater, and not much more. Ecology as we conceive of it—as a web of interactions among organisms—did not yet exist. And consequently, for those lovely but dull Ediacaran creatures, at rest in a quiet and mostly empty ocean, evolutionary pressures had more to do with physical variables—the temperature of the water, the availability of light, the supply of current-borne nutrients—and less to do with the ways and deeds of the animals around them.

And then—maybe—what emerged in the Cambrian was the possibility of interaction. For the first time, animals hunted and fled; they sniffed and searched and hid. Thread by thread, the web of ecology came into being. And with each new ligature, natural selection could apply new pressures: The predator drove innovation in its prey; the prey did the same for its predators; and every encounter between competitors was likewise an occasion for selection. Even without making direct contact, animals in a crowded, dynamic world could impact one another. When a creature dug in the mud, it collapsed some niches and created others; it opened up a new dimension of biological space—maybe more than one, if we consider fractals; it mined sequestered nutrients and returned them to biological systems. The basic idea, then, is that animal interaction entailed a sudden elaboration of evolutionary pressures, resulting in a sudden expansion of animal diversity. Or, to put it differently, when animals themselves became a meaningful part of the environment, the number of ways of being an animal was suddenly multiplied.

But wait, you may be thinking, haven't we just shifted the bump in the rug? Don't we now have to ask why and how animals started interacting? Yes, we do. But what's intriguing about the evolutionary scenario we've just outlined is that it may unfold in much the same way regardless of which particular contact kicks it off. An interaction—almost any interaction—complicates the environment for an organism; evolutionary change in that organism complicates the environment for other organisms; and the process cascades. If that's correct, then we may never know which ecological links came first—which two Ediacaran organisms happened to find themselves in a particularly favorable benthic setting, where their densities could climb to levels at which competition would finally set in. But our ignorance of such matters may not be so grave, because the process could have been initiated by many different pairs of organisms, capable of many different kinds of interaction, and the result would have been basically the same: a sudden surge of evolution. When you have a chunk of fissile material, it doesn't matter which particular nucleus splits first, because whichever one does, it's going to start a cascade. By one path or another, there's going to be a big explosion.

But even if we cannot expect to reconstruct the earliest events—to track those first few cases in the outbreak of ecology—we still, of course, must find evidence for our evolutionary story. And we do see, toward the end of the Ediacaran, the first clear signs of predation. A small Ediacaran animal called *Cloudina* left a number of fossils with perfectly round holes bored into their sides; and in today's ocean, when you find such a tunnel in the shell of a clam, you can be fairly sure it was the work of a predatory mollusc. Then, in strata that come immediately after the first pierced *Cloudina*, we see a different kind of evidence for predation: animals start donning all sorts of thick protective armor. And more generally, the simple oval of the Ediacaran body begins to sprout new ways of reaching out: sensory appendages; eyes; new forms of mobility. Even the microscopic plankton join the fray: new spikes and plates make them look like medieval mace heads. And they grow suddenly larger, too, which is probably a case of what ecologists call *size escape*: nobody can eat you if you get stuck in the gullet.

Which brings us back, interestingly, to another hypothesis we considered—that rising oxygen levels unleashed the flood of animal evolution. When plankton die, their fate depends on their size. If they are very small, they get bounced around in the water like dust motes in a drafty room, and consequently do not sink. If they are larger, they filter downward, taking with them all the nutrients bound up in their carcasses. Oceanographers call this form of transport the biological pump, and they consider it an important part of the earth's circulation of carbon. If the descent of nutrients is somehow interrupted or counteracted, grave consequences follow: The suspended compounds react directly with oxygen, and what's more, in the upper reaches of the ocean, where much light is available, the nutrients feed microscopic organisms, which in turn suck up even more of the available oxygen. Eventually, much of the water—not the layer right below the surface, but everything deeper—becomes so depleted of oxygen that it is uninhabitable to anything that needs to breathe. It is a noxious and fetid state, which may be familiar to anyone who has seen and smelled a bog.

And so, when we see that plankton became suddenly larger at the start of the Cambrian, we have to suspect that this shift would have brought about the very first pushes from the biological pump. And this is an intriguing hunch, especially in connection with those recent studies of ancient rust, which revealed that much of the ocean was low on oxygen until well into the Cambrian. For it seems that, as we contemplated animal evolution and oxygen levels, we may have had our causation backward: it was not that rising oxygen unleashed animal evolution, but rather that the flood of evolution—the rise of predation, the enlargement of plankton—inaugurated an oceanographic process that allowed oxygen levels to climb for the first time since the early Ediacaran. If that's true, then it may have been the discovery of one animal by another—the advent of interaction—that ultimately transformed unimaginably great ponds of muck into the clear blue expanses we know today.

PART V

Canis familiaris

ENDURING YOUTH

1

Whenever the evening seminar is mine to teach, as it is tonight, I visit the station kitchen shortly after dinner to make myself a cup of strong black coffee. Stepping from the dusk, which descends abruptly here on account of the shadow of Mike's Mountain, I find the dim room so deliciously inviting I could laugh—for it seems ridiculous, in a way, to take such immeasurable pleasure in a simple rectangle of stone masonry. But there is much that is right here—remarkably and unexpectedly right. Alejandrina keeps the place immaculate and orderly, despite the formidable forces of entropy that swirl through. The wind blows grit through the doorways—beach sand from the east, desert dust from the west. The ants form rush-hour highways across the concrete floor. Yukon enters, soaked and panting from his gleeful pursuit of seagulls. He exits, lies down to rest in a dusty pit he has dug just outside the kitchen door, then changes his mind and returns once more to visit his beloved Alejandrina, who rewards such incorrigible mischief with bits of huevos or tortilla. And after all that, when I walk in to make coffee, there is not a grain of sand on the concrete; bottles and cans of food, tanks of filtered water, pots and pans are stored away as tightly and rationally as they would be in a ship's galley; the stainless-steel surfaces of the kitchen equipment— wash station, industrial stove, refrigerator—gleam in the white light; and there is a very slight smell of lemon juice and bleach.

This rectangle of stone resides in conditions that have succeeded many times in sweeping away civilization. Through the steady application of irritating pressure—abrasion, heat, corrosion, more heat—this place wears the trappings of humanity down to an austere minimum, a bleached backbone of human existence, and sooner or later, that too is scattered in the desert or relinquished to the tide. Even the mining company abandoned its building and returned home. The mining company, though, like a number of ventures that came and vanished before and after it, was a thoroughly male and foreign affair; at least, I know it was run by men who had been sent here from the United States. Alejandrina was born in Bahía de los Ángeles. She is first cousin to Samy Diaz. Perhaps she knows ways of holding the forces of disorder at bay—ways that American mining executives did not understand. In the time that I have known her, Alejandrina has taught me how to rescue a truck that is caught in shifting and unsteady sand, how to use toothpaste to mend a tank of propane, how to dig Venus clams in a tenth the time it takes the uninitiated, and how to numb the sting of a Cortez round ray. And all of that is well outside the area she considers her real dominion, which is here, inside these four stone walls.

Besides the clear signs of Alejandrina's hand, there are other things to admire about this room. The open doorway to the east, which is at present a tall rectangle of eventide blue and gray, like a misty Japanese watercolor of the bay and its angels, will glow orange tomorrow morning, when the rising sun presents itself at the center of the door frame. Whether the flawless composition of this view is fortuitous, or was achieved through painstaking calculation—as in an Aztec temple where sunlight penetrates at a holy hour to strike a stone icon—I cannot be sure, though what happens here for the rest of the day strongly suggests that something other than luck was involved. Shortly after dawn, the eastern doorway will fall into shade, because the kitchen, which is at the northern end of the station, was built with smaller dimensions than the rest of the structure, and is therefore tucked behind the taller and wider walls to the south. And a portico of heavy stone archways further protects the other doorway, the western one, from the late-afternoon sun. The two doorways, east and west, are precisely positioned such that the doors may be propped open to create a tunnel that amplifies even the slightest sea breeze. I'm not sure how this strange local weather pattern is affected by the windows on the north side of the room, behind the wash station and the stove, but I have a vague sense that they play an important role. All of which seems to testify to a great deal of design and engineering behind this room's triumphs. But when you are standing in here, with Alejandrina, on an August afternoon, and the temperature outside is ascending the ladder of one-twenties, and four of the industrial stove's burners are making the sound of a jet beneath large pots of stock and beans, and still the air feels fresh and comfortable—eighty-five, perhaps—the capabilities of such a simple stone rectangle beggar comprehension, and in that moment, the space seems beyond human contrivance and no less mysterious than that Aztec temple to the sun.

In any case, through engineering or by conjury, the room was added on to the old mining headquarters by Lane McDonald, whose rightful title with respect to the Vermilion Sea Field Station would be a word that combines the meanings of proprietor, guru, and creator. Before he moved to Southern California to pursue graduate degrees in marine sciences, Lane was a farmer in Missouri. And it occurs to me, as I admire the mark of his work, that perhaps the mining company would have needed someone like him, too.

With my coffee in hand, I cross the gravel lot between the kitchen portico and the staff house. The ground here still emanates the afternoon heat, and will do so, like a sunburn, well into the early morning hours. The staff house windows are dark, but on the left-hand side of the concrete porch, where Graham often lies on his canvas cot to read, there is a single bright point of halogen blue: his headlamp. As I climb the short concrete stairway up to the porch, Graham, mostly invisible behind his bright light, says, "Barracuda, baby! This is gonna be good!" Evidently, he has briefly set aside *The Faerie Queene* and taken up the course-reader. Tonight's seminar is to be on the limitations of natural selection—the inability of the evolutionary process to engineer optimal solutions—and one of the articles I've assigned is about metabolic proteins in barracuda; their structure, it seems, represents a compromise, a trade-off between rapid function at normal temperatures and stability on exceptionally hot summer days. Graham has not read the article before, because there is an implicit agreement between us that while the general themes of our seminars may remain unchanged from one summer to the next, we will assign new reading every year. It is essentially a promise to keep teaching each other.

"I channeled you today," I say, pausing on the porch before I go inside.

"Really? Was I smart as hell?"

"We'll see. I'll tell you about it after seminar."

"Just give me the ten-second title. I can't stand the suspense."

"We were talking about German Idealism and embryology."

"Uh-oh."

"Uh-oh what? I could handle it. And besides, you were up here with the fairy queen. I had no choice."

"So what did you say?"

"I have to prepare, Burnett. I'll tell you later."

"Come on. Just the sixty-second summary."

I take a seat on the knee-high wall that bounds the porch. Graham remains on his back but props up his head, so his headlamp, which is canted downward, no longer shines in my eyes but instead illuminates a wedge of his supine form: his prominent nose, the tip of his chin, a slice of his white tank top, and a widening swath of his red-and-yellow sarong; his bare feet are backlit. The tank top and sarong compose Graham's daily uniform here in Baja. When we're likely to encounter locals, Veronica makes him put on sleeves and pants. She tried to impose a similar rule for seminars, but Graham bucked authority and, to the students' delight, he often teaches in what they call his skirt and muscle shirt. When we're out on the boats, he adds some layers for sun protection: a clean white dress shirt, a baseball cap, a red bandana that hangs around his face like a sheik's headdress, and, of course, the usual hornet-eye sunglasses. With his bandana, shirttail, and sarong streaming behind him in the wind of the panga's speed, he looks madcap, harlequin, deranged.

Now, beneath his glancing stripe of bluish light, he waits to hear what his medium—or impostor—has said on his behalf. So I take my first sip of coffee and begin to tell him about my conversation out on the patio. I had pulled up a chair, I explain, with five students—Allie, Chris, Anoop, Rafe, and Cameron. They were talking about the cucumber that gutted itself this morning on Anoop. Kind of out of nowhere, Allie asked me why the cucumber doesn't have the fivefold symmetry of other echinoderms. And when I said it does, but it's obvious only in the juvenile, Anoop raised his hand and asked, *Does that mean ontogeny recapitulates phylogeny?*

"Anoop," Graham says, with a smile that is invisible but clearly audible, a mixture of amusement and affection.

"So that's when you got channeled," I say, and I tell him what I said this afternoon about the German Idealists—that their universe was supposed to be unified by fundamental laws and striving toward perfection; that naturalists brought this worldview to the study of embryos; and that they found in those embryos a superb instantiation of their own philosophical commitments. But what was sort of surprising this afternoon was that, when we tried out the theory of recapitulation on our cucumber, it actually kind of worked. The succession of different forms of symmetry in the cucumber seemed to bear it out.

"Whoa," Graham says. "That's weird. Tell me about that."

"I have to prepare." I say, standing up. "But I did okay?"

"The Germans aren't really my thing, Hirsh."

"But did I say anything patently wrong? That's all I want to know."

"No, not wrong, I don't think. But it seems a bit weird to represent German Idealists mainly as evolutionists."

I sit back down. "Ernst Haeckel?" I ask.

"Haeckel came late in German Idealism. The *Naturphilosophen* who came before him—and I'm not just talking about von Baer—they didn't believe in evolution at all. In fact, they firmly rejected it." As Graham sits up to continue, I am struck with an odd sense of revolving roles. Just as I had hoped, out on the patio this afternoon, to take several steps backward from Anoop's question in order to give the others a sense of where his peculiar phrase had come from, Graham is now leading me back further still, filling in a vital, earlier chapter of the story.

It begins with Lorenz Oken, a German embryologist who preceded Haeckel by a generation. In Oken's view, there were five classes of animal, defined by their relationship to the five senses. The Dermatozoa, or skin animals, included almost everything we would call an invertebrate; these were, according to Oken, creatures of touch, the sense he considered most primitive. The Glossozoa, or tongue animals, were mostly fish, which Oken believed to be relatively lowly creatures, possessing only two of the senses in perfect form—the most basic one, touch, and the slightly more advanced sense of taste. Yes, he knew they had eyes, but eyes lacking proper lids were too primitive to qualify as perfect. The Rhinozoa, or nose animals, included lizards, snakes, and some other reptiles; Oken had taken their open breathing holes to be indicative of a sense of smell. The birds were Otozoa, or ear animals. And the mammals, finally, were Ophthalmozoa, or eye animals. The classes were arranged along an ascending ladder, on which each step upward was supposed to represent an improvement in the precision and sensitivity of perceptual ability: smell, for instance, was deemed superior to taste, but not quite as worthy as hearing.

It is an intriguingly weird, if scientifically doomed, system of arranging animals. More to the present point, though, Oken described a parallel between his five-rung ladder and the process of human embryological development. It was, in a sense, a theory of recapitulation. However, of the two correlated series, only one, ontogeny, was thought to unfold in time. The other, the sense ladder, was a purely static hierarchy of organisms.

Oken's successors would eventually relinquish his system of five sensory classes, but not his notion of recapitulation. In fact, they went on to describe in much greater detail how embryonic development presented a lesson, in fleeting time, of an order that existed outside of time—how ontogeny revealed the ideal chain of God's creations, from organic muck at the bottom to European Man at the top.

In a rather poignant episode in the history of biology, it was the Harvard zoologist Louis Agassiz, a man committed to the idea that God had created every species separately, who inadvertently brought the theory of recapitulation into the service of the theory of evolution. A student first of Oken, and later of the great anatomist Cuvier, Agassiz combined the Idealist's philosophical bent with a Cuvierian ardor for empirical work. He was an exceptional paleontologist, and it was in his reconstruction of the historical order of fossilized fishes that he first showed how the theory of recapitulation could be extended. Where Oken had perceived an ascending parallel between human development and the hierarchy of existing organisms, Agassiz intended to add a third rising progression, consisting of the historical succession of species on earth. As deeply convinced as Agassiz was that each species had been divinely created, he could not see in his own work what many others, including Charles Darwin and Ernst Haeckel, immediately did: the historical succession of species on earth was not just a third correlated series, to be arranged alongside development and the Great Chain of Being. It was, rather, the deepest series of the three, the one that gave rise to the other two, and thus caused them to run in parallel.

I stand up, feeling vaguely appalled at Burnett's erudition.

"Where are you going?" he asks.

"So the Germans aren't really your thing, eh, Burnett?"

"Nope, not at all," he says, and he sounds vaguely amused or mischievous, but says nothing more.

"What?" I ask.

"Nothing. You'd better prepare."

"No. What is it?"

"Anoop and Rafe told me about your conversation. I've been thinking that one through."

"You dog!"

His headlamp shakes as he cackles.

"I have forty-five minutes to prepare a seminar, Burnett. And you'd better do the reading, because I'm going to be asking you *a lot* of questions."

"Barracuda, baby!"

2

Inside the staff house, I seat myself at an old wooden desk, in front of one of the large open windows. A breeze has just begun moving off the bay and over the field station roof, rustling the edges of my notes and chilling my bare arms. It feels like a beneficence, a kind gesture from the Channel of the Whales, but even so, I cannot quite hush the vague suspicions at the back of my mind: the air is supposed to be still at this hour; every evening, it is perfectly still; so what is the meaning of this strange gift? Through the lines of my notes—through the sequences of letters representing the amino acids of the enzyme lactose dehydrogenase—a peculiar word percolates into my consciousness. *Horripilation* is the word. *Horripilation. Horrere,* as in horror; plus *pilare,* to grow hair. So are those goose bumps on my arms from the chill of this wind, or the chill of my uneasiness about it at this hour, in this place? I don't want to seem thankless—the animist in me is certain the channel will revoke the breeze in response—but neither can I receive the gift without reservation. I could swear it carries the scent of deep, cold water.

"Burnett," I say out the window, onto the dark porch, "what's with this wind?"

"What?"

"Nothing."

"What's Coomassie blue, Hirsh?"

"It's a dye for visualizing proteins. But why are you reading the methods? You don't have to read the methods."

"I've read everything else."

"Well, read it again, Burnett."

I have to focus.

"Hi, Yukon," Graham says, "how's it going?" Unlike most people, Graham doesn't have a different tone or language for interaction with dogs. His voice doesn't rise in pitch; he doesn't pat a leg or make clicking sounds. Rather, he addresses them as if they were his peers, and he would no more reach out and pet a dog than he would do so to a human being. But this last consequence of species equality is one that Yukon can't seem to accept, so although the two of them are out of view from where I sit, I am certain that Yukon is now testing Graham's tendencies for the thousandth time. Thirty seconds later—about the duration of Yukon's patience in requesting physical contact—he walks into the staff house, as expected, and rests his head on my lap. I wait him out, and he lies down by my chair. Like the cool breeze, this strikes me as a puzzling sign, since he would ordinarily find a place outdoors. But maybe I'm just being edgy. Maybe it's just that he too is a little disoriented by the strange wind.

I have less than forty minutes now. *Sphyraena lucasana*, the species of barracuda here in the gulf, has a slightly different version of lactose dehydrogenase than *Sphyraena argentea*, the Pacific species, perhaps on account of the extreme fluctuations in temperature that *lucasana* must endure. Extreme fluctuations—that's for sure. Like right now. But something else is troubling me, too. Something besides this illegible message from the channel. Part of Graham's story, I realize now, makes no sense to me. No sense at all.

The Idealists, from Oken to Agassiz, posited a hierarchy, a Great Chain of Being. I've known this for some time, I suppose, though certainly not with the coherence Graham has just given it. But what suddenly strikes me as mysterious is this: when Agassiz and others finally went out and uncovered the fossil record, it lined up quite nicely with the hierarchy they already had in mind. Why would that be? They'd known nothing of evolution—and indeed, as Graham has just explained, most of them rejected it outright. And yet, there they were: the Accidental Evolutionists, their Great Chain of Being anticipating the very succession of beings they found embedded in the earth.

Sifting through my notes, I search out some white space to scribble in—just for a minute or two. First I need to pin down the pattern that wants explaining, because it is wavering for me now—between seeming mysterious one instant and utterly self-evident the next. So here's a go at it—the thing that may, or may not, require an explanation: Groups of living species that the Idealists called "lower" did in fact originate *earlier* than groups they deemed "higher." In their ranking of vertebrates, for instance, they generally placed fish near the bottom, reptiles above them, and mammals at the top. And when you consult the fossil record, fish do in fact appear first, followed by reptiles, and lastly mammals. Is it totally obvious that there should be such a correspondence between the Idealist's lower and the evolutionist's older? For a moment, it seems maybe so: *Of course,* the thought goes, *animals that are ancient are also primitive, and therefore appear lower. Humans, by contrast, evolved recently, and with all their advanced capacities, they appear highest.* But no sooner do you think that thought than it seems a little suspect.

Animals that are ancient are also primitive . . . and therefore appear lower . . . But if the first fish existed hundreds of millions of years before the first mammal, then the fish have had that much longer to produce fish-specific evolutionary adaptations and innovations. Why, then, would the fish look "lower" rather than "higher"? *Humans appear higher . . . because they evolved recently . . .* But there are plenty of species that are newer to the world than humans, including, in fact, a great many species of fish. Why, then, don't these newest of species look highest of all?

So, no, this matter is not as straightforward as it momentarily seems. Our initial *Oh, this is easy, lower means older,* does not hold up under closer inspection. And yet there was, undeniably, a real correlation between the Idealists' Chain of Being and the actual succession of species on earth—at least, enough of one for Agassiz and Haeckel to perceive it. So the question stands: *Why?*

The answer might have something to do with the way Idealists formulated their hierarchy. They must have had in mind a list of the attributes belonging to their favorite organism, which happened to be *Homo sapiens.* The list may have been explicit, and their procedure for using it a conscious one, or the entire affair may have been largely implicit and subconscious—but either way, when they wanted to determine where in the hierarchy a certain organism belonged, they must have done so basically by checking how many attributes from their key list that organism possessed. The more they found, the higher up they would slot the creature. And I believe this procedure—in itself and irrespective of the particular species one believes is ideal—will generate a hierarchy for which a correlated historical succession of fossils can be found. Just to test this notion, I want to try out the procedure, but without reference to any real species. I think some doodles might help in thinking it through, so here goes:

Let's say a certain Idealist, a peculiar fellow named Helmut, has a favorite organism that is not *Homo sapiens*. (It could be anything; just don't think human.) And this beloved creature of his has three attributes he finds noteworthy, which we'll represent as ●, ◆, and ■. It's not hard to imagine an evolutionary tree that could have led to Helmut's ideal:

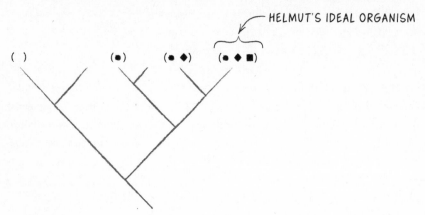

At the top of the tree, I've written which attributes four living species possess. The number of noteworthy traits these creatures share with Helmut's ideal ranges from zero { } up to two {● ◆}, with closer relatives of the ideal sharing more with it. This pattern may seem fairly intuitive, but it might prove helpful in a moment to state now why it is so: the attributes themselves have an evolutionary history; they have come about in the long line of descent that leads to the ideal. In fact, I can just draw onto the tree where in the history of these species the attributes first arose:

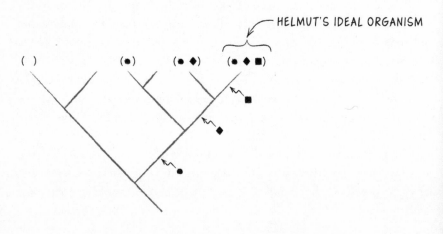

When Helmut draws up his hierarchy, he ranks these species according to the number of attributes they share with his ideal. So his Great Chain of Being would look like this:

And what will Helmut find when he now heads out to look for fossils? We can imagine fossilization taking snapshots of the lines of descent shown in our tree. And if I just draw them in—these moments fossils are deposited in the earth—it becomes clear, I think, that the historical succession of species will reaffirm Helmut's hierarchy:

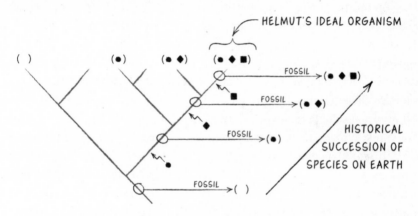

"Look at that," I mutter as my pen streaks an arrow from the bottom of the imaginary fossil record up to the top.

"What's that?" Graham says.

"Nothing. Sorry. Just thinking."

"Do you realize, Hirsh, that I'm in my sleeping bag?"

"I know—this wind is freezing—and does it smell like deep water? Like squid or something?"

"Squid? I don't know, Hirsh. Ask Yukon."

I'm being irresponsible—I realize that. I've got to get back to lactose dehydrogenase and the limits of adaptation. But now that I have the basic model sketched out, I can't resist playing with it a little. What seems somehow revealing, pleasingly iconoclastic about it, is that it shows that you can actually choose anything at all as your ideal and still have your choice nicely borne out by the evolutionary record. Let's say, for example, that Yukon here is an Idealist. And he takes *Canis familiaris* as his ideal species. (He might, in fact, be more inclined to choose Alejandrina, or maybe seagulls—but anyway, for the sake of argument, let's just imagine he thinks very highly of his own kind.) His list of meaningful attributes is: 1. supremely sensitive olfaction; 2. mammary glands; and 3. a backbone. Dogs possess all three traits, so they, obviously, belong at the top of the hierarchy. Most other mammals, including humans, fall well short of *Canis familiaris* in the nose department, but they do possess the other two traits, so Yukon slots them one rung lower on the ladder. And finally, reptiles have only one of the three attributes, a backbone, so they're at the bottom.

Yukon now dons his pith helmet and heads out to do some digging. The earliest appearance of *Canis familiaris* is 20,000 years ago (about 80,000 years more recent, Yukon points out, just incidentally, than *Homo sapiens*). The first appearance of the next tier down, the mammals, is around 200 million years ago. And, as expected, the reptiles appear even earlier in the fossil record, about 300 million years ago. Yukon has documented a perfect correspondence between his great chain—no, make that the Great Chain—and the succession of organisms in the fossil record.

To really push this—to scramble the Idealists' hierarchy even more drastically—we could take as our favorite species a creature that any good Idealist would have considered echt low. We could pick, for instance, the brown cucumber, *Isostichopus fuscus*. With our reverence for him running high, we might write this list of ideal attributes: 1. defense by the toxin holothurin; 2. mutable connective tissue; and 3. deuterostome embryological development, which means the mouth develops after the anus. At the top of our Great Chain of Being, with three attributes out of three, we place only those species of sea cucumbers that are capable of synthesizing holothurin. One rung lower, we have all the other echinoderms, which have both mutable connective tissue and deuterostome development. At the bottom, we place the non-echinoderm deuterostomes, like humans, which never managed to evolve mutable connective tissue. Turning to the fossil record to test our hierarchy, we find the earliest deuterostomes at the base of the Cambrian, around 540 million years ago. The echinoderms don't make their first appearance until about 20 million years later. And the Holothurians show up another 10 million years after that. Once again, then, the succession of species proceeds from lowest to highest on our Great Chain of Being.

I think it's becoming clear now why this always works. As we move further away from our ideal, toward species that possess fewer of its attributes, we are moving outward in the evolutionary tree. In the process, and without quite meaning to, we are also taking into consideration progressively broader groups of species: we move from *Canis familiaris* (one species), to all mammals (thousands of species), to all vertebrates (tens of thousands of species); or, in our last example, from cucumbers that synthesize holothurin (just a handful of species), to echinoderms (an entire phylum), and finally to deuterostomes (a few phyla lumped together). But consider what this means for the ancestry of each group. In a tree, a pair of adjacent twigs come together just a short ways down from the canopy, whereas a broader patch of twigs will reach deeper into the tree before they all join in a single branch. Whenever we widen our group of species, we will necessarily find the ancestor deeper down in the tree—or, more accurately, in the earth. This is why we keep finding that the further we go from our ideal species, the deeper the ancestor sinks in the fossil record. It is an unavoidable consequence of our algorithm for making a hierarchy.

It seems so evident now, I have to wonder: How could the Idealists have failed to recognize that they were considering progressively wider groups of species as they moved away from their ideal? But of course they didn't even think in such terms—did not envision a tree of life, or the concentric circles that can be drawn in its canopy, circumscribing successively wider groups of organisms. They thought rather in terms of ladders, great chains, ascending lines of one sort or another. And therefore they had no notion that each group, each tier in the hierarchy, could have been nested within, rather than positioned on a line with, the groups below it. They were thinking with a different geometry. And what reason had they to doubt it? The beautiful correlation between the Great Chain and the fossil record merely confirmed a conviction that was already strong: *Homo sapiens* stood at the top of a clear cosmic order. How could it even occur to them to wonder what perfect correlation might have been discovered by their dogs?

"Burnett," I say out the window, "you're off the hook."

"What do you mean?"

"I'm not going to ask you about lactose dehydrogenase in front of all the students."

"Why not? I've been studying. Go ahead, ask me something, right now . . ."

I confess that I got myself involved with a certain puzzle, and have been sitting here playing with it instead of preparing for seminar.

"Strong work, Hirsh. What are you going to do?"

"Talk about something else, I guess."

"Like what?"

"Dogs and cucumbers, I think."

Graham must hear the worry in my voice, because he doesn't complain about having studied my assignment only to have it pulled off the agenda. And I can tell he's trying to sound encouraging as he says, "Sometimes improv works. Sometimes the kids are game."

"Right," I say, feeling the wind again. "Let's hope they're game."

3

Holding my coffee and notes—the wrong notes, on the backs of which I've scribbled some hurried and hopeful thoughts about where our conversation might take us—I enter the brightly lit seminar room. Most of the students are still out on the terrace, waiting to be called in, but Ace, Haley, and Miles are already here. They have folded themselves into writing desks that look terribly undersized beneath their three long bodies, as if they were adults visiting an elementary school. They would have more space at one of the wooden benches beside the room's long central table, but I think each of them is the kind of student—and together, they form the kind of group—that is drawn, by some complicated amalgam of humility, independence, and pride, to the periphery of a classroom, far from its front and center. And besides, by taking those undersized seats, I now see, they are able to sit together directly in front of one of the two open window frames on the windward side of the room.

Ace and Haley greet me with beaming smiles—they are both a bit sunburned from our time on the water, so their eyes and teeth appear luminous. Miles doesn't see me enter, because he is doubled over his tiny desk, laughing so hard there's no noise to it, only the occasional gasp for air. His blond curls, drooping from his head, bounce like loose springs with the rhythm of his laughter. He looks as if he would collapse from his chair were it not for his hand grasping Ace's arm. Glancing up at last, he sees me through teary eyes and attempts to say hello, but doubles over again. He can only lift a long, lean arm in greeting and, I gather, apology for his helplessness.

Ace says, "Did you know that Miles swam the fastest two hundred meters ever by a seventeen-year-old?"

As startling and remarkable as I find this revelation, I also recognize that it is being offered mainly as a form of cover: Ace is directing us to a new topic, so they won't have to disclose the real reason for their mirth. But I'm happy to play along, because there are many conversations I should not be a part of—conversations that would collapse what distance remains between me and the students. I am perhaps not so detached from them as Graham resolves to be, but I do agree with him that it is better, for many reasons, not to behave entirely like their companion and peer. Veronica, for her part, possesses certain qualities of character— instinctive reserve, formality, maternal solicitousness—that win for her a role that is warmly caring but clearly not collegial. It is an enviable balance, but those qualities of hers are not ones I naturally share, and consequently, moments like this can be sort of tempting, and therefore hazardous.

"I had no idea," I say. "That's incredible, Miles. Why didn't you put it in your application?"

Ace looks down at Miles's bowed head. "You didn't?" he says.

"No," Miles says, sitting up now and leaning back in his small chair, as if to catch his breath after his fit. "I guess I didn't."

"Why not?" Ace asks. "I'd tell everyone. I'd make T-shirts with my name on the back, and the record time where the number would be."

Miles seems to shake his head—a gentle refusal. "I'm not even on the team anymore," he says, and there is something closed off in his voice. So instead of asking him why—why the fastest seventeen-year-old in the world would be off the swim team at age twenty—I ask him about something he did choose to mention in his application: the triathlons he races.

He sits up, enlivened, and speeds through his recent results: Santa Cruz; Wildflower; several other top-ten finishes. And he's new to the sport, he says, so he still has a lot of room for improvement. In his first race, he completed the first stage—the 1.5-kilometer swim—so far ahead of all the other competitors, including professionals and race favorites, that the spectators did not at first understand that the young man emerging from the water was actually part of the race. Instead of cheering, they looked on in puzzled silence as he jogged up the sloping beach. And poor Miles was terribly certain he had somehow violated triathlon etiquette— maybe he'd skipped a part of the swim course, or maybe he had an embarrassing hole in his wetsuit. On the next two stages—running and biking—he was comparatively slow, and had to watch many competitors whiz by him before he reached the finish line.

"But I'm getting faster," he says, "and it's more fun than five hours a day in the pool . . ."

So maybe he does feel a need to return to the question. "Is that why you left the swim team?" I ask.

"I had shoulder surgery," he says. "It made me sit out a season. And when it was time to go back, I didn't really want to." He explains that a few months away from the pool had allowed him to devote more time to his schoolwork, and he had found it unexpectedly rewarding. He had started working in a lab in the chemistry department; his grades had improved; and now that he was keeping up with the reading, his classes seemed more interesting. On the advice of his coaches, he had started running and biking to stay in shape while he couldn't swim, and though he knew perfectly well he was not as talented in these new sports, he had discovered that he enjoyed them—maybe a bit more than his coaches might have hoped. Now he can't really bring himself to contemplate going back to practice, because he knows how much he would sacrifice.

"You just can't be as good a student—" Miles is saying, when he seems suddenly to catch himself and fall abruptly silent. Only when he leans forward in his chair and looks over at Haley do I realize that he has held his words on her account. She too, of course, is a collegiate athlete of the highest caliber; the women's basketball team, like his own former swim team, is one of the best in the country, and I am sure the demands it makes on its players' time and energy are no less consuming. Shaking his head slowly, Miles says only, "I don't know how you do it." It feels as if some pain is distilled in those words, and I can't but admire Miles for the way he has elected to characterize his own decision. He permits himself no contempt for athletic monomania, and says only that he himself cannot sustain it.

After a moment of silence, a moment that feels taut, Haley says, "I guess I just like being busy," and her tone is entirely untroubled, as though she has felt nothing of the aching tightness in the air. As she shrugs her shoulders, indicating nothing more to say—it's just that, she likes being busy—I feel terribly impatient with her, but then it occurs to me that there is a reason for such bluntness. Maybe she can't really contemplate the choice Miles is making, because doing so could open a crack in the ramparts of her own commitment. Maybe the only way to keep spending five hours a day in practice is never to let yourself wonder what it would feel like not to. Still, even if there is necessity behind her answer, the seconds that follow it nonetheless feel awful—up until the moment Ace opens his arms wide, puts one hand on Miles's back, the other on Haley's, and, leaning back in his seat, says, "Fortunately, I myself never had to struggle with these issues."

"Not much of an athlete?" I ask.

"Loved baseball. Truly *loved* it. But you know how, when you're little, you're kind of waiting for your personal superpowers to be revealed to you? Waiting to find out which of the X-Men you're going to be?"

"Totally," Miles says, sitting up. "I was Aquaman."

"Right. Well, for me, the moment I knew I'd never go pro actually came *before* the moment I understood the superpowers weren't coming."

"Oh, Ace," Miles says, laughing, "that's rough."

"I was *bad* at baseball," Ace says. "Really bad."

Graham appears in the doorway, arriving from the staff house. "What's going on?" he asks.

"Nothing," I say. "What do you mean?"

"Did you cancel seminar?"

"No. We were just talking."

He turns around to look at the clock above the doorway. "Seminar was supposed to start five minutes ago."

"Well, you're late."

He exhales and walks past me, out the other side of the room to the terrace. We can hear him as he says, "What are you all doing?"

"Just talking," says Veronica's voice. I didn't know she was out there—and I'm surprised by it, because I know Becca is: even in here, I've heard her shrill voice rise now and again in domineering salvos.

"Well, get in here," Graham says. "You're five minutes late."

They enter carrying their notebooks and course-readers. Many of the readers are folded open to the article about lactose dehydrogenase in barracuda, and the sight of it causes me an inward wince. What will they all think of this last-second swerve from Veronica's detailed syllabus? What will Veronica think? She comes in last, still talking with Allie and trailed closely by Taiga. Then, just behind Taiga, another dog, a small cream-colored stray, appears in the doorway. She has large black eyes and a short snout, like a puppy's, though I would guess she is an adult. One ear hangs down beside her face, while the other has just enough lift in it to stick straight out to the side, as if she were attempting to signal an imminent right turn. She stands tentatively at the threshold, cautious in the way of strays, as she watches Taiga follow Veronica to the other side of the room. Veronica takes a seat on a bench against the wall, and Taiga settles into a regal, watchful pose by her feet. With Taiga down, the little one appears to decide that she's going to go for it: she lowers her head to its most submissive posture, averts her gaze from Taiga, and moves with quick small steps across the threshold, pausing just a few feet into the room. Taiga has tolerated the entry, and the little one's sidelong gaze seems to fix itself now on Miles, at his little desk by the room's seaward wall. He extends a hand, and she makes a decisive move—settles in quickly at his feet and directs all her attention to manic, submissive licking of his shins. He places his hand on her small head, and she ceases the licking.

Of the various ways life in this town can confront you with the uglier aspects of human habitation—Guillermo's generator roaring late into the night, the dusty litter tumbling past your ankles on a windy afternoon, the boozy gringos excreting piles of aluminum cans onto the beach—it is, I think, the stray dogs that are most troubling of all. Every year, a half dozen or so will attach themselves to our group. They will lurk on the terrace and loiter near the kitchen door, awaiting scraps. One or two might appear surprisingly healthy; I remember a certain blue-eyed husky who wore a shiny coat of silver and black, and could stare at you with the intensity of a wolf. Most of them, however, look to be suffering. That very same husky, just a year later, had open sores on his back, and something had infected his ear, causing him to hold his head sideways and shake it often in irritation. All of them, healthy and sick alike, will respond to the slightest signals of affection, or even of forbearance, with overwrought enthusiasm—whining, tail wagging, trembling of the entire body—and when the spastic fits of joy at last subside, they are replaced by absolute fidelity. Your new companions will wait for you on the beach while you're in the water, and they will trot behind your car as you drive out of town.

Most years, we warn the students about incurring devotion. If they are not prepared to take a stray home, we say, then it would perhaps be better not to invite unwavering fidelity. This year, however, the strays have been mysteriously absent. They weren't prowling the roadside on our way into town, and they have not shown up by the kitchen door. I shudder to imagine the story behind their disappearance from the area, but then, I shuddered at their suffering presence, as well, so I'm not sure what to think. In any case, because this shifty little one is the first to appear, we have not warned our group about the supplicant pack, and now, evidently, it is already too late.

"Who's that?" Graham asks.

"Millie," Miles says.

"Millie?" Graham repeats.

"As in Vermilion," Veronica says. "They named her for the sea."

"You're in on this?" Graham asks.

"No, no," Veronica says. "I just met her—out there on the terrace."

"She showed up this afternoon with Yukon," Miles says, "so we gave her some water and granola bars."

"Oh, Jesus," Graham says.

Yukon and Taiga have always worked at cross-purposes in relation to strays. Yukon befriends them all, invites them in for hot dogs from his patron saint, Alejandrina. Taiga meanwhile threatens the mongrels, especially when Veronica's around, and seems to say, *No, Yukon—there are limits.* But somehow this little one has placated her. Perhaps she's exceptionally adept at canine diplomacy.

Looking around now for Yukon, Veronica asks if anybody knows where he is.

"Up at the staff house," Graham says. "I asked him if he was coming, but he didn't move."

"He's acting weird," I say. "I think it's the wind."

4

"I have two reasons," I say, "for changing the topic of tonight's seminar. The first is a conversation some of us had this afternoon, out on the terrace."

Rafe is seated on the seaward side of the large seminar table, with his back to Ace, Haley, and Miles in their miniature desks. He still has white toilet paper springing from his ears, but I've gotten so used to it over the course of the day that it's no longer much of a distraction—just another part of his look, like the gold hoop. When he realizes he's just been mentioned, he sits up straight and looks around, like a guest of honor who is momentarily uncertain whether he should perhaps stand and gesture to the crowd, acknowledging the speaker's kind words. If I'm not mistaken, he nods slightly to Becca, who is directly across the table from him: Is he perhaps confirming for her that he, and not she, was among the five involved in the conversation?

Anoop happens to be sitting beside Rafe, and he looks pleased as well, though not in a prideful way. He is peering up at me through the peepholes of his lenses, and his expression, which I would describe as thankful, or perhaps even *relieved*, makes me suspect that he has been wondering, ever since this afternoon's conversation, when we would return to the topic of recapitulation. For there was a moment out there on the terrace when I left something unsaid—I failed to mention von Baer's theory of embryology, the main alternative to Haeckel's recapitulation—and I know Anoop noticed the omission, because he looked nearly alarmed that we might let such a misconception stand. And it must have been lastingly troubling to him, because—I am sure of it now—he looks distinctly relieved that our misdeed is soon to be rectified.

"And the second reason," I say, "for skipping the barracuda is that Millie has just joined us, and I'd like to talk a bit about her."

"About *Millie*?" Graham says, as if he may have misheard. He has assumed one of his habitual seminar stations—standing in the door frame on the inland side of the room—and he is smiling with what looks to be authentic surprise, though I just told him half an hour ago that I would be winging it tonight, and might even talk about dogs. Maybe he thought I was joking—or maybe he's intentionally joining in early, helping me drum up interest and energy.

"Yes," I say, "about Millie. I want to try to explain why she's so darn irresistible."

Since the students know that I met Millie only about a minute ago, I may have just confessed that my plans for tonight's seminar aren't exactly well-laid. But maybe that's for the best: maybe they'll take it as an invitation to help me out—to be game, as Graham put it.

Becca raises her hand. Her brown eyes look wider, more protuberant than ever. When I gesture to her, she says, "So are we totally skipping George's work? I mean, will it be on the exam?"

Evidently, Becca is on a first-name basis with the Stanford professor who wrote the barracuda paper.

"Barracuda might still be on the exam," I say, trying to smile. "But Millie will for sure."

As if she were aware that she has been invoked, Millie stands from her place at Miles's feet, balances on three legs, and uses the fourth to perform a standing, twisted scratch of her own scruff. It's a distinctly cur-like posture, a move perfected only among the flea-bitten, and it prompts Graham to suggest that maybe we should put her back outside.

"Come here, Millie," Miles says, tucking her under his long legs.

"Okay," Graham says, with resignation. "Okay."

Now—how to begin? It feels to me like there are two ways this could go very wrong. We could spend much of the next hour rehashing this afternoon's conversation, which would be tedious, if also flattering, for the five students who participated in its first rendition. Or, the other way to fail, the precise complement to the first, would be to proceed from the end of that conversation, holding the interest of the five, but almost surely losing everyone else. The only path between adjacent pitfalls, I think, will be to dash swiftly through Haeckel and the cucumber, and then build from there as a group. And yet—here's the next problem—if I gloss Haeckel myself, I'll end up lecturing, and that's exactly what I can't do. I'm too ill-prepared. I'll land us all in a purgatory of awkward silences. But I think there may be another way—a safer way forward . . .

"So, Anoop," I say. "How about giving us a speedy primer on Haeckel's theory of recapitulation?"

"All right," Anoop says. He touches his hair, adjusts the platinum temple of his glasses. Turning toward him to listen, Rafe appears surprisingly content to be just outside the spotlight; in fact, I think he looks deferential.

As Anoop explains Haeckel's theory, he adds a certain point of emphasis that we did not touch on this afternoon. What the developing embryo reiterates, he says, is the evolutionary succession of ancestral *adults*. Several more times he says it—adulthood, adults, it is the adult morphologies that are recapitulated—until I have little doubt that he is thinking ahead to von Baer, consciously laying the groundwork for our conversation to move in that direction. He could easily lead us there himself, but instead he pauses and looks at me, as if to inquire whether he should proceed. As a student who is used to knowing far more than he has time or space to say, he is practiced in the self-restraint that makes discussion possible.

"Thanks, Anoop," I say. "And Rafe, what evolutionary mechanism would generate such a parallel?"

As if the ideas were his own, coming to him spontaneously, Rafe guides the group through the concept of terminal addition—the idea that evolution has more success appending new stages at the end of development than it does making changes somewhere in the middle. At an especially important and enthusiastic juncture, he yanks the paper out of his ears—perhaps to hear himself more clearly—and deposits it, as a small nest of tissue, on the table in front of him. His usual impatience with uncertainty seems to have been swamped, for the moment at least, by the sheer satisfaction of articulating a theory for his peers. Nearing the conclusion of his short lecture, he stands from his seat, steps confidently past me to the whiteboard at the front of the room, and draws a diagram there: an evolutionary tree, which he labels with pathways of development. With a final flourish of his marker, a sweeping green circle around a line of descent, he concludes: "So you see, if evolution does happen by terminal addition, then it's got to be that *ontogeny recapitulates phylogeny*."

Just as I'm beginning to fear that he might actually remain at the front of the room to field questions, he places the marker on the table in front of him and returns to his seat. He is looking at Becca across the table, but she has her gaze fixed on a page of her course-reader—a page with illustrations of the different species of barracuda. Her manifest lack of interest prompts me to scan the group anxiously, but I find most of them intently engaged: Miles is leaning forward, one hand frozen, mid-scratch, on Millie's head, as he studies Rafe's diagram; seated at the seminar table, Isabel is copying the figure into her notebook; and beside her, Cameron is typing rapidly on the brailler he uses to take notes. Though he was there, on the terrace, this afternoon, I'm not sure he was really with us for this part of the conversation, because it was at about this point that I first noticed him looking awfully dreamy.

I think most students, even the very good ones, find it more engaging to listen to one another than to one of us. There is a kind of magnetism there, an added element of plot or suspense, whenever they are listening to a peer. For some of them, it is no more than the edacious edge of competition—*How well is he doing? Can I, will I, do better?* But for most, I believe it is something quite different from that, even opposite to it—something closer to empathy. Now and then I will see an attentive student bite her lip, another clench his fist around a pen, at the moment their peer seems to get stuck on a word or tangled up in a difficult concept: they are urging him forward, willing him through. Sometimes it's as if the speaker were giving voice to thoughts that are not entirely his own—as if the class were a single yet multifarious mind, working its way toward an articulation or understanding that it, or rather we, can settle upon.

It might be a maxim for teaching: Never say something yourself when one of your students could say it just as well. Heeding that rule, I'll turn next to Allie, who first led us into this conversation with her question about the cucumber's pentamery. She is seated directly across from me, at the other end of the seminar table, and her remarkably legible, silent-movie-star visage is presently expressing anxiety about rising conflict: she is looking back and forth between Rafe and Becca, as if it were up to her to broker amity. But as soon as I say her name—"Okay, now, Allie . . ."—her expression transforms itself abruptly and she projects only clear-eyed attentiveness. "How could all this apply to that brown cucumber?"

She gives a short, lucid summary of our series of symmetries: the cucumber starts as a radially symmetric embryo, develops into a bilateral youth, remakes itself as a pentamerous teenager, and finally returns to being bilateral—but only outwardly so—as an adult. And we can trace the same series in the cucumber's deep evolutionary history.

So far so good: we've dashed through Haeckel's theory, and now we're poised to call it all into question by invoking the alternative. Anoop will breathe a sigh of relief, but right beside him, I'm afraid, poor Rafe is going to lose his mind: no sooner has he relinquished his hard scientist's superiority and embraced Haeckel's story than we're going to turn around and remind him of its essential uncertainty. And it's not just his epistemological sensibilities that stand to take a bruise, but also his status in what appears to be an ongoing tournament with Becca. With his avid presentation at the whiteboard, he has just climbed atop the shoulders of Haeckel's monument and stared down victoriously at his rival. If we now dig in our grappling hooks and bring the monument down—and it does seem likely that our conversation will take us there—Rafe will find himself in the rubble, looking up through the dust. But I mustn't concern myself with their competition. The room feels good—the group is thinking together now—and I believe I can introduce the missing scientific figure without tipping us irretrievably into a lecture.

5

Karl von Baer, German embryologist, preceded Ernst Haeckel by four decades and made several fundamental discoveries. He identified the mammalian ovum; he found that the earliest stage of the animal embryo is a hollow sphere of cells; and he was a major contributor to the theory that a developing vertebrate partitions itself early into three distinct kinds of tissue—endoderm, mesoderm, and ectoderm. He rejected the theory of recapitulation, both as it was advanced by Haeckel's predecessors and as it was reformulated, in the last decade of von Baer's life, by Haeckel himself. He also rejected the theory of evolution by natural selection, though Darwin, with his characteristic omnivory, invoked the work of the anti-evolutionist von Baer as often as he did that of the evolutionist Haeckel. But we'll get to that later.

For von Baer, the deepest problem with the theory of recapitulation—the reason it simply could not be right—was that, in his view, embryological development was the gradual emergence of form, and certainly not a succession of well-defined forms. In other words, an embryo doesn't look like much of anything at first, and it acquires clearly defined features—the morphologies that will show you what, exactly, the embryo is planning to become—only as time passes and development proceeds. How, then, could development possibly reprise the adult forms of evolutionary ancestors, as Haeckel maintained, when embryos don't look like adults of any kind whatsoever?

Haeckel and his followers were aware of this problem. They acknowledged that embryos are not equivalent to full-blown ancestral adults, but maintained that most of the attributes embryos do exhibit materialize at just the right moment in development—that is, at a moment corresponding to the evolutionary era in which those attributes were appended to the developmental program, yielding a new adult morphology. When a troublesome attribute slipped its place in development and appeared out of order with the parallel evolutionary succession, Haeckelians simply set it aside as a case of *heterochrony*—evolutionary change in the timing of an attribute's development, resulting in an exception to the rule of recapitulation. But the rule, they argued, is indeed recapitulation, for how else could they—and, for that matter, others before them—have detected recapitulation in the first place?

For a clear and forceful answer to this question, we cannot turn to von Baer himself; he could never really get started on it, since he repudiated the very possibility of evolution. So we must look, instead, to those biologists of the early twentieth century who heartily embraced both von Baer's embryology and Darwin's evolutionary theory. They argued that the gradual emergence of form in the embryo—the process von Baer emphasized—does indeed generate a parallel between development and ancestry, but it is distinct from Haeckelian recapitulation. Because the attributes that distinguish one species from another emerge gradually over the course of development, the immature stages of a descendant species are simply bound to look like the immature stages of its ancestor. Yes, this is a kind of recapitulation, but it's not the kind Haeckelians talked about, because the ancestral adult never gets recapitulated.

This distinction—between Haeckelian and von Baerian versions of recapitulation—may seem rather subtle, but it has important implications. The two patterns emerge from fundamentally different mechanisms of evolutionary change. So in principle, if we can figure out which sort of recapitulation we are witnessing, we will know also which mechanism of evolution has been in play. As Rafe has just explained, Haeckelian recapitulation would arise from the process of terminal addition. But what evolutionary process would produce von Baerian recapitulation?

When we were thinking about terminal addition, we started with an organism that develops from immature form A to adult form B, and we imagined that this creature gave rise to two new species. Let's start there again, in order to see, this time, what would have to happen differently for von Baer's pattern to emerge:

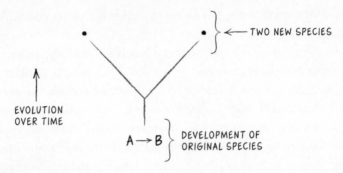

You'll recall that, as our metaphor for development, we invoked the challenge of navigating from point A to point B in an ancient and labyrinthine city. And to extend this metaphor into the realm of evolutionary change, we imagined that, while we held in our hands a set of directions from A to B, we now wanted to reach destination C, which was just two blocks away from B. The only truly safe solution, we said, would be to travel first all the way to B, and then to go from there to C. But perhaps we took the argument a bit too far. It is surely true that making the two-block adjustment very early in our set of directions would be a big mistake, leading us into an unfamiliar maze of streets. But would we really have to go *all the way* to B? Perhaps we could, for instance, tweak our route when we were just a few blocks away from B. That would probably allow us to find our way to C, and do so rather more directly than if we had first gone all the way to B. Similarly, perhaps evolution need not append every new stage to the *very end* of development; perhaps it can instead tinker with the final steps of the existing developmental pathway, changing morphology B, bit by bit, to morphology C:

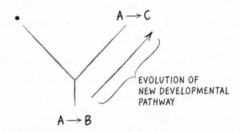

As you can see, under this evolutionary scenario, the developing descendants never exhibit B, the adult morphology of their ancestor. But they do reprise A, the ancestor's juvenile form—and that, as we've said, is von Baerian recapitulation.

Turning around now from the figures I've been drawing on the whiteboard, I find the students looking befuddled: Allie's nose is crinkled, as if she were enduring a pain in her gut; Ace's head is canted back, and he is peering at me through narrowed eyes—a posture of musing resignation to the universe's state of confusion. Even Millie seems to sense an atmosphere of disconcertion: she is holding her head low, eyeing me warily. If the group, just a short while ago, felt like a single mind alight with activity, then I've somehow delivered a sturdy knock to the head, causing many bright connections to wink out. Not knowing what else to do, I search anxiously for attentive eyes—something, anything, to focus on.

First I find Veronica and Graham, and though it's clear they are not confused like the others, they cannot help—in fact, they only make matters worse, because I can see that they are trying hard to look wildly interested, which means that I am indeed in trouble. And in any case, staring at them as I go forward would be to euthanize the seminar; it would be relatively painless, but also effectively over. There is Anoop, of course, who still appears gratified that we are finally talking of von Baer, but looking to him is almost like relying on Graham and Veronica—similarly easy, and a pedagogical surrender. So my best chance, I think, will be with the other students who were there this afternoon. But Chris, who was such an important part of our conversation on the terrace, is presently showing me only a head of glossy black hair: he is hunched over his notebook, examining phylogenies he has drawn there. Cameron doesn't look lost, but neither does he appear entirely comprehending: he is typing something, his blue-green eyes of glass stationed straight ahead. And then there's Rafe—who actually looks quite calm and attentive. Not the least bit frustrated or disturbed. Whether this means he does not feel toward Haeckel's challenger the way I thought he would, or rather that he has understood so little of what I've said that no such challenge is apparent to him, I cannot say for sure. But at this point, I can't be picky. So be it: I'll rely on Rafe.

"Rafe," I say, "did you see the needlefish today?"

"Needlefish? Those silver eely things?"

"There were a million of them," Miles says. "I swam with them for a while. They were eating stuff off the surface."

The room murmurs about needlefish; others saw them, too; someone says, "Did you see those nasty teeth?" The recollections seem to awaken them, but even so, as I venture back into recapitulation, I keep my eyes mostly on Rafe.

The needlefish's closest relative is called a halfbeak. And when a school streaks past, the way that one did today, it's sometimes hard to tell which species we are glimpsing. The main difference is in the jaws. As its name suggests, the halfbeak has only half—the lower jaw, or mandible—of the needlefish's elongated snout. The upper jaw, or maxilla, is short and inconspicuous, and looks much like that of the halfbeak's other next of kin, the flying fish, which has a rather normal-looking fish mouth. A few years ago, I read a superb paper that determined the phylogeny of these fishes. It looks like this:

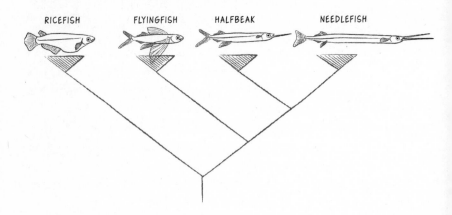

Above the tree, I've sketched the basic form of each lineage. Based on the arrangement of different jaw forms at the top of the tree, we can infer fairly easily where in the evolutionary history of these various species the jaws underwent important changes. And what's intriguing is that, when we compare our reconstruction of evolutionary transformations to the development of just one species, the needlefish, we find nicely parallel sequences of change:

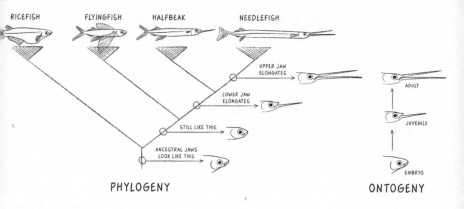

Perfect! proclaims the Haeckelian: *Ontogeny recapitulates phylogeny, just as we expected!* But the von Baerian stops him right there. *Just you wait!* he says, and then points out that if you watch a halfbeak develop, you see it go through these steps:

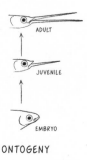

ONTOGENY

So how do we know, the von Baerian asks, that the developing needlefish is recapitulating the morphology of the *adult* halfbeak, and not that of the *juvenile* halfbeak? Perhaps what we are seeing is not Haeckelian recapitulation at all. Perhaps the fishes share their early developmental program, but each follows its own late developmental program.

But what's the big deal? I mean, why does it really matter which half-beak stage is recapitulated during needlefish development? Well, if it's the adult stage, then evolution of the needlefish happened by terminal addition: elongation of the upper jaw was appended to the very end of halfbeak development. But if it's the juvenile halfbeak that's recapitulated, then evolution happened not by terminal addition, but rather by modification of the late stages of development: the adult halfbeak stage was effectively replaced by the adult needlefish stage. And in the history of the study of evolution, this distinction—appending new stages versus replacing old ones—turns out to be very important indeed.

You'll recall that Darwin made room for a variety of mechanisms in his own view of evolution. Among these was the mechanism he called *use and disuse of parts*: if an animal's persistent efforts modified its own form—enlarged an important muscle, for instance, or stretched out a limb—then the modifications could be inherited by the animal's offspring. Consequently, the members of a lineage could strive their way, generation by generation, to a substantially altered form. We have come to associate this idea with the French naturalist Lamarck, who preceded Darwin by sixty years, but it was also embraced by a number of Darwin's contemporaries, including his mentor Robert Edmond Grant.

Terminal addition fits comfortably into a Lamarckian view. Toward the end of their own development, the fortunate offspring of industrious parents get to progress effortlessly through the bodily improvements that Mom and Dad worked so hard to achieve. And that, of course, makes for perfect Haeckelian recapitulation, which therefore turns out to be a natural embryological complement to a Lamarckian theory of evolution. A von Baerian process of gradual differentiation, on the other hand, fits less easily with Lamarckism. That embryological process, we've said, emerges when evolution revises late stages of developmental differentiation. But since many of those stages only ever appear in an embryo—not in an active adult—it's hard to see how they could be transformed by effort and striving. After all, the embryo isn't *doing* anything; it's just developing. So it's harder to make von Baerian embryology, and the evolutionary mechanism that leads to it, harmonize with a Lamarckian story.

Graham raises a finger—a Socratic finger. But why am I looking at Graham anyway? I'm supposed to be looking at the students—at Rafe, Chris, maybe Anoop for the really difficult parts. But I see now what's happened: I've gotten involved in the argument—this claim that von Baerian embryology didn't jibe with Lamarckism—and my attention has shifted to Graham simply because I'm eager to see how he'll respond, what he'll say to a claim I've never actually tried on him. Wisely, however, he says nothing. The Socratic finger tips over, directing my attention across the room, to Anoop, who is, I see now, waiting patiently with his hand in the air. And despite my lapse, my abandonment of the students, they appear to be with me now: Allie's nose has uncrinkled; Ace no longer looks like he's philosophically exhaling cigarette smoke; Rafe, remarkably, remains calmly attentive; Becca I choose not to check on.

"Yes, Anoop?" I say.

"Do you think this is why Darwin was enthusiastic about von Baer, even though von Baer rejected Darwin?"

"You mean because von Baer could fit with Darwin's own theory of natural selection, but not with Lamarckism."

"Correct," says Anoop. "It was a clever strategic move on Darwin's part, embracing von Baer."

I was indeed heading in Darwin's direction, but to make a different point. Anoop is certainly right that rhetorical considerations alone would have instructed Darwin to embrace von Baerian embryology, the story that could count against Lamarck's mechanism of evolution. But that is not, as best I can tell, the way Darwin thought. Perhaps because he was personally familiar with an extraordinary variety of embryological observations, some apparently Haeckelian in character, others more von Baerian; or perhaps because there was so much uncertainty still in the evolutionary science he himself was just founding; or maybe just because he was a humble sort of man—whatever the reason, in embryology as in other areas, Darwin proved himself a theoretical pluralist, a both/and sort of thinker, never much inclined toward either/or. In the last paragraph of the *Origin's* long section called "Development and Embryology," Darwin somehow managed to incorporate both von Baerian and Haeckelian views into his final synthesis. To me, it's a beautiful instance of the way he remained broad-minded—perhaps to meet the measure of the wide range of natural history he knew.

"Hang on," says Graham, just as he vanishes from the door frame he's been standing in. From the room next door—the station library—he continues to say, "Hang on, just hang on, two more seconds," until he reappears in the door frame, holding now a thick volume. He flips through the pages to find his place, then holds the book open in one hand as the other hand exhibits, once again, the Socratic finger, and he reads with stentorian verve:

"Thus, as it seems to me, the leading facts in embryology, which are second to none in importance, are explained on the principle of variations

in the many descendants from some one ancient progenitor, having ap-peared at a not very early period of life, and having been inherited at a cor-responding period. Embryology rises greatly in interest, when we look at the embryo as a picture, more or less obscured, of the progenitor, either in its adult or larval state, of all the members of the same great class."

How does Darwin pull that off? How does he make room for two theories that just about everyone looks on as incompatible? Novel varia-tions arise *at a not very early period*—so it could be in late stages, as von Baerians would have it, or it could be at the very end, by terminal addi-tion, as Haeckelians would have it. And sometimes it's the larval form that's recapitulated, sometimes it's the adult. It's as if he is remind-ing us all of the enormous amount of theoretical room afforded by the great variety of living things. You can almost hear him saying it: *There are more things in heaven and earth, Horatio, than are dreamt of in your philosophy.*

6

"But what about Millie?" asks Miles.

"Millie?" I ask.

"You said you were going to talk about Millie, and why she's cute."

"It's her sideways ear," Haley offers.

"Her ear is cute," I say, "but I meant ultimately, not proximately."

"The ear is ultimately cute," Ace says.

"Right, but what's the ultimate cause of her ultimate cuteness?"

"As in, how did she evolve to look like that?" asks Anoop.

"Right."

"Hey, mates?" Rafe says. "Before we move on to Millie, could I just get this straight? We're saying, sometimes recapitulation is Haeckel's kind, and sometimes it's von Baer's kind?" He appears intently focused, but not upset about Haeckel's demotion, as I had feared he might be. Nor does he seem impatient with what is, I would have to agree, a not-so-simple story. I do not understand how we've been delivered this new Rafe—the student I rode with down the peninsula, even the one I spoke with on the terrace this afternoon, would have been terribly tense by now, demanding a clear and *scientific* answer, signaling *enough already with the dead Germans!* But for reasons that remain mysterious to me, he has adopted a new style, a new mode of discourse. I don't know where he's gotten it, or how he's snapped into it so abruptly, but what I do know is that I'm very grateful for it. And there's a chance, I think, that it might even help us retrieve Becca from her walled resentment: maybe Rafe's surprising admission of uncertainty will make a little room, inviting her to drop the silent boycott with which she began the seminar. But when I finally bring myself to check on her, she is still looking down at her reader, which is open now to an article about the history of whaling, the assignment for Graham's lecture tomorrow night.

"That's right," I say to Rafe. "Either kind of recapitulation—von Baer's or Haeckel's—can happen, depending on the mechanism of evolution."

"Darwin's or Lamarck's," Rafe says.

"No, no!" I say urgently—maybe too urgently, for it occurs to me that Rafe could take it badly, that I might refute him right out of this wonderful new sensibility he's found—but this is a profound misunderstanding he's just expressed, and I have to correct it as clearly as I can. "Lamarck's theory is ninety-nine percent wrong," I say. "When an animal's form changes because of its efforts or experience, those changes are not passed on to the next generation."

"Right," Rafe says, "'cause those changes aren't encoded in the genes."

"Right," I say, relieved. "But *Darwinian* evolution can produce either form of recapitulation. If selected mutations add on to the end of development, we get Haeckel. If selected mutations alter the latter stages of development, we get von Baer."

"Got it," says Rafe, as he smiles and leans back in his chair. "Now, what's that mongrel got to do with all this?"

Millie has fallen asleep. Her black nose and furry white muzzle are pressed against the tile floor as if it were goose down. Above her, Miles's tan legs shelter her like a lean-to. She looks harmless, adorable, and extremely trusting of her new clan. And since we're thinking about Darwinian evolution, we've got to wonder how that process could possibly create an animal that deserves those three particular descriptors: harmless, adorable, trusting. I mean, how did nature red in tooth and claw give us this?

"It didn't," Miles says.

"It didn't?" I ask.

"Darwinian evolution didn't make dogs."

To clarify the point, Anoop adds, "Dogs were domesticated—by humans."

7

I would agree with the first part of that statement, but not the second. Dogs were indeed domesticated. But it's probably wrong to assign human agency in this most ancient case of domestication. Dogs began their descent from Eurasian wolves sometime between 15,000 and 35,000 years ago. That's at least 5,000 years before the Neolithic Revolution, when the first plants and several other animals—sheep, goat, pig, and cow—were domesticated. And the dog was also something of a zoological anomaly: it was the only one among the ancient domestics that had descended from a carnivore, and it therefore shared few behavioral or ecological traits with the rest of humanity's menagerie. For a number of reasons, then, one might not be surprised to find that the process by which the dog entered our fold is a story all its own, distinct from the other cases of domestication.

Contemplating the earliest suite of doglike skulls—those older than 12,000 years, let's say—you could characterize their relation to the archaeological remains of human settlements in this way: dogs are sporadically associated with Paleolithic sites, but they are not predictable or especially abundant residents; and dog bones do not seem to occupy a place of special or distinct status in human settlements, but rather seem to be heaped together with other animal skeletons. The earliest skull— the age of which is still somewhat befuddling, as it appears to be 32,000 years old, or about 18,000 years older than its closest successors—was recovered from a Belgian cave, where Paleolithic tools reveal intermittent human residency for at least the past 40,000 years. The skull was found among bones from mammoth, lynx, and red deer. The next skulls, dating to the neighborhood of 12,000 to 15,000 years ago, were likewise found among skeletons of other species, including rhinoceros and mammoth. Many of the bones, including some of the canine ones, show cut marks or perforations, indicating that the animals were eaten.

Then, sometime around 10,000 years ago, the relationship seems to shift. Dogs begin to appear more commonly in close association with human settlements. Most tellingly, they begin to show up in carefully laid graves, sometimes with other dogs, and sometimes with humans. They assume, it seems, the special status they are to maintain for millennia to come, leading right up to Taiga keeping watch over Veronica, and Yukon perpetrating all sorts of mischief without fear of serious punishment, like getting eaten.

The evidence is piecemeal, but even so, what is worth focusing on in this fragmentary picture is the relative timing of two different kinds of change. The skeletal transformations that mark the descent of dogs from wolves appear to *precede*, by thousands of years, the cultural changes that would seem to indicate a shift in the relationship between dogs and humans. Morphologically, at least, wolves became dogs while they were still lurking at the edges of the Paleolithic camp, and well before they were welcomed into Neolithic life. To put the point differently, I would note what we do *not* find: evidence of captive wolves kept in or near settlements; a gradual morphological transformation, from wolfish to doglike, among canids already living in close association with humans; or any other signs of Paleolithic taming or breeding of wolves. Rather, it would appear that, before wolves could join the village in any stable or intimate way, they had to become dogs.

And what exactly did that require? What sorts of changes had to be made? There are several diagnostic measurements that distinguish the skulls of ancient dogs from those of wolves: among the ancient dogs, the snout is proportionally shorter and broader; the total length of the skull is smaller; and the braincase is wider. And the ancient dogs also seem to have had relatively larger and rounder feet. Interestingly, there is a very simple way to summarize all of these differences from wolves: adult ancient dogs looked like oversized wolf puppies. And that may in fact be what they were. For there may have been a straightforward way for evolution by natural selection to extract a dog from a wolf: simply slow down or postpone developmental maturation relative to growth, so when the animal is full grown, it remains, in shape if not in size, a puppy.

But even if it was just one developmental dial that had to be turned, why did natural selection actually turn it? What special advantage accrued to an overgrown wolf pup prowling the hem of the campfire's light? Here we are speculating, but once we start telling it, a certain story does seem to gain momentum. *Homo sapiens* was already an extraordinarily successful species, expanding its numbers and range like no other mammal before it. For wolves, therefore, scavenging scraps at the settlement's edge was probably a recipe for an ecological expansion of their own. But how tolerant would Paleolithic humans have been to a predator in their midst? Wouldn't they have driven off any prowling animal that appeared to pose a threat? Perhaps so, in which case, the best strategy for the scavenging wolf would have been to disarm, in both senses of the word: to divest itself of threatening weaponry, and thus to relieve its host of suspicion or hostility. And one way to do that would have been to look a lot like a puppy.

Perhaps, then, it was not we who domesticated the wolves, but the wolf pups who disarmed us. And in all likelihood, it wasn't a trick of mere appearances. Turning that developmental dial in the direction of youth probably affected not only the animal's morphology, but also its mind and character. By the standards of wolves, adult dogs do not always act their age, but rather behave more like pups in certain ways. For one, dogs continue to engage in play for the duration of their lives, whereas wolves eventually grow out of the fun and games, becoming more serious adults. And, with a few exotic exceptions, adult dogs are generally less aggressive than wolves, even those that are reared among humans. So it seems likely that ancient dogs not only looked less dangerous, but really were.

And so we are joined by puppy-faced, sweet-natured Millie, who also happens to give our theories of recapitulation a surprising twist. In Haeckel's strong-program version of the theory, the descendant passes through a developmental stage in which it looks like the adult ancestor. But consider carefully what happens here, with the wolf and the dog. In this case, the ancestor, the wolf, passes through a developmental stage in which it looks like the adult descendant, the dog. In other words, recapitulation is exactly *inverted*. And it is not only the dog, with its unique history of self-domestication, that exhibits such an up-ending of recapitulation. In fact, there is a technical term for evolution that turns back the master developmental dial: it is called *paedomorphosis*, as in, taking the form of a child. And it has played a role in the evolution of a wide variety of species, both domestic and wild. Among amphibians, for instance, paedomorphism is not uncommon: whenever a species finds itself in an environment where gills would be useful throughout life, it has the evolutionary option of dialing development down and thus becoming a lifelong tadpole. And whenever evolution happens by paedomorphosis—be it in a dog, a pig, a newt, or a cockroach—recapitulation will go all topsy-turvy. Ontogeny will recapitulate phylogeny upside down. So if von Baer blurs the edges of Haeckel's theory, calling into question, for instance, whether it is the juvenile halfbeak or the adult that gets recapitulated during needlefish development, it is really Millie—with her large eyes and round paws, her short fuzzy snout, her right ear struggling to stand up—who reminds us how unwise it is to venture universal theories in the area of evolution. It's best instead to bear in mind Darwin's pluralistic bent.

Stephen Jay Gould, who wrote so compellingly about many topics in evolutionary biology—including, come to think of it, the limits of adaptation, my long-lost seminar topic—argued in a number of books and articles that human beings were derived from great apes by a process of paedomorphosis. Or, as he liked to put it, that adult humans are essentially *juvenilized great apes*. I can certainly see why the idea is attractive. Most obviously, adult humans share a number of physical features with very young apes, including reduced body hair, flat faces, and a large ratio of brain to body size. But there are also deeper reasons, I think, for the argument's appeal. As we all know, we like to play, even as adults and, perhaps most enduringly, with our minds: crossword puzzles; poetry; non-Euclidean geometry—from a certain grown-up and utilitarian point of view, it all seems rather ludic and useless. And I think we harbor a suspicion, too, that we are a uniquely callow species: petty; irresponsible, at least in comparison with the single-minded parental figures of the wild; and, most terrifyingly, inclined to resentments and temper tantrums, which might one day take the form of a thermonuclear explosion. In what senses this notion of special puerility is accurate, I'm not sure, but I do think it was part of the reason Gould found such a receptive audience for the juvenilized ape argument. And the eager reception does need explaining, because the argument is actually quite wrong.

As we saw in dogs, paedomorphosis involves reducing the total extent of development. Natural selection may achieve this end by two different means: by decreasing the rate of development or by shortening its total duration. In dogs, paedomorphosis results mainly from a decrease in developmental rate, though some toy breeds do have abbreviated gestational terms, as well. But when we consider that definitively human trait—our large and complex brain—we find that it is formed not through any retraction of development in general, but rather through an extension of brain development in particular. The part of the brain that is most different between humans and other primates is the cerebral cortex, that crumpled outer quilt, which is composed of billions of wildly interconnected neurons, like so many leggy spiders holding hands, interspersed with their microscopic life-support systems, or glial cells. The generation of new cortical neurons occurs almost entirely during a key phase in early fetal development. Among humans, that phase is many days longer than it is in other primates. As a result, the total amount of neuron production is dialed up—not down, as it would be in a case of cerebral paedomorphosis. Subsequent phases of brain development also have longer to play out in humans: the process of myelination, in which neurons are sheathed in a layer of electrical insulation, goes on for years longer in humans than it does in any other primate; dendritic and synaptic growth—the extension and interconnection of those spidery legs—carries on for decades longer. When we really look carefully, then, we find that our brains have not evolved by paedomorphosis at all, but rather by the contrary process of *peramorphosis*, or overdevelopment.

Perhaps you are now wondering if this means dogs have underdeveloped or small brains relative to wolves. They do not. Development is slowed in dogs, but because brain growth happens early and bodily maturation late—in canines as in primates, juveniles have relatively big heads—the dog brain does attain wolfish proportions. But there is a subtler point to make here. Adjusting the extent of development—turning the master dial, and thus tuning many variables at once—is a potent mode of change available to natural selection, but it is by no means the only mode available. Some genetic changes affect the master dial, but others affect development in more circumscribed ways. For instance, dogs are more inclined to form emotional attachments to human beings. And though this might strike us as a rather childlike attribute, we can be fairly sure that it did not evolve by paedomorphosis, because wolf pups—even those that are bottle-fed from birth—do not bond as deeply as dogs with their human caregivers. More concretely, when you walk to the front door to leave, your dog will show all sorts of distress; your wolf pup won't worry overly much.

And dogs have certain cognitive attributes, too, that do not seem to have come about by paedomorphosis. If you've hidden a cookie somewhere in your living room, a dog will read your gestures—anything from obvious pointing to more discreet but clearly directed gazing—to find his way to the prize. You can train a wolf to do the same, but it takes some work, and even then, he is unlikely to reach a dog's level of proficiency. It's tempting to interpret this as a sign of greater intelligence in the dog. But that would be to make an error precisely analogous to that of the Idealists who found a way to see anything akin to human beings as higher, and anything less like humans as lower. Because in fact, the dog's skill in reading our gestures is probably not a single facet of a generalized intelligence, but rather a highly specific attunement to human beings—a perceptual focus that, in certain circumstances, can produce remarkably daft behavior. Say you tell a dog to watch you—*Fido, look!*—as you hide a cookie, four times in a row, in location A. And then, in Fido's plain sight but without the same verbal instruction, you place the prize in location B. The dog, remarkably, will perseverate in searching location A, the one you seemed so intent on teaching him about just a moment ago. The wolf, meanwhile, will simply go get the cookie.

In this last experiment, the dog resembles not young *Canis lupus*, but infant *Homo sapiens*: a ten-month-old baby will show precisely the same misguided persistence. What we see, then, in this particular cognitive attribute, is not a case of paedomorphosis, but rather one of evolutionary convergence—between *Canis familiaris* and *Homo sapiens*. Both species appear to be prioritizing social information—the enthusiastic communication they have received from their mentor—over their own direct observation of the prize being hidden in the other location. And yet, as is generally the case with convergent evolution, attributes that appear similar at one level of resolution prove different upon deeper inspection. Say you hide the cookie in place A four times in a row, generally doing your best to look pedagogical—*Look here, I'm hiding it!*—and then you leave the room. A new person takes your place for the next round, in which the cookie is deposited—quite obviously, but without all the pedagogical fanfare—in the other location, B. This time, the baby once again perseverates at A, but the dog now goes for the cookie at location B. It's as if the baby trusts all members of its species, whereas the dog trusts one person in particular. The wolf, meanwhile, elects to go it alone.

8

Years ago, I saw a solitary coyote stretched out in the crescent of shade beneath a sandstone boulder. I had just cut my panga's engine and was letting her glide quietly toward the ivory sand of El Rincón. Stingrays skittered from the boat's shadow, like dark paint splattering under clear water, but my eye caught movement somewhere unexpected—at the back of the beach, where the steep rocky slope disappears into the smooth shoulder of tide-swept sand. The coyote was there, with stone at his back and sand beneath his belly, and he had just lifted his head at the sight of my approaching boat. The insides of his tall triangular ears were gray, and his eyes were golden, but everywhere else he was dark, like an etching of fine lines on copper. As the hull eased into the sand with a hollow whisper, he rose on long, ready legs. It must have seemed strange and threatening to him that a large creature of the sea had just become a large creature of the land. In a smooth dark swirl, he rounded the boulder and came up its back, to stand atop it and view me. His golden eyes were somehow balanced on the thin edge between unseeing and all-seeing, like the strangely unfocused gaze of a cat. His torso and tail were long and tubular, and dark—so surprisingly dark. In three more rising curves—an inky strand of oil smoke—he was up the hill and gone.

I would remember him, because encountering a wild animal feels always to me like a silent benediction, and when the animal is unusual—I had never seen a black coyote—the moment of grace feels especially inscrutable and potent. In an openly superstitious age, we took such sights for omens, and though we now resist—wisely, I believe—the impulse to read them that way, such appointments with the unique feel nonetheless distinctly meaningful. So I was not quite sure how to revise my memory of that moment, how to edit its meanings, when I later happened upon an article on the genetics of melanism—dark coat color—among North American canids. The researchers showed that blackness of coat in the wolves and coyotes they investigated was due to a mutation that had originated in domestic dogs, and had subsequently crossed over into the other species, via bridges of hybrid animals. The technical term for this kind of movement of a genetic trait from one species to another is *introgression*, and suddenly now I heard in the term certain connotations I had not noticed before, and they felt apt—for this seemed an intrusion, and also a transgression. It felt to me like my encounter had been tainted. What I had witnessed seemed suddenly less like a being turned black by the touch of fate, less like a glimpse at that deep source of variation that has fueled evolution, and less like the wink of an eye from nature. It seemed more now like a shard of smoky mirror, discovered deep in a forest where we had been so naïve as to dream we were the first ones to tread.

But then I remembered his eyes—those all-seeing and unseeing eyes. They were not the eyes of a dog: not the eyes that look to you with bonding affection, and certainly not the eyes that look to you for a cue. They were the eyes of a wild animal. He was a wild animal, wreathed though he was in our creeping domestication. And now I do take him to be a sign— not an inscrutable omen, but a symbol for which I know certain meanings. He is another reminder that the boundary between the town and the wilderness is porous; that we are already out there in profound and unavoidable form; and that, despite all that, there is still something there that is other, something that is not our own reflection.

PART VI

Mobula lucasana

THE END OF NATURE

1

It is disorienting to wake to a state of incredulity. Are we really going to flee? Evacuate? Abandon ship? None of the words makes it seem any more real. I count the mornings we've had on the water: three—only three— though it feels like more, because the group has already eased into the rhythm. And now, all of the sudden, that rhythm is to be halted? But this morning looks just like the others: the bay is brimming with placid fire; two pelicans slide effortlessly across the brilliant, glassy surface, stringing behind them a pair of slight, parting wakes. If something is different, it is only that the gulls are strangely quiet. And where Veronica and Graham should be, there are only vacant cots, mutely admonishing me for being the last to wake.

Folding the cots, I survey the south bay: Estero La Mona; the thin strip of green behind the beach; the flowing gown of sand that sweeps up to the higher peaks, which look exceptionally rough and variegated in the sidelong morning light. And no sign of weather. Not even a wisp of cloud. So how am I going to make myself think seriously, on a morning like this, about grave messages and impending threats? How is one supposed to feel the gravity of such warnings, when they are delivered from a satellite somewhere in outer space? And when they arrive so unconvincingly, in the abstract form of words and cartoons on the screen of a laptop. And when you can just as easily look to the south and see for yourself only clear sunlight, shattering into rays across sharp and jagged peaks.

Like the gulls, I am moving slowly. Last night, when I finally closed my notebook, checked the laptop one last time, and walked from the staff house down to the cots, where Veronica was long since asleep, I decided to leash Yukon to one of the legs of my cot, because I feared that his anxiety about the wind might cause him to run off into the desert and lose his way. Joyful, affectionate Yukon has a dark side, an alternate personality, in which he becomes monstrously strong, wickedly intelligent, and maniacally bent on escape. It is brought on, this terrible mixture of Hulk and Hyde, by anything Yukon associates with thunder. I learned about it the first time I met him, at the animal shelter, where I was handed a notecard written by his previous owners. They gushed about his playfulness and unbounded love for creatures great and small, but then their voice swung precipitately from warm and personal to detached and clinical: *Warning*, they wrote. *During thunderstorms, this dog may cause EXTREME DESTRUCTION.* The final two words were underlined three times. When I looked up from the card, the attendant who had handed it to me—an ashen-faced man in his forties, wearing a black T-shirt under green scrubs—looked bored and smug, as if he knew we were wasting our time. He then dutifully informed me that Yukon was the only animal that had ever escaped from that particular animal shelter, and he had done so twice.

I did take these warnings seriously, but I also registered them, at some level, as a kind of sign, because I remembered that my very first dog, the dog of my childhood, had also been terrified of thunderstorms. So when the attendant said that those underlined words on Yukon's notecard amounted to a death sentence, since they would stop anyone in his right mind from adopting him, I suddenly felt that I had little choice in the matter. Since then, Yukon has crashed through second-floor windowpanes, deciphered or destroyed countless latches and locks, and on one occasion, when he was unable to shatter a double pane of glass, removed the entire window frame from the wall of my house, leaving a gaping rectangular hole. So last night I knew perfectly well that if Yukon wanted to escape, he would, and therefore I had to tie him to my cot—rather than to the post of Melissa's porch—so that he would wake me with his struggle, and I could tumble from my cot and clutch him. Which I did, over and over again, before the wind finally exhausted itself at four in the morning.

By the time I am crossing the gravel lot on my way up to the staff house, the sun is already hot against my back, and Veronica and Graham are there, sitting on the veranda.

"Hirsh," Graham says, "what the hell?"

"Yukon kept me up. He was freaked by the wind."

"Clever dog. We gotta break camp."

"Any update?"

"Yep, this just in: San Diego, here we come."

"Same red line," Veronica says, morosely.

She is referring to what the National Hurricane Center calls the *path of highest probability* for Hurricane John. When I refreshed the center's website for the final time last night, its map of the eastern Pacific and the coast of Mexico emerged slowly from a mist, as if I were skydiving onto it through high clouds. Our Internet signal comes to us through a satellite dish on the roof, which seems to invent new and creative forms of interference whenever a connection is truly needed. And this time, even after I had penetrated the clouds and was gazing on a clear view of the map, no sign of the storm appeared. No red line showing where the hurricane would go. No little dingbat slowly rotating, calling to mind a small boat propeller more than anything else, but supposedly marking the precise location of the storm's eye. None of it. Just a brilliant image of blue ocean and green continent, and for a moment, I thought the storm might never appear. Maybe it had been a big mistake. Or maybe it had dissipated at a miraculous rate.

But then, like a final wicked flourish from our satellite dish, it all appeared: The dingbat was 180 miles west of Acapulco. And from that point, the path of highest probability traced northward, straight for Baja. Just before it hit land, though, it swerved left, heading out harmlessly into the Pacific Ocean. So at least there was that—that merciful turn to the west, like a last-second reprieve. But there was also a little button, at the top of the Hurricane Center website, which promised to show *possible trajectories*. Clicking on this caused swaths of transparent color to appear around the red line, and all of a sudden, the dingbat looked less like a boat prop churning northward than a projectile hurtling southward, streaming behind it a widening tail of red, orange, and yellow fire. But what the fiery tail actually represented, of course, was not where the meteor had streaked across, but where the hurricane might go. The narrowest swath of color, which was a slightly paler tint of red than the line down its center, showed the region that had a seventy-five percent chance of experiencing the hurricane's strongest winds. A slightly wider swath was orange, and covered the places that had a fifty-fifty chance. The widest swath was yellow, and showed the area with a twenty-five-percent chance of seeing the storm's worst. Last night, Bahía was on the outside edge of that widest swath.

"Are we still in the yellow?" I ask as I unfold a chair to sit beside Graham. Veronica is seated on the veranda's low concrete wall, where she has a view past the north end of the station, at a slice of the bay. Her binoculars rest on the wall beside her.

"Just barely in the yellow," she says. "We're almost outside it."

"So there's a chance—"

"Wait," Graham says. "What are you guys trying to say?"

"We're just thinking it through," I say. "Right now, we're in a zone that has a twenty-five-percent chance—"

"A twenty-five-percent chance of what, Hirsh? Flash floods? Wind tearing off the roof? Mud slides off Mike's Mountain?"

"Mud slides, Burnett?"

"I don't know, Hirsh. That's the whole point. We have no idea. And we've got these kids with us. And if just one of those kids' parents starts freaking out and calling deans and chairs of departments and—you know, we could be in big trouble. So I say we go."

"I'm getting that sense."

"I mean, I'll defer to you two. V's the boss, so it's her decision in the end. I'm not bucking the structure. At least not yet. But I don't want these kids on my conscience."

"Any more parental e-mails?" I ask.

Veronica shakes her head, and at the same time lifts her binoculars to peer out at the bay. It's not the quick, focused movement she would perform if she'd just spotted something; it's more of a perfunctory scan of the horizon, and the gesture somehow suggests to me that she feels the same numbing incredulity that I do. All morning, something has been conspiring with the stillness of the bay and the clarity of the skies to heighten the sense that this hurricane, this immense storm supposedly coming our way, is of dubious reality. And I think I know now what it is, what's been helping to conjure such doubt. It is the sheer contingency of our own awareness of the storm. After last night's discussion of cucumbers and Millie, the three of us gathered in the staff house, feeling rather jovial. Graham pried caps off of beer bottles, saying the wind was so cold we ought to make hot toddies instead. Veronica absentmindedly opened her laptop for our daily glance at e-mail.

"There's a message here from Allie's father," she said, and suddenly we all fell silent—I think because our first assumption was that there had been some sort of emergency in Allie's family. Then Veronica read aloud: *"I am sure you are already taking all appropriate precautions in view of the sudden upgrade of Hurricane John to category four."*

"Who?" Graham said, setting his beer on the table beside Veronica and leaning over her shoulder to look at the screen.

She read on: "*I just wanted to offer to do anything I can, on this end, to assist in timely evacuation, should that become necessary.*"

As Veronica tried to click through to the Hurricane Center website, our Internet connection went down. She tapped her track pad again and again, harder and harder, as if the impact might make the satellite dish feel her urgency.

"Fuck," she said.

"It'll come back," Graham said, still hanging over her shoulder.

"Right after the hurricane," she said, without turning to look at him. "It's a fucking sea cow."

"A what?"

"A Hansen sea cow."

Graham looked at me for translation. "Steinbeck's outboard," I said. "It ran perfectly on nice days, when he was happier rowing, then died when it could strand him in a bad spot. He called it a sea cow."

An image began to materialize on the screen, but then stalled again.

"Wait," Graham said, "it was working. Maybe you just have to say the words *sea cow.*"

"What?" I asked. Veronica, staring at the frozen screen, was holding her head, or maybe covering her ears; I wasn't sure which.

"Sea cow, sea cow," Graham was chanting. "See. It's working. Sea cow, sea cow . . ."

When the image finally came into focus, we were looking down on a vast churning galaxy of clouds. It looked terrible and huge, about a third the size of Mexico. And yet, even in that first moment, the photo from space felt not just staggering and unignorable but also unreal, and therefore it was hard not to feel like our main mistake had been to check e-mail. That's what Veronica is feeling now, I think, as she gazes out at the bay through her binoculars. She is thinking, *If only we didn't know.*

"How many days do we have?" I ask.

Without putting down the binoculars, Veronica answers. "They say three till it hits the peninsula. But then it's supposed to go west. So they don't really tell us when it would get here."

"So let's say four," I say. "That's conservative, right? And that means we have three days, at least, until we have to make a decision."

"What about the roads?" Graham says. "Did you see how huge that thing is? Even the edge would wash out the roads. Then we won't have a decision to make at all."

It's a fair point. There are many *vados*—seasonal river crossings—on the roads we take home. A few hours of hard rain would fill them with a foot or two of water. We'd have to turn back at the first one.

"Okay," I say, "let's say we have a day and a half. Tomorrow evening, we have to decide."

Veronica's spine straightens and she leans forward into her binoculars. She has seen something.

"Bryde's whales," she says quickly. "Mom and calf, just off the nose of Cabeza. Come on—we gotta get out there."

"Out there?" Graham says. "Out there? A huge fucking hurricane—that's what's out there."

"We decide tomorrow," I say. "Agreed?"

"Agreed," Veronica says, standing and walking quickly into the staff house to get our things.

"I can't believe this," Graham says as he follows us inside. "It's your conscience, Hirsh. These young souls—they're on your conscience."

2

As we approach the squat concrete pillars and slack catenaries of cable that mark the edge of the Diaz boatyard, I can see that it is almost empty: there are only three pangas left, and two of them are our own.

"The fishermen are all out," I say happily to Veronica.

"Why wouldn't they be?" she asks with a touch of reproach. Her feelings of doubt about the storm seem to go beyond mine. In fact, I think she might be determined to doubt it right out of existence—as if it feeds only on our credence; as if we will be fine so long as we don't say the word *hurricane*. And without Graham here to balance such headlong commitment to denial, suddenly it feels important to get Veronica to acknowledge the meaning of that image—that huge circular sawblade of storm that materialized on the screen of our laptop.

"Well, there is a hurricane," I say. "A real live hurricane."

"It's four days away."

"But it's huge. There could easily be storm surge in the channel."

"Okay, Burnett," she says, stepping over the cable ahead of me.

"The young souls," I declare, "are on your conscience!"

She points at the panga sitting on its trailer beside ours. "Look," she says, "the Islas del Golfo guys thought it was too dangerous to go out."

This panga is almost always parked here. On its side is an official-looking insignia that reads ISLAS DEL GOLFO DE CALIFORNIA, which is the name that a certain department of the federal government has given to the islands out in the bay. Representatives from this department first showed up in Bahía several years ago, shortly after the announcement of Escalera Nautica. They came from the city of Ensenada, and they called themselves park rangers. They were here, they said, to protect the islands. This was all very puzzling—not just for us, but also for the locals, especially those who already considered it their own responsibility to look after the islands. Several years earlier, Antonio Reséndiz—ejido leader, environmental activist, and tireless protector of sea turtles—had organized a small association of townspeople to serve as the islands' guardians. They had erected small signs on all the beaches, asking visitors not to litter or disturb the wildlife. They made monthly trips out to the islands to clean up. And they kept a little office in town. It was right behind the Diaz boatyard, and whenever we arrived in Bahía, we would go there to pay a small visitors' fee in support of the association's work.

But when the people from Ensenada arrived, they opened their own office, on the main street into town. They put an air conditioner in the window, moved in office furniture, and hired several locals to serve as their secretaries. The original office, behind the boatyard, was soon shuttered, but whether this was done merely out of a sense of bureaucratic inferiority—they didn't have an air conditioner, office furniture, or paid secretaries—or rather because they were directed to desist from their activities, I've never really known. But I do know that Antonio himself did not resent the new office, because the first time we saw it on our way into town, we went to Antonio to ask his advice. How should we view the people from Ensenada? As well-intentioned officers of the Mexican park service? Or as interlopers who had seized an admirable local effort?

"Go see them," he said. "I think they're good guys."

We were greeted inside the cold office by a pudgy-faced young man with tightly curled black hair. He wore khaki cargo pants, which would have been unbearably hot outside, and a collared shirt emblazoned with the Islas del Golfo insignia. He pointed us to a dark veneer table and told us to have a seat. Then he went away, and for a while moved some documents around on his desk. When he finally returned, he spread out two glossy brochures on the table in front of us, much as one might place laminated placemats in front of a pair of children before a meal. Then he started to tell us about the ecology of the islands—nesting seabirds, sea lions, that sort of thing—and as he spoke, he pointed at maps and diagrams and photographs in the brochures, using them as his instructional aids. After about ten minutes, Veronica interrupted him as politely as she could. She said we were familiar with the islands because we taught a science class here. But the young man wanted to finish his presentation, and he carried on. It took about half an hour. When he finished, there was a long, awkward silence, perhaps because we were not able to think of any good questions for him. Finally, he ended the silence by saying, "How may I assist you?"

I said we had just wanted to introduce ourselves.

"Do you plan to visit our islas?" he asked.

Veronica said yes, we did so every year.

"Then you will need a permit."

"A permit?"

"Excuse me just one moment." He went to his desk, and after a few minutes, came back with a form for us to fill out. I wrote our names, affiliations, and reasons for visiting the islands. Then I handed it back to him.

"You do not submit it to me," he said, leaving his hands decidedly by his side. "You submit it to the office in Ensenada, where you receive your permit and your bracelets."

"Bracelets?"

"Every individual visiting the islands must be wearing a bracelet."

"Ensenada is two days away."

"It is about twelve hours, but best to drive during daylight hours."

"Can you fax our form instead?"

"Yes. Of course we can fax it."

"Good. Thank you."

"But then you must go to Ensenada to pay your fee and receive your permit and bracelets."

"Ensenada is two days away."

"Next time," he said cheerfully, "you can simply stop on the way down."

We thanked him and told him we'd better get on the road, since we had such a long drive ahead of us. The next morning, when we went to the boatyard to launch the *Eagle* and the *Angel*, we saw the Islas del Golfo panga for the first time.

"Uh-oh," Veronica said, pointing at the insignia on the side of the shiny white panga.

"You think they really patrol?" I asked. I couldn't imagine the man in the cargo pants outside of his icy office. Veronica asked Samy if the panga would be going out soon.

"Not till Wednesday," he said.

"Wednesday?"

"They patrol on Wednesdays. They hired Old Paco to run the panga."

"I guess we don't set foot on an island on Wednesday," I said to Veronica.

On Wednesday, we were preparing for a boat dive on Ventana reef when the shiny white panga motored slowly by, getting a good look at us. Old Paco was at the center console, running the boat. Two other towns-people were sitting behind him. The pudgy-faced man was stationed in the bow. He was wearing a safari hat with the sides turned up and the Islas del Golfo insignia on the front.

The office in town lasted about a year. Then they closed it up, leaving behind the air conditioner and the office furniture. Still, every once in a while, they drive into town in a nice white pickup truck with the insignia on the side. They unlock the office and sit there awhile, and sometimes they even take the panga out for a cruise around the islands, just to make sure that everything is in order. For the rest of the time, which is, essentially, all the time, the panga is parked here, giving up a bit of its white sheen with every passing year. So when Veronica points out that the Islas del Golfo people decided not to go out today, what she really means is that, with our anxiety about the hurricane, Graham and I are revealing a certain kinship with the pudgy-faced fellow in the cargo pants and safari hat.

We walk past the pangas, on our way up to the men seated in folding chairs at the back of the yard. Old Paco is there, and two other leathery dignitaries, whom I recognize but don't really know. One chair is empty.

"No Samy," I say quietly to Veronica.

"Hola," Veronica yells to the men.

"Hola, Veronica," they reply in concert.

"Está Samy o Octavio aquí?"

"Sí," they say, smiling with amusement. Then they point at our boats behind us.

"I'm over here," Samy says, standing up in the *Sea Eagle*. "I'm putting oil in your engine. You shouldn't let it get so low."

"He's putting oil in your engine," one of the men repeats, smiling. It is clear that they find this unusual and amusing. Paco adds that Samy already replaced the spark plugs on the other boat.

Samy's solicitousness has not wavered since he startled us by climbing into the pickup truck to launch our pangas. If anything, he has become only more helpful, more generous, though it is now much less mysterious to me. Yesterday morning, I was talking with Alejandrina in the kitchen, and I mentioned Samy's strange behavior—how he'd been helping us himself, instead of delegating; how he'd seemed interested in chatting with us; how he'd laughed with me about my Spanish, about Graham, the loco professor who wears a skirt, and about my having a wife who is a mejor capitán de panga than I am. Alejandrina laughed about it all, especially about Veronica being a superior boat driver, but then she became serious and looked me in the eye.

"Pero, profesor," she said, "el es mi primo."

I said I knew, of course, that Samy is her cousin, but she seemed to be waiting still for me to catch on.

"Y ustedes," she said, finally, "ustedes me ayudan."

She was referring, with characteristic discretion, to a loan that Veronica and I gave her last summer. It was meant to help her build her restaurant—the place we saw, half-constructed, on the way into town— but I'm not sure how much of the money really ended up there. Most of it, I suspect, went to medical bills, and the remainder to the lawyer who represented her in the dispute over her land. For me, the experience was a lesson in the delicacy of dreams and plans in a less secure and prosperous society. And it was a reminder, more generally, of the strong gravitational pull of poverty. By Mexican standards—and by those of most of the world—*poverty* is perhaps too extreme a word for Alejandrina's situation. After all, she owns an automobile that can be driven on the highway. Her family runs a simple but pretty hotel here in town. She wears glasses, eats three meals a day, and is always nicely dressed. And yet, I have watched Alejandrina exert herself, strenuously and unceasingly, not to better her situation, but only to resist a powerful downward draw.

Alejandrina has worked many different jobs during the weeks and months when no one needed her here at the station. For a while, she was employed at a cannery in Ensenada. She would pick a fish from a mountain of ice, slice open its belly, yank out its innards, and throw the hollowed animal in a hopper. What made the work hard was that the room was kept extremely cold, and her hands, which were supposed to wield a gutting knife, were constantly plunging into ice. Many coworkers had lost fingers. Still, she did not mind the job, because it allowed her to be near her son, whom she had sent to secondary school in Ensenada, since Bahía's schooling goes only through the eighth grade. Eventually, though, her hands froze up—simply refused one morning to close around the knife's handle—and so she had to quit. I remember her seeming very disappointed about this.

But her next job was better. Capitán Muñoz, who worked for many years as a pilot flying gringos down to Bahía, first met Alejandrina at Las Hamacas, the family hotel. He must have perceived what a warm and sturdy woman she is, because when he fell ill, toward the end of his life, the capitán and his wife asked Alejandrina to become his caretaker, in San Diego. Somehow, they secured a work permit for her. And for the next few years, Alejandrina drove the fifteen hours between Bahía and San Diego twice a week, returning home on weekends to help at the hotel. When the capitán died, Alejandrina continued to make the journey now and then to help Señora Muñoz. She also started working construction in the San Diego sprawl, swinging a hammer alongside teams of men. She laughs about this, saying she's muy fuerte and showing me her fist. But it is good work, she says, and she'll return to it when our group departs.

When we gave Alejandrina the loan for her restaurant, an outcome I did not foresee was that she would one day be working construction in order to send us a check. This feels untenable, so yesterday morning, when the topic of the loan arose, I tried to tell her she could repay us whenever she wanted—sooner, later, it didn't really matter. But she waved me off, saying she would surely repay us very soon, and not only the entire amount, but also some interest, on account of Escalera Nautica. With some hesitation, I said I thought Escalera Nautica had been canceled.

No, no, Alejandrina said. The project was still very much alive. Just last week, in fact, two men in very nice clothing had come to the general meeting of the ejido. They represented a group of investors from *Texas*.

The way she said *Texas*, raising her eyebrows and pausing to let it sink in, suggested that the word carried weight and magic. *Texas*, I gathered, was a place of untold wealth and power. And Alejandrina went on to explain that the men at the meeting had not advertised the fact that they came from Texas, but everyone knew they did, and their reticence on the matter only confirmed it, because of course they would not want the ejiditarios to know, as it might cause them to demand a higher price.

"A higher price for what?" I asked.

"For the ejido," she said.

"The whole ejido?"

She nodded gravely.

I tried to understand what this could mean. The ejido around Bahía is the largest in Mexico; it was meant to provide a workable ranchero for each ejido family, and in such a dry desert, that requires truly vast acreage. Anyone buying the entire ejido, then, would be buying great expanses of empty desert. But did the investors perhaps deem this the most efficient way to acquire the entire coastline?

"How much is the group from Texas going to pay?" I asked.

"Mil millones dólares," she said, slowly. A giddy smile flashed through her seriousness, but she quickly suppressed it.

A billion dollars. There are just over a hundred members of the ejido, so that would be, what? Just shy of ten million dollars per person. So someone had lied to Alejandrina. Because no investors, not even those from Texas, would pay each member of the ejido ten million dollars when they could easily pay them a hundredth as much. But I didn't know what to say. Would it be more hurtful to tell Alejandrina what I thought of this offer, or not to? It wasn't like I had a clear understanding of what was afoot, either. All I knew was that two well-dressed men had come to the ejido meeting. They were there for something. And somehow this huge number had come up. Was Mike's Mountain maybe full of bauxite or lithium or something? But even if it were, why would anyone pay ten times more for it than they had to?

"Es mucho, sí?" Alejandrina said, awaiting my reaction. Maybe she thought I'd been struck dumb by the figure she'd just given me. In a way, I had been.

"Sí," I said. "Es mucho."

I would very much like to ask Samy Diaz, who is now climbing down from our panga with an empty oil can in his hand, what he thinks of these Texans and their impossible offer to purchase the ejido. But something stops me. Partly it is the way Alejandrina spoke—as if she'd been telling me a secret. Perhaps ejiditarios are not supposed to talk about official business with outsiders. And if that's the case, I would not want to get Alejandrina in trouble with her cousin Samy. Nor would I want Samy to get the wrong idea about my own interest in these affairs. Now that I know he is aware of the loan—and even seems to find it so meaningful that he hosts me as his guest, whereas before I was a distant outsider—I would not want him to think that I am, in some indirect way, asking after my money. And so I say only, "Muchas gracias, Samy," as we meet beside the pangas, and I ask him what he thinks of Hurricane John.

"No big deal," he says, and he waves vaguely toward the bay, as if to say, look for yourself.

"You'll tell us when we shouldn't go out?" I ask.

He nods as he turns away and walks toward the cab of the old gray pickup truck, which is already hitched to the *Sea Eagle*'s trailer—prepared for our departure even before we arrived. Veronica climbs into the truck-bed. "See you out by Cabeza," she says. And then, glancing at the calm bay, she adds, "Don't get lost in the storm."

3

No bottom in sight. Just my own toes, pale and foreign in the watery light, and suspended above a depth that is impossible to judge. The sun is still low in the sky, so the light has not yet arranged itself into steely rays slanting downward, but instead dissolves into a bright but slightly cloudy confection of radiance and seawater. Around me, here and there, are more human forms, most of them in the curved, prone position that the floating body takes of its own accord: faces in the water, arms out to the side, legs drooping lazily from the surface. Flying slowly beneath them, and behind them, and even among them, are great numbers of devil rays. They are three or four feet across, and they are nothing but wing—as if every other part of the body had been smoothed away, eroded over millennia of slow soaring though the ocean. Even the edges of their flat, diamond-shaped bodies are beautifully rounded. Their backs are black like honed granite, their undersides white like milky quartz, and their movement as steady and timeless as weathering stone. Each rise of a wing reveals a glimmer of quartz, and the flashes count a slow, pulsing rhythm: Rise, white. Fall, black. Rise, white. Fall, black. And because animals close together appear to be keeping time with one another, synchronizing their stately flight, the flashes form brief strings of pearls, appearing and winking out all around us, as if the ocean itself moved to a perpetual metronome.

But where is it going, this vast shoal? And where was it coming from when it first appeared? We had come to the edge of Cabeza de Caballo, just as Veronica had suggested, to look for the Bryde's whale and her calf. Waiting in our pangas, Veronica, Graham, and I tried to keep everyone—especially Becca—relatively quiet, so that we would hear the pair of breaths: first mother and then, before she had finished her long inhalation, the awkward little splash and gasp of calf, who must push his entire head out of the water because he has not yet understood the placement of his own blowhole. On a morning like this, when all is still, it is much easier to find whales by sound than by sight. In fact, vision starts to feel distinctly limited, because in any given instant you can see only a small segment of horizon, but you can hear everywhere around you at once. And with the water such a perfect plane, the red cliffs of the islands serving as sounding boards, and the air so very still, sound seems to travel farther than light: You hear pelicans and boobies hitting the water in pursuit of fish, and the two sound very different—pelicans a clumsy harrumph of a splash, boobies a sharp pop of streamlined penetration—so you can even say which species it was that you just heard; but when you look in the sound's direction, there is no sign of the birds, because they are too far off. The same is true for whales: various species sound distinct, and you hear breath that is too distant to see. So we knew that if mother and calf were anywhere nearby, we would hear them, and we would hear them very soon, too, because such a young whale cannot hold his breath for long. When you are looking for large fin whales, you have to wait ten minutes before you can be sure they are not there, somewhere far below you, but with this little Bryde's, we knew after just a minute or two that the pair must have made their way into the sound shadow of one of the islands. Sitting on the rail of her panga, Veronica looked at me and very slightly shook her head.

As I stood from the rail of my own panga, I quietly asked the students if they wanted to slip in to cool off. They started shedding T-shirts and hats, seating themselves on the rails so they could ease in without making a loud splash. But just as they were about to go I heard myself say, "Stay there—just a sec." Something had grazed the surface to the north of us, where the arch of Isla Ventana stood in the background. Whatever it was, I had hardly perceived it. But the impression still lingering somewhere between my retina and memory looked to me like a dark fin, and for a moment, I thought it might be the very tip of the tail of a whale shark. They swim like that sometimes: a fish thirty feet in length and ten feet in draft, with four horizontal feet of open mouth, discloses its substantial presence with approximately one and a half inches of tail barely tracing a serpentine line along the surface. So when I glimpsed that black tip slicing delicately through the glass, I held my breath, seeing in my mind's eye the entirety of the animal below.

But then another fin rose, and another, and it was apparent then that the fins came and went in pairs, like fish swimming two by two, and then suddenly there was a whole field of them, riffling the surface like a wind and moving toward our boats. They sounded like rainfall on the sea.

"What is it?" Isabel asked, pulling her toes up from the water and swinging her feet into the hull.

"Just a big school?" Chris ventured. He was reaching down to pull his dive mask out of his bag. "Can I get in?"

"Mantarraya!" Veronica yelled from the other panga. She was standing now on top of one of the bench seats, getting a better view into the water. "*So* many devil rays."

"I'm getting in, 'kay?" said Chris, putting on his dive mask.

"Vica," I yelled, "are you sure?" I had seen devil rays before, but only one or two at a time, when they had hurtled themselves inexplicably from the sea, flapping spastically in thin air until they returned home with terrific, smacking belly flops. I had no idea that shoals like this even existed.

"Okay?" Chris said again, peering at me through his dive mask.

"Vica," I yelled again. She had climbed up onto the center console.

"Definitely *Mobula*," she yelled.

"Go ahead," I said to Chris.

"They're not dangerous?" Allie asked, reaching now for her own mask.

"Not if they're what she says they are."

Chris lowered himself into the water, but continued to grasp the rail of the panga with one hand. Allie and Cameron did the same; she would tell him what the rays were doing. The rest of us stepped up onto the bench seats to watch the wide front of gentle disturbance as it approached us across the water. Then, just as the rays' dark bodies were becoming clear, the middle of the phalanx disappeared—went quiet—leaving two arms of textured water, left and right, with a glassy stretch in between. But of course they were not gone; they had only dropped from the surface. And a moment later, they were flying beneath us. They came in chevrons, like fighter planes or migrating birds, and they passed silently beneath Chris, Allie, and Cameron, into the shadow of our hull, and out again to the brightness on the other side. Then they angled upward, and the phalanx reassembled itself.

Then, mysteriously, it circled back. It traced a large round arc across the surface and approached us once again—this time from the west, where the town lay low in the distance and Mike's Mountain rose up behind it. That was when the rest of us pulled on our masks and slipped from all sides of the pangas into the water. Again the rays passed beneath us. And again they circled back. They have come four times now, and meanwhile our group has drifted slightly apart and away from the pangas, so we are now a sprinkling of human forms and a pair of boat hulls, haphazardly arrayed on a placid sea and vaguely marking the crossing point, the waist, of a large figure eight that is steadily described, over and over again, by a great shoal of rays.

With each pass, the time between their visits has shortened, so I think their turns must be getting tighter—as if they were thinking of joining us, or were wondering, perhaps, why we have not moved to join them. And with each pass, too, they are venturing closer to their floating spectators—these strangely passive creatures with four long appendages and a single glassy eye that tracks them as they go by. Have they been testing us? Gradually proving our harmlessness before moving in for closer inspection? Surely that is more cognition than I ought to attribute to the ancient kin of sharks and skates. And yet I am certain I have read somewhere that the brains of the mobulid rays are exceedingly large in comparison with those of their evolutionary relatives. What that really means, I'm not quite sure, but their present behavior is too far from utilitarian, too purely pointless, to look like anything but a kind of curiosity.

Now that they are willing to come within a few feet of me, they look different. From the air above, and then through the filter of aqueous distance, they appeared so abstract—black diamonds glinting white; moving gear teeth of an oceanic chronograph. But now, much closer, they are not shapes in formation or pieces of a machine, but individual animals. They even have faces. Strange ones, for sure, but definitely faces. The mouth is flanked, left and right, by flaps that reach out in front of the head: seen from above, they resemble a pair of horns jutting forward, and that, along with the bullwhip tail streaming behind, must be why these graceful creatures are called devil rays. But it's unfairly menacing, that name, because in fact the mobulid rays are harmless. Their tail has lost its sting, and the mouth flaps serve only to funnel water into the throat, where gills take oxygen and catch planktonic food. The mouth hardly has teeth.

At the base of each flap, on the side of what you would call the animal's head, and just where the black above meets the white of the underside, there is an oval of cloudy brown. I am certain now that it is an eye, and I keep catching myself peering intently into it. I feel a bit silly for this, but I can't help myself. As each one passes, I try first to look at some other detail—the wingtip, which is very slightly falcate; the base of the tail, where there rises a low, hydrodynamic vane—but invariably I end up looking stubbornly, expectantly, into that eye. There's something about the way I can't quite tell if anyone is there—if anyone is home, looking out at me—that keeps me trying again and again.

When I have watched the last of the albatross silhouettes vanish into the aqueous haze, I lift my head from the water to locate the *Sea Eagle*. Veronica might be near her panga, and I want to find her so we can watch the shoal go by, at least once, together. The white hull with the red rail is not far, but I swim hard for it because I know I do not have long—just twenty or thirty seconds, I'd guess—before the rays circle back. And sure enough, just as the underside of the hull comes into view—a large white wedge settled into the bright surface of the sea—I glance to my left, where I think they may appear, and there they are: innumerable stroking wings, materializing from the haze. Then I look back at the hull, and from behind it, someone dives straight down.

It is Miles. I can tell because of his shape—he is long and sleek, unencumbered by baggy trunks because he wears a Speedo. And I can tell, too, by the way he swims—two powerful dolphin kicks and he is fifteen feet down, where he pauses, upside down, in the water column. The first rays are upon us now. To my left, I can look them head-on, straight between the points of the crescent moon. Under me they are close—almost close enough for my fingers to graze a wingtip. And to my right, where I expect to see them vanishing once again into the distance, they are instead flying a steeply banked turn to their left, sweeping like a streamer in the hand of a spinning child, one ray following on the next, and they are flying now straight toward the panga, toward Miles, just ahead of me.

He is still upside down. And though I am watching him now through a stenciled pattern of black wings, I can see that he is watching them come. Unlike the rest of us, he wears not a dive mask but swim goggles, mirrored droplets like quicksilver eyes. His arms have fallen overhead, like a diver's in midflight, so he is now a long, pale jet in the shade of the panga's hull. He appears motionless, but somehow, through the subtleties of a gifted swimmer, he is slowly rising, feetfirst, toward the bottom of the boat—a dive in reverse slow motion. When the rays are about to reach him, and it looks as if some may actually hit him, they bank right, and then they bank left again, sweeping around behind him. For an instant they are coming back toward me, but then they bank to their left again, closing the loop of the skein, following the rays that just a moment ago were flying behind them, and thus creating a whorl, a great scroll of rays, slowly winding itself in a counterclockwise direction around the pale spindle at their center, the spindle that is Miles.

His feet have settled now against the underside of the hull, so he is truly motionless, standing there inverted with his arms overhead. He and the rays closest to him, the ones flying the tightest circle, are in the shade of the panga. The outer whorls of the scroll, including the rays that are just now banking in, are flying in bright water. I do not understand what is going on, and therefore I cannot quite quell my apprehension. But I can still see Miles's face now and then, through the rotating screen of black wings, and though the silvery eyes make his expression difficult to read, I think he looks calm. And I know he is choosing to stay where he is, because just a few of his powerful dolphin kicks could propel him out the bottom of the spool, where he could easily exit in any direction and return to the surface. So he is fine. He is not afraid. He's fine. But what are they doing, these devil rays? Is it perhaps just the shade of the panga that has attracted them? Has their minuscule food, the tiny crustaceans and cope-pods, gathered there, out of the sunlight, where the rays are now circling through especially rich water? Even as I try to think this through—try to explain in some plausible fashion what is happening right in front of me—still the best explanations I am able to produce seem to be brushed aside by the image itself: a lithe human form with quicksilver eyes, at the center of a rotating cylinder of rays. It is an image so eerily iconic that it feels as if I might see it again someday—etched, perhaps, among the hieroglyphs we sometimes discover on the undersides of boulders in Cataviña.

A ribbon of rays peels off, tearing the outer layer of the scroll and leav-ing in the direction they were flying when they came. More rays follow, and the scroll unwinds itself that way, as a ribbon that streams into the distance, and is gone.

4

Afterward, I am sitting in the *Sea Eagle* when Miles pulls himself easily from the water up onto the rail. He shakes his head like a wet dog, flinging seawater from his blond hair, then lifts his goggles onto his forehead, where they look like stumpy silver horns. His right thumb and forefinger press the water from his squinted eyes, sliding together into a firm pinch on the bridge of his nose. When his hand finally falls, he looks directly at me, awaiting my comment. But I have no idea, really, what we've just witnessed. Again he presses his eyes, as if there were something besides seawater he needed to clear from them.

A hand appears beside Miles's hip and drops a dive mask into the hull. Ace struggles up onto the rail. And from the way he looks back and forth now between me and Miles, it is clear that Ace was watching too: he must have been just on the other side of the *Sea Eagle*, where I saw Miles dive down. Finally, Ace laughs and nods, affirming our collective speechlessness.

As others arrive, I watch to see if they look meaningfully at Miles, but no one does. It must have been just a section of the phalanx that peeled off and wrapped itself around him, because Veronica, Graham, and the others have been occupied, all along, with other parts of the shoal. Everyone is climbing into the *Sea Eagle*, heeding an instinct to congregate after such an experience. Stories of rays coming *this close*, wingtips so near to grazing fingertips, vie with all the noises of snack time: a bag of nuts is passed around and playfully fought over; Rafe says the devil rays were "so badass, like fighter planes"; the overloaded panga rocks precariously to one side and people shriek and tip clumsily off the rail, into the water; Allie asks about the two dots on each side of the head, and Veronica says yes, one is indeed an eye and the other, she thinks, a spiracle, a vent to the respiratory system; standing in the bow, Anoop looks like an awkward heron as he imitates the devil ray wing-flap. Neither Miles nor Ace says a word about what happened. I will tell Veronica later, I decide, when we are on our cots and everyone else is asleep.

Chris is the last out of the water, pulling himself up onto the rail beside me. On every dive, he has stayed in until the last possible moment, as if all his experience in the ocean, his hundreds of hours of snorkeling and diving, has yet to sate his appetite. His pale skin has darkened some over the past few days, so its contrast with his black hair and eyes has lessened, and he is looking more and more like a local.

"Holy shit," he says, smiling, and everyone seems to start over again with their excited stories. Anoop repeats his wing-flaps.

"How young were they?" Chris asks.

"How young?" I ask.

"I saw a manta once, in Thailand, and it was like three times as big." Smiling, he adds, "Just one, though," as if to clarify that he's not trying to trump the experience we've just had.

"Those weren't mantas," I say.

Turning to Veronica, Chris says, "Didn't you call them mantarraya?"

"That's what the locals call them," Veronica says. She explains that the rays we've just seen are members of the genus *Mobula*; it's hard to say which species, maybe *lucasana*. What we call mantas are related, but belong to a genus of their own. And yes, they grow to be much larger than the rays we've just seen.

"What do locals call actual mantas?" Chris asks.

"Mantarraya," Veronica answers, shrugging a bit.

"Really?" says Chris. "They call them all, just, mantarraya?"

The other conversations wash over this one. Anoop and Rafe are trying to estimate the number of rays in the shoal by guessing at volumes and densities. Becca is talking loudly about something her roommate once saw or did. And Chris turns his attention to tracking down the bag of nuts.

5

Later, I kept going back to that passing exchange, because it had made me uncomfortable. My glance had flashed apprehensively to Allie and then to Isabel, the two students whose disposition and background had by now made them seem like arbiters of what's considerate and sensitive. No sign of judgment or disapproval had come, but nevertheless, I worried, especially later on, that someone might have heard in that brief conversation an indictment of the locals of Bahía. What Chris called "an actual manta" is *Manta birostris*. It grows to be fifteen or twenty feet across and to weigh well over a ton. And it does occur in the waters around Bahía. So when Chris asked what locals call it, and Veronica replied that they call it by the same name as the species we had just seen, which she thought was *Mobula lucasana*—in fact, quite different from *Manta birostris*—the next question, the one Chris did not ask but which felt dangerously nearby, had to be something like "Don't they know the difference? The locals really think the various species of ray are all the same?" And the reason that question, even merely broached, made me writhe a bit is that it unavoidably casts us, the visitors, in the role of knowing scientists, pronouncing our Latinate names and carefully distinguishing among the species we see, and just as unavoidably casts them, the locals, in the role of rather ignorant and careless inhabitants of their own seascape. Aside from being unseemly, such a characterization turns out also to be incorrect. But to be honest, it took me a while to understand what that surprisingly loose usage really shows—what it reveals, that is, about us, the visitors, and them, the locals.

When we returned to shore that day, I started asking around about rays—what they're called, who had seen them, how big they get, and so on. And my little survey has continued, intermittently and opportunistically, to this day. It seems that most of the older citizens of Bahía have seen huge manta rays—mature *Manta birostris*—at some point in their lives: the fishermen have seen them in the bay, in the Channel of the Whales, or between the peninsula and Isla Coronado; the shorebound, like Alejandrina, may have seen a huge diamond-shaped shadow in the shallow water, or they may have seen an immense limp slab, towed ashore by the team of fishermen who had managed to harpoon a manta. The entire town, Alejandrina remembered, would go down to the beach to see the monster. But what would have been done with such a catch? The meat, she said, is similar to angelito, angel shark, which is often used in fish tacos, so probably the animal would have been butchered and sold.

The younger citizens of Bahía have not seen huge manta rays, and neither have most recent visitors, including Veronica and me. In the early nineties, a friend of hers named Clay was sitting on the station terrace when he saw a great commotion just offshore: men on two pangas were yelling and pointing, leaning over the rails and madly swinging a boat-hook at something in the water. When Clay realized it was a manta, he scurried down the rocks and swam out to the boats. Pretending he meant to help the fishermen, he drove the giant away.

In one sense, it is not surprising that only the older inhabitants of Bahía have seen huge manta rays. Such is the pattern, after all, for many species: sharks; openwater game fish like yellowtail and sierra mackerel; huge reef dwellers like goliath grouper and giant sea bass; even shellfish like pearl oysters and lion's paw scallops—for all of these creatures, I suspect, you would find a very strong correlation between the age of a fisherman and the size of the largest specimen he remembers seeing. In Bahía, I've observed this trend only in a rather casual and anecdotal way, but several biologists who work farther down the peninsula, in La Paz, have applied more rigorous methods to confirm that older fishermen indeed remember larger fish. You might raise an eyebrow and suggest that the trend has something to do with the growing nostalgia of old age, but that hypothesis seems to be refuted by old photographs and, for the shellfish, certain oversized ashtrays and other curios. Nor is the correlation merely a result of older fishermen having had more years to catch the big one. If that were the case, at least a few fishermen would be getting lucky every year, but in general, they're not. In other words, the trend is just too clean to interpret as a consequence of the truism that fishing takes patience.

The reason for the correlation is pretty clear. When a population is fished too hard, it disappears from the top down: the bigger cohorts go first, followed by progressively smaller segments of the population. In part, this occurs because fishermen tend to direct their efforts at the largest individual fish; only when the big ones are gone do the smaller ones start to look like they're worth the trouble. But that's not all there is to it. Even among species that are fished without attention to size, overfishing still causes a shift in the population toward smaller sizes. Think of it this way: every year, an individual fish has a chance of getting caught. Consequently, the odds of still being alive decline geometrically with age and, by extension, with size. Here's a simple example. If fishermen catch twenty percent of the population every year, paying no attention whatsoever to size, a fish has an eighty percent chance of making it to age two, but only a thirteen percent chance of making it to age ten. And ten-year-olds, of course, are a lot bigger than two-year-olds.

Why, then, would we be at all surprised that large mantas, too, were last seen many years ago? Well, in at least one respect, mantas are a bit different. Each of those other animals I listed, from goliath grouper to lion's paw scallop, was at some point a mainstay of fishermen's work. That is to say, fishermen went out looking to catch that species in particular, or perhaps that one and a few others very much like it, and they tailored their gear, scheduled their outings, and mapped their routes accordingly. To the best of my knowledge, the same cannot really be said about manta rays around Bahía. Rather, mantas have always been killed, in one way or another, on the side.

The greatest toll, I suspect, has been taken by drift nets. Four to eight miles long, they hang as curtains of nylon mesh, draping from about thirty feet below the surface down to a hundred or so. They are set mainly for billfish and sharks, but their loose mesh of heavy-duty monofilament can also ensnare the occasional manta ray, a second-tier prey item that might be kept and passed off in pieces as angelito, or might be pitched over the bulwarks as waste. A few years ago, such nets were being set just offshore from Isla Ángel de la Guarda, and it seems like they've been positioned somewhere in Bahía's general neighborhood for much of the past decade or so.

Aside from drift nets, the main cause of the manta's decline in the gulf has probably been opportunistic, helter-skelter pursuit, much like the incident Clay witnessed. When fishermen were still going for big animals that were hunted with harpoons, pangas were better prepared for pursuing the occasional manta than they are these days, when hulls are filled with small fish traps or fine-mesh gill nets. But even in the era of harpooning, effort was focused elsewhere—on the sea turtles, for instance, or the grouper—and in a way, this makes the manta's decline especially disturbing. People weren't even really *trying*, and yet, in a single generation, from Samy's youth to Octavio's, the mantas have taken a noticeable step toward extinction. And you have to wonder, then, how many generations before Samy's had already exacted a similar toll—how many steps downward the population had taken before this last.

There is some interesting evidence on this count. The earliest intensive fishery in the Gulf of California was focused on the pearl oysters *Pinctada mazatlanica* and *Pteria sterna*. The pearl of *mazatlanica* is black, its nacre a lustrous gray; in *sterna*, pearl and nacre alike are fabulously opalescent. Europeans have coveted both species for a long time. In fact, the handful of mutineers who were fortunate enough to survive their discovery of La Isla de California brought back word of pearl oysters. Cortés himself confirmed their claims on his own ill-fated expedition. And a century later, in 1632, Nicolás de Cardona, a Spanish explorer and early entrepreneur in the pearl business, wrote this about the shores of the gulf: *Along the seacoast of the interior region, over a distance of one hundred leagues, all that one sees are heaps of pearl oysters . . . They are the size of a small plate and full and complete they would weigh from one to two pounds.*

Other explorers offered their own astonished accounts of the pearl oysters, but the most telling indication of the beds' former richness is their sheer endurance: they withstood three centuries of sedulous extraction, from Cardona's time until the 1930s, when the fishery at last declined into commercial extinction. Today the beds are gone, even from the particular bays and island coasts where they were most abundant. We occasionally come across a lonely specimen of *Pinctada mazatlanica* or *Pteria sterna*, and when we do, we show it to the students and try to imagine how the shore would have looked coated in their shells. The entire coastal ecology must have been different: The seafloor would have had a very different texture—oyster shells instead of sand or rock—and therefore must have sheltered a different suite of fish and invertebrates. The water itself probably looked different in some bays, since shellfish beds are capable of filtering enormous volumes. And in many places the shoreline must have been different in both color and topography, due to the accumulation of hills and beaches of shell. In a way, then, the transformation wrought over those three centuries of extraction should be thought of not just as an ecological change, a matter of plants and animals, but as a geological one, as well—a matter of water and rock.

But what do centuries of pearl extraction have to do with mantas? Strangely, from a very early time, the pearl industry seems to have considered the manta ray its particular enemy. Miguel del Barco, a Jesuit missionary who traveled to Baja California in 1738, recorded in his chronicles what he had learned from pearl divers about the manta ray: *when it discovers a diver under water, it tackles, holds and embraces him with its own body, without letting him return to the surface.* A bizarre image, to be sure, with its hints of dark intimacy in deep water, but it proved remarkably persistent. In the late 1860s, J. Ross Browne, a travel writer, was hired by the United States Treasury to investigate lucrative resources in Lower California. In his final report, Browne assessed threats to the pearl industry:

> *The manta raya is an immense brute, of great strength, cunning, and ferocity, and is more the terror of the pearl-divers than any other creature of the sea . . . The habit of the animal is to hover at the surface over the pearl-divers, obstructing the rays of the sun, and moving as the diver moves, and, when he is obliged to come up for breath, hugging him in its immense flaps until he is suffocated, when the brute, with his formidable teeth and jaws, devours him with a gluttonous voracity.*

This is perfect nonsense, of course. The manta, like its smaller relatives of the genus *Mobula*, feeds only by filtering seawater for tiny prey. And far from being ferocious, it is intriguingly peaceful and curious in the presence of human divers. Such facts of the matter aside, however, it seems very likely that pearl fishermen, in their terror, endeavored to kill every manta they happened to encounter. Indeed, in the nineteenth century, pearl ship owners were required by Mexican law to carry *a harpoon, a hook and a chain* to kill any *marine monsters* or *enemies of pearl divers* that might pose a threat.

But if the story of the manta's decline is one part cultural, hinging on certain myths and habits, it is also one part biological. Species vary greatly in their resiliency to the losses humans impose. Animals that produce huge numbers of offspring, investing very little time or energy in each one, often endure human impacts with remarkable aplomb. They are, in a sense, set up to absorb the losses: their reproductive strategy naturally entails making far more offspring than the environment can support, so if humans skim some of the excess, there are still plenty of fry around to fill out the ecosystem's capacity. Of course, we ought not forget the pearl oysters. They practice an extreme version of minimal investment, spewing their millions of gametes into the open water, and still they succumbed, eventually, to human diligence. So even the most fecund species are not entirely immune to overexploitation. But their decline is generally long in coming—the pearl oysters' three hundred years represents quite a run— and it often involves dips and recoveries before the final plummet of no return. By contrast, fate is fast and decisive for animals that raise few, carefully tended young. They've got hardly any offspring to spare, no margin for humans to shave. And consequently, their declines often trace a steep and uninterrupted slope down to zero.

On this continuum of resiliency to human impacts, mantas occupy an unfortunate position. They sit in the range of elephants and whales, far away from oysters and anchovies. For the manta is what zoologists call a viviparous matrotroph: mothers do not lay eggs, but rather give birth to live young; and they nourish those young directly, via uterine fluid, supplementing whatever nutritional reserves they were able to store away in the yolk of the egg. Such a reproductive system allows female mantas to invest heavily in each offspring: they bear only one or two pups at a time; the gestational period is close to a year; and the pups are born big, about four feet across. All of which is to say, mantas are not equipped to compensate gracefully for their losses to harpoons, drift nets, or swinging boathooks.

6

This returns us to that sensitive question: What do the locals, and what do we, really know about the various species of ray? For it was not from a scientific publication—not from a technical work using Latinate species names or terms like *viviparous matrotrophy*—that I first learned that *Mobula*, like mantas, give birth to large live young. Rather, I learned this fact from the first fisherman I happened to encounter after we'd seen the rays. It was three in the afternoon, and not quite so furiously hot as usual, since the wind coming off the water was again strangely cool—the day's first sign of the great storm churning toward us. Veronica had asked me to fill our gas cans—as if she knew we'd be out on the pangas again—so I'd tossed a half dozen empties into the back of the pickup and driven to the shiny new gas station. Like the traffic circle and the lofty steel sculpture, the Pemex was one of those occasional rumblings from distant federal agencies, a reminder that FONATUR was still out there, and had not forgotten Escalera Nautica.

I was carefully measuring out marine engine oil and pouring it into my gas cans when an old fisherman named Pablo pulled his rust-eaten truck up to the pump next to mine. Climbing stiffly from the cab, he greeted me with a faintly amused expression. There was no telling what was funny—my fastidiousness with the oil, perhaps? The way I had all my empty gas cans all lined up in front of the pump? It could have been anything, really, or nothing in particular, because Pablo was always looking at me this way. Evidently there was just something about me, or about my presence here in Bahía, that Pablo found endlessly silly. I didn't mind, of course, because it meant that he was always up for a conversation, and was sure to enjoy himself thoroughly. The only difficulty, on my side of such conversations, was that Pablo tended to talk very quietly, keeping one corner of his mouth closed around a wet, fraying toothpick. Along with his burnished, ruddy skin and darting black eyes, which were stuck in a permanent squint from a lifetime on the water, his habit of mumbling made him seem quite demure. Though in truth, I don't think he was shy at all; I think he was only clever—clever and highly entertained by my frequent misunderstandings. And so he happily went on mumbling, even when I was peering hard at his half-closed lips, struggling to make out the words. And it wasn't just that I found the words hard to recognize or the sentences hard to parse. Even when I'd gotten them mostly right, I would for some reason get his meaning largely—and amusingly— wrong.

When I saw that Pablo had come to the Pemex to fill a hundred-gallon plastic drum that was sitting in the bed of his truck, I figured he'd be there awhile, so I decided to ask him about the word *mantarraya*. Did it really mean any and every ray? Did Pablo and his peers really consider them all the same animal?

"Pablo," I said, "if it's okay, I have a question for you."

"If what's okay?"

"Pardon?"

"If what's okay?"

"If it's okay with you."

"If what's okay with me?"

He was enjoying himself already. The corner of his mouth with the toothpick had crept up into half a crooked smile. Behind him, I could see the level of yellowish gasoline in his plastic drum. Rising slowly, it was to me like the sand in the bottom of an hourglass, indicating how many minutes I had left to glean what Pablo thought of our rays. Exacerbating my sense of haste, the oversized timer was not so peacefully silent as an actual hourglass, but rather thrummed with the hollow sound of liquid rushing into a barrel, making it even harder than usual to discover the meaning of Pablo's sounds.

"Anyway, Pablo, today we swam with a really big school of manta-rraya."

"Hm," he said, "qué bueno," though the look on his face was balanced between bewilderment and mirth.

"They were beautiful," I added, feeling suddenly that I needed to explain.

"Qué bueno," he repeated.

"Anyway, Pablo, my question for you is this: Is there more than one kind of mantarraya?"

He thought for a moment. It was a sign of Pablo's generosity, I thought, that he was willing to consider the question so carefully, even as he appeared to consider the person asking the question quite addled. "No," he said, finally. "A mantarraya, it's a mantarraya. Just like a truck is a truck."

I was surprised, and also, I admit, vaguely disappointed. If an old fisherman like Pablo really took all rays to be the same, then I doubted I would find anybody in town who thought differently.

"So a mantarraya," I said, perhaps a little incredulously, "it's just one thing."

"Like a truck is a truck," he said.

"Like a truck is a truck," I repeated.

"Sí. You have Toyota trucks and Ford trucks, Four-Runners and Rangers—but still, a truck is a truck, right?"

"I see," I said. "And a mantarraya is a mantarraya—in the same way."

"Sí," he said, "you don't have these mantarrayas and those manta-rrayas. Any mantarraya is a mantarraya."

"Pablo," I said, "you're a philosopher, you know that?"

He shrugged and fiddled with his toothpick, indicating, I thought, that he did not really know or care if he was a philosopher.

The plastic drum was already half-full, but I felt that my only hope of understanding what a mantarraya was or wasn't would be to enumerate species and get Pablo to tell me whether or not they belonged to the category mantarraya. The problem was, I had no idea what to call each species; for many I had a common name in English, for some a Linnaean designation, but I didn't think either of those would do me much good with Pablo. So my only option, it seemed, was to describe the distinctive appearance or behavior of each species. Fortunately, Pablo seemed to enjoy this charade. Leaning back against his truck, he watched me struggle urgently with one animal after another. Sometimes it seemed he would wait longer than he really had to, allowing me to carry on gesticulating about body size or wing movement or the way a certain ray's body might rest on the sandy bottom, before he would finally raise his hand to interrupt me, and pronounce his curt judgment—sí or no, as in, mantarraya or not—at which point I could finally move on to the next animal. In this way we firmly established that anything with a devil ray's horns, regardless of the animal's size, was mantarraya, whereas anything with a round body for resting on the bottom, instead of wings for flying through the open water, was certainly not mantarraya. And there were two species—the spotted eagle ray and the bat ray—that seemed to mark the boundary of the concept, for Pablo made me impersonate both of them, twice over, before at last deciding "probably no."

By the time I'd completed my series of ray charades, Pablo's plastic barrel was full, and the gas pump had clicked to its automatic stop. But Pablo did not move to draw the pump from his barrel or walk into the Pemex to pay. He stayed right where he was, leaning against his truck, fiddling with his toothpick. So I started to tell him about our shoal of mantarraya—each one about three feet across, black on the back, pearly white on the belly.

More than once, when I've mentioned a certain animal to a fisherman in Bahía, his response has been to tell me the most bizarre thing he knows about it. I have heard strange tales about many animals: the Bryde's whale that breached onto a boat full of drunk American fishermen; the giant sea bass that was heaved onto the deck of a fisherman's panga and forthwith vomited a yellow-footed gull, which flailed around, spread its wings, and flew away; the orca who had a habit of spy-hopping in perfect silence behind unsuspecting fishermen, until the humans finally sensed something and spun around to discover the whale peering at them over the rail. I actually believe that many of the stories are true. (Okay, maybe the gull didn't fly away.) But their veracity doesn't change their nature; they still feel arcane, mysterious. And so I sometimes feel like the keeper of a vast *kunstundwunderkammer* of zoological narrative, a cabinet in which any sufficiently weird or spine-tingling anecdote about animals merits a niche.

Leaning against his truck, Pablo undertook now to add another item to the cabinet. "In the summer," he began, "I sometimes find a baby mantarraya in the net, right up against its mama."

I became excited. On the one hand it seemed preposterous—fish just don't do that sort of thing—and yet our encounter with the shoal had suggested a level of intelligence and sociality I never would have imagined in a ray, so maybe, just maybe . . . "You really think," I said, "the little ones stay with their mothers?"

"No, amigo," Pablo said, laughing, "I don't think the kids swim with their mamas." Saying it out loud made him laugh even harder. And when I asked how, then, he would explain a big mantarraya and a baby one right next to each other in his net, he only shrugged and feigned bafflement, trying to get me to offer another terrific hypothesis. But I held out in silence, and eventually, he explained: the mother must have given birth in the net.

It didn't seem impossible. After all, extreme stress can induce delivery in mammals. So why not in a viviparous fish?

Pablo explained his theory further: a few times, he had cut open an adult mantarraya and found a fetus inside, and it was the same size as the little rays he'd found in the net.

"How big?" I asked. "Over a meter?"

Pablo started laughing again. At what, specifically, I wasn't sure. Then he said, "My nets would not catch a mantarraya gigante."

Mantarraya gigante—what quantile was this, I wondered, of the broad category mantarraya? Anything larger than five feet? Ten? No. As I would realize later, Pablo had just given me my first glimpse of his precise zoological vocabulary: gigante signified not a size, but a species—the precise synonym, in fact, of *Manta birostris*. And once I'd understood that, I would see also why Pablo suddenly abandoned the word *mantarraya* in our conversation and started using a different name in its place. He was taking pity on me, trying to avert further bewilderment by dropping the general term and using a more specific one instead: "Bueno," he said now, "a few times, when there was a really fat cubana in the net—"

"Cubana?" I said. To me, that word meant only "Cuban woman."

"Sí," Pablo said. "Una cubana de lomo negro. Una *gordita*."

A black-backed Cuban woman. A fat little Cuban woman with a black back. Of course: this was the *ray* he was talking about. And this much was clear: The cubana was not gigante (whatever that meant), because Pablo had caught it in his net. The animal was black on its dorsum. And presumably, in Pablo's estimation, it was the very animal we'd seen that morning, which was why he was sharing with me this bit of esoterica.

"Okay," I said, "comprendo," though that was perhaps an overstatement.

Pablo continued: "Right next to the big fat cubana, in the net, there would be a very little one."

"How little?" I asked.

He held up his hands—"Sixty centimeters," he said.

"Sixty centimeters?" That was nearly the size of the rays we'd seen—but Pablo was talking about individuals that hadn't even been born yet. "How big was the mother, the gordita?" I asked.

He stretched out his arms. "Two meters," he said. "Bigger than you see them now."

"How long ago was that?"

"Ten years ago, when we first started catching them."

"You didn't fish them before that?"

He shook his head.

7

Several months later, I came across the papers of Giuseppe Notarbartolo-di-Sciara, an Italian zoologist who had spent three years in the early eighties hanging around the base camp of a fishing cooperative at Punta Arena de la Ventana, just south of La Paz. Some of the fishermen there were going for *Mobula*, and whenever they brought their rays ashore for processing, Giuseppe would borrow the animals, determine their species, sex, and size, and open their stomachs to see what they'd been eating. His publications about this work are fascinating, partly for the knowledge he gleaned from the fishermen, who sometimes had more to say about rays than did the scientific literature Giuseppe had been reading. When he'd first arrived in Punta Arena, for instance, Giuseppe's studies had led him to expect two species in his group of interest: he hoped to find *Manta birostris* and *Mobula lucasana*. But the fishermen told him that in fact their waters offered four different kinds of devil ray, in addition to the giant manta. And Giuseppe soon realized they were correct.

The most abundant of the devil rays was *Mobula lucasana*, which, I was intrigued to read, the fishermen of Punta Arena referred to as "cubana de lomo azul"—the blue-backed Cuban. Evidently, Pablo's puzzling term, cubana de lomo negro, had not been one of his elusive jokes. Giuseppe described the dorsum of *lucasana* as blue-black in color, so I assumed that what fishermen near La Paz deemed azul had come to be called negro farther north. And Pablo's size estimates were also corroborated by Giuseppe's papers. The largest individuals Giuseppe measured were six feet across—just shy of Pablo's round two meters. And newborns, Giuseppe reported, were between two and three feet, which seemed to corroborate Pablo's estimate for the rays he had taken to be late-term fetuses.

But there was a dismaying implication here: If the black-backed rays we'd seen were in fact the same species as Pablo's cubana de lomo negro and Guiseppe's *Mobula lucasana*, then our entire shoal must have been composed of adolescent rays. As I noted before, overfishing erases populations from the top down, from the oldest to the youngest. Therefore, that astonishingly vast shoal, which one might have taken as a sign of abundance or health of the population, could just as well have been the opposite: a clear signal of decline.

But was it possible, I wondered, that the rays we'd seen were actually a different species, one that matured at a smaller size? The second most abundant ray in the hauls Giuseppe studied was *Mobula japanica*, which the fishermen of Punta Arena called "cubana de lomo blanco," white-backed Cubans. The *blanco* part of the name seemed as puzzling as *cubana*, since Giuseppe described *japanica* as being exactly the same blue-black color as *lucasana*. Despite that similarity, however, I was fairly certain our shoal could not have been *japanica*: that species, Giuseppe reported, has a spine at the base of its tail, and I had seen no such thing on our rays. In any case, there was no optimism to be found in such a mistaken identity, anyway, because Giuseppe reported that *japanica* grow to be even a bit larger than *lucasana*.

Another species Giuseppe had seen, however, seemed to offer a more hopeful possibility: it was smaller than the others, reaching maturity at about three feet. The fishermen of Punta Arena called it "tortilla"—etymology, once again, obscure—and when they first brought one to Giuseppe, they handed him a rather remarkable discovery. For while the fishermen were perfectly familiar with it, the scientific literature had neither name nor record for such an animal. Giuseppe christened it *Mobula munkiana*—etymology less obscure, since Guiseppe's papers acknowledge the support of the renowned oceanographer Walter H. Munk. Much that Giuseppe reported about *Mobula munkiana* sounded like the rays we had seen—size, shape, and, not least, what he called *social habits: This is the only mobulid species in the Gulf of California*, Giuseppe wrote, *that was consistently seen in schools.* But there was one troubling detail. Giuseppe wrote that *Mobula munkiana* was *mauve gray.* Our rays had been black, maybe even blue-black, just as Giuseppe had reported for *lucasana* and *japanica*.

It was at this point that I spoke with another group of fishermen in Bahía. I found them at the fish cleaning station, a concrete table beneath a tin awning, just inland from the Diaz boat ramp, so our conversation was conducted around busy hands wet with blood and glittering with silver scales. To me, the keen hands swiping steel blades and tugging red innards were mesmerizing, like a campfire, but I was the only one intermittently dazed, as the fishermen had no difficulty talking as they worked.

Only one of them had ever heard the term *tortilla* applied to a ray. Yes, he said, the species was present here, around Bahía, but they had a different name for it. He paused for a split second as the tip of his blade slipped into the soft silver belly of a yellowtail. "We call it 'cubana de café,'" he said.

Ah, the others said, of course, cubana de café—but who called it tortilla?

I was diverted by the yellowtail's guts flying through the air and alighting on the sand, where they disappeared beneath a great mob of gulls. "The fishermen of Punta Arena," I said, "near La Paz." And everyone around the bloody table nodded and frowned at the interesting new name, tortilla. Another name, I said, was *Mobula munkiana*, and everyone nodded and frowned some more.

As the next silver fish was pulled from the plastic cooler below and slapped onto the table, I ventured to confirm my hunch that what Pablo had called "cubana de lomo negro" was the same species that the fishermen of Punta Arena referred to as "cubana de lomo azul." But the men around the table said no: they knew cubana de lomo azul well, and it was different, they said, from cubana de lomo negro. I asked which species had a spine at the base of its tail. Negro, they said. Azul has no spine. But this meant that "cubana de lomo negro" to these men was "cubana de lomo blanco" to the fishermen of Punta Arena. When I mentioned this, the men around the table laughed and work ceased for a moment. Wasn't it just like the guys down south, someone said, to say blanco when they saw negro? Someone else said something about cubanas negras in La Paz, and they all laughed even harder.

To this day, I am not certain what species of ray it was that encircled Miles as he stood, inverted, on the hull of our panga. And, not knowing exactly what it was, I am not sure what to make of it—how to interpret their numbers, their size, and their remarkable behavior. I suspect it was *Mobula lucasana*, cubana de lomo azul in Punta Arena and Bahía de los Ángeles alike. And if that's correct, then the rays were young. Forebodingly young.

8

Another mystery Pablo left me with proved less knotty. When the fishermen at the cleaning table had finished laughing about lomos blancos and negros, I asked them why they'd been catching rays only for the past decade or so. "Other things to take," an older man said, tersely. "Expensive things."

While the demographic shift from old, large fish to young, small ones is an ominous signal, the old man's words disclose an even higher order of change. Imagine a stairway of many flights and landings. With each season that the fishermen pursue an overexploited species, they take a step down in the economic returns of their work: The fish they catch get smaller and therefore less valuable, and what's more, the fishermen have to spend more time and money to catch them. Areas close to home get fished out first, and the fishermen respond by traveling farther afield, which requires more gasoline for their pangas and more hours of their own labor. After enough steps down in pursuit of a given species, the fishermen make the rational economic decision to switch to the next most precious animal. Soon, however, this species too begins to decline, and so, after a brief pause on a landing, the fishermen start their way down the next flight of stairs. And so it goes: stair by stair, floor by floor, the fishermen march further into poverty. When they were up amid the higher floors, taking valuable species like grouper, they would not have considered fishing for a ray. But once the grouper were just about gone, the fishermen switched to sharks; and once the sharks disappeared, the rays didn't look so bad after all. Yesterday's trash fish had become today's prime target.

Admittedly, the image of a staircase is too simple in certain respects. The order in which species are removed is determined at least partly by developments in the marketplace, which is intriguingly complex and fickle: witness the run on Bahía's cucumbers, triggered, in all likelihood, by new statutes in Ecuador. And lest that seem an exotic anomaly, here's another example: In the early nineties, a merchant from Ensenada started showing up in villages on both sides of the gulf. He said he represented an importer in Japan, and offered to purchase as many murex, black or pink, as the local hookah divers could provide. The shells of these snails are gorgeous plump coils: the black murex is strikingly marked with zebra stripes; the pink is china white on the outside, luxurious pink within. But it was not for their beauty that the Japanese importer had taken an interest in murex. The meat is something of a delicacy in Japan and parts of China, and, more strangely, the murex operculum—that little round door the snail pulls closed behind itself whenever it retracts into its shell—is a key ingredient in ancient recipes for incense. The year after the importer first appeared, the total catch of murex in many villages was ten times higher than it had ever been before. Five years later, the total catch was lower than it had been in decades, though the total number of hours that hookah divers were spending in search of murex was probably at an all-time high. In short, the population had collapsed under the sudden surge of extraction.

This story somewhat complicates the metaphor of a stairway, since the divers did not switch to murex at the moment they depleted another, slightly more precious species of shellfish. Rather, the price of murex suddenly rose, causing divers to abandon their previous targets in favor of the newly precious species. Even in situations like this, however, the metaphor is not too far off, because buyers are smart about the price they offer. They do not pay more than they must, and what they must pay is just a tad higher than the going price for any other species the divers might choose to pursue instead. So perhaps, on occasion, the switch to a new species allows fishermen to take a step up. But the step is generally quite small, and the time short—as short as a single season—before the fishermen plunge once more down the next flight of stairs.

It is important to see why they descend so swiftly. The fishermen of Bahía are essentially artisans: their gear is light and inexpensive; their boats are mere pangas; their crews, just a few men. However, their fishing grounds have been visited often by much larger vessels from Ensenada, from port cities on the Pacific coast of Mexico, and even from Japan. Large drift nets, those indiscriminate curtains that have been draped lately near Guardian Angel Island, are not deployed by small pangas, but by ships with large payloads and heavy winches on the rear decking. Similarly, to scrape the seafloor with weighted nets that gather up small mountains of varied life in every sweep; or to lay miles of line dangling thousands of large baited hooks; or to drop a voluminous purse of netting around a shoal, then cinch the purse closed—these too are activities that require vessels, gear, and crew on a scale that exceeds Bahía's local fleet. So the artisans of Bahía are not the sole culprits in their own economic descent; in part, at least, they have been chased down the stairway by better-armed fishermen.

The critical question now is whether the stairway has crumbled behind them. Is ascent still possible, or did each tread turn to dust as the last fisherman's foot passed to the next stair down? One form of decay is ecological, and it is evident in populations that don't recover the way you expect them to. In most populations, a decline in the number of individuals alleviates competition for important resources, such as food or places to lay eggs. And, under such favorable conditions, individuals are more likely to reproduce successfully, allowing the population to bounce back. In some cases, however, the dynamics are not so simple and homeostatic.

Imagine a marble on a hilly surface. As long as you don't move it too far from a certain swale, it will always roll back to the same spot. But as soon as you get it atop an adjacent ridge, it could roll down the other side, coming to rest in a new basin. An ecosystem can behave in just the same way: disrupt it too severely, and its own dynamics will send it careening the rest of the way into a whole new condition.

I encountered a dramatic example of this about ten years ago, on a moonless night in late summer. The town's generator had been doused for the night, so the darkness was nearly perfect. The stars were reflected in a placid bay, and the islands were long, dark forms floating in space, like the clouds of cosmic dust that make dark bands along the equator of the Milky Way. When I heard a clamor of engines and urgent voices coming from the direction of the Diaz boat ramp, I rose from my cot, picked my way cautiously down the rocky slope, and walked the length of the pale beach. At the ramp, I came upon a great commotion. Pickup trucks were backing trailered pangas into the sea; fishermen, standing astern, shouted commands to boys who scrambled in the shallows, shoving and guiding the crowded pangas as if they were reluctant livestock. The scene was dimly lit by red taillights and brakelights, but whenever another truck would swing in from town, its headlights would sweep through, unfurling a reel of images: rock lice skittering across the granite breakwater that flanks the boat ramp; a boat rail of chipping red paint; a great bundle of thick fishing poles and gaffs, bound together in the bow of a panga; an outboard engine of gunmetal blue; a boy's brown face, worried and shiny with sweat.

A few outboards coughed and revved. The pickup trucks began to depart. I distracted a boy long enough to ask what was going on, and he said they were going for diablos rojos. More outboards crashed into the uproar, props tore at the sea, and the pangas headed out. A moment later, the engines' separate hollers were blending into a single dissonance somewhere in the reflected stars. And soon, the drone faded into the Channel of the Whales.

Curious to see a red devil, I decided to settle on the beach beside the boat ramp. I watched the southern skies for falling stars, found a satellite instead, and then woke, several hours later, to the sound of a lone panga coming in across the bay. They must have radioed ahead, because a pickup truck converged with them at the ramp. As the truck turned itself around, its headlights flashed, ever so briefly, across the idling boat: three faces, an old fisherman and two boys, their eyes wide and jaws drooping open in hollow masks of exhaustion; in the bow, visible over the rails, a high and glistening heap of tentacles, eyes, and bloodred flesh. Los diablos rojos. For a moment, I might have believed that the man and his boys really had been dropping their weighted hooks clear down to the underworld.

Even in bright daylight, the Humboldt squid is an unnerving creature to behold. It is about five feet long. Half of that length is a torpedo-shaped body of muscle, which looks slippery and mucosal, as if the animal had been stripped of its proper skin. The other half consists of eight arms, which are lined with suction cups the size of nickels, and two longer tentacles, which terminate in lanceolate leaves coated with still more suckers. Toward the torpedo's tip, the body is flanked by wide triangular fins. Aligned with these fins, but at the other end of the torpedo, where the arms meet the body, are a pair of eyes like large nacre ashtrays filled to their brim with liquor. On the base of the torpedo, at the center of the eight arms, is a large sharp beak, exactly like that of a macaw; it seems zoologically misplaced, and when you first see it, you might briefly think something is trapped inside the squid, trying to hatch its way out. When the animal is dead, it is bloodred, just as its epithet would suggest. But when it is alive, its color is unstable, flashing cadaverous white, then red, then white again. Squid, like octopus, have large brains, and many scientists believe they use their flashing skin to communicate. As the panga pulled up to its trailer, one of the boys shined a light on the pile to show me. Waves of white and red passed over the wet mound as if it were an enormous heart, still pumping outside its body.

That same night, around the gulf, more than a thousand pangas were making similar outings, and each one was bringing back, on average, a thousand pounds of squid. Several hundred shrimp boats would have been out, as well, and on average each of those returned to port bearing six thousand pounds of squid. It's worth meditating on the arithmetic for a moment, because it is staggering: the quantity of Humboldt squid pulled from the Gulf of California in the course of a year is somewhere around 250 million pounds, which puts the species in the running to be the gulf's single most productive fishery in terms of the biomass it yields each year. And these numbers are even more surprising when you compare them with the annual commercial catch of Humboldt squid prior to 1977. Official statistics were not kept, but the quantity is not too hard to estimate: It was zero.

Like murex and cucumber, squid is sold mainly in East Asia. In this case, however, the fishery's expansion was awaiting not only the arrival of the right buyer, but also the appearance of the animal itself. Neither Miguel del Barco, who was in Baja from 1738 until 1768, nor J. Ross Browne, who was there in the 1860s, had anything to say about schools of predatory squid. And given the enthusiasm with which they described the bloodthirsty mantarraya, it seems unlikely that they would have neglected to mention a grotesque creature that pulsates bloodred and, unlike the manta, actually is a ravenous predator.

When John Steinbeck and his friend Ed Ricketts took a chartered fishing vessel up the gulf in 1940, they trolled lures behind their boat at night, shined lights into the dark water, and scooped through it with nets. But in the detailed and thorough list of the species they found, the Humboldt squid does not appear. In travel and fishing books dating from the 1960s, you find the occasional flabbergasted description of a jumbo squid hoisted from the depths, but certainly no one mentions great schools of them; you rather get the feeling that catching one was a weird and surprising event. But then, sometime in the late 1970s or early '80s, the strange experience became routine. And a few years ago, when a small team of scientists retraced Ricketts and Steinbeck's voyage, they reported that the waters just south of Bahía de los Ángeles were busy with Humboldt squid every night.

What happened? How did a creature that was hardly seen before 1975 come to provide a wildly prolific fishery by 1999? The squid that pangas pursue at night have risen from depths of six hundred to a thousand feet, and the reason for their ascent is almost surely to hunt. They grow at an astonishing rate—when they hatch, from a floating mass of eggs, they are about a millimeter long; two years later, they are pushing five feet—and to fuel such growth, they must eat voraciously. Mostly they feed on great schools of lanternfish, dark-water creatures that make their own nightly migration from depth. But Humboldt squid are not especially discerning eaters, and they will consume most anything in their path, including a variety of fish, crabs, and, quite often, one another. To a large extent, this diet overlaps with those of many predatory fish, such as yellowfin tuna, skipjack, albacore, yellowtail, and mackerel. All of those species, of course, are highly desirable on Mexican and American markets, and were therefore among the first to go when fishermen began descending the stairway of marine resources. Interestingly, the schools of tuna and jack were fished hard in the fifties and sixties, and catches appear to have taken a dive in the early seventies, just a few years before squid populations boomed. This timing, together with what we know about similarities in diet, would certainly seem to suggest that the removal of large predatory fish opened the way for an ecological expansion of large predatory squid. And, now that the squid are dominant, they may hold that position not only by competing with predatory fish, but also by eating their youngsters.

Still, I should not neglect to say that there may be more to the story. The scientists who retraced Ricketts and Steinbeck's journey wrote this: *Fishing pressure on tuna, a warming climate, and increased agricultural runoff from the Yaqui Valley in Sonora may have acted in concert to alter pelagic food webs in the Gulf in ways that favor jumbo squid over competitors, such as yellowfin tuna.* Personally, though, I'd put my money on the fishing. Time and again, in one place after another, fishing has pushed ecosystems hard enough to tip that metaphorical marble over the ridge, sending it down into a new basin below. And that makes me believe the rise of squid was, in all likelihood, a direct consequence of fishermen descending several flights in the stairway of marine resources. You many be wondering, however, if the story of the Humboldt squid doesn't altogether contradict the metaphor of a stairway. Fishermen were swiftly descending when—what luck!—the ecosystem quite suddenly thrust up a landing for them to stand upon. And, unlike the murex, which disappeared in just a few seasons, the squid keep bouncing back. Have the fishermen perhaps overfished their way into a *better* situation?

I doubt it, for two reasons. First, Humboldt squid is cheap. A fisherman today dumps back into the sea the squid's arms, tentacles, and head— pretty much everything but the mantle, that tubular torso of muscle—and for the mantle, he gets about five cents per pound. That works out to be about two or three cents for every pound he pulls out of the water. By comparison, the predatory fish are precious. For every pound of yellowtail jack or yellowfin tuna, a fisherman will get at least a few dollars, probably more. Here is another way to see the problem: By weight, Humboldt squid ranks within the top five slots on the list of the most productive fisheries in all of Mexico. But by value, it ranks somewhere south of twentieth place, squeezing in just above sea urchins.

The second serious problem with the squid fishery is its volatility. The predatory fish that have suffered under the squid's recent dominance make up a relatively diverse assemblage, which includes yellowfin tuna, black skipjack, albacore, bonito, and probably several species of mackerel, as well. Diversity in an ecosystem often confers the same benefit as it does in a stock portfolio: when conditions are bad for one part of the system, they are good for another, and thus the system as a whole is less prone to spikes and crashes. The lesson has been harshly reiterated about once per decade since the Humboldt squid displaced diverse predatory fish at the top of the food chain. In the most recent collapse, the catch plummeted from 75 million pounds in the month of November 1997 to 6 million pounds in December 1998. By April, the population had bounced back, but even when they are fleeting, such fluctuations can be devastating for fishermen and their communities.

So, no, I do not believe fishermen have fished their way into a new and improved ecosystem. And sadly, at this point, the reduced system seems to have staying power: the predatory fish haven't bounced back, while the squid, even after their crashes, keep resurging. The stairway of marine resources, then, has indeed begun to decay behind the fishermen who descended it. And the decay is due—at least partly—to the ecosystem's own dynamics. I say partly because it is not clear that the reduced system would be so tenacious without some ongoing help: fishing for tuna, mackerel, and the other formerly dominant species has never completely ceased. The reason for this has to do with a different sort of rot in the stairway. It is a decay in ideas of the sea. And although that may sound mistier and more abstract than the ecological trap we've just touched upon, its impacts are felt in the same definite terms: in stubbornly small populations of fish and missing sources of income for fishermen; more briefly, in kilos and pesos. But to see the direct connection, we need to think for a moment about managing a fishery.

9

A critical question for any manager: At what size does a fish population yield value at the highest possible rate?

The answer is not, as one might initially think, the population's greatest possible size. As the population grows toward the ecosystem's capacity, limited resources become scarce, constraining fish growth and reproduction. On the other hand, when the population is very small, there just aren't enough parents around to produce new offspring. Somewhere in the middle range, then, the population produces offspring most rapidly, and is therefore most valuable to fishermen. Still, while the population's maximal size is not economically optimal, a good manager must nonetheless base his estimate of the optimum on what he *thinks of* as the maximum. If he thinks the population, left to its own devices, would grow to a million, then his rational plan for management is to fish it down to around five hundred thousand. But if he thinks the population, left alone, goes to a hundred, then his management plan is to fish it down to around fifty. In short, low expectations lead directly to overfished populations.

It is important to add that the manager has no reliable way of knowing how big a population, left alone, could become, if a population is never left even remotely alone. And therefore the manager's estimate of maximum size must be based mainly on memory. This manager, of course, is a bit of a figment, a conceptual stand-in for the broader network of individuals who have some degree of influence over the way fishing is regulated. In the case of Bahía de los Ángeles, that network includes not only the officers of CONAPESCA, the federal agency that regulates fisheries, but also the fishermen themselves, the merchants who buy their fish, and even, in some diffuse way, any citizen who takes an interest. But if management decisions depend directly on memories, and the manager is not really a person but a broader community of relevant parties, the vital question is what this community remembers, and what it forgets, about each species fishermen have passed on their way down the stairway of marine resources.

The commercial potential of a species is hardly ever forgotten, even after it has gone commercially extinct, because knowledge of the commodity is embedded in a wide marketplace: merchants know where to buy it; fishermen know where to sell it; industrial operations know Bahía as a place to find it. And yet the community's idea of a species does evolve in other ways. In Bahía, the only kids familiar with specific terms for mantarraya are the children of fishermen. And even these seafaring youngsters, when asked how big the various species get, offer estimates that effectively turn all cubanas into tortillas, and all mantarraya gigante into large cubanas. It seems possible, therefore, that in another generation, the giant manta ray will be gone not only from the waters around Bahía de los Ángeles, but from the community's idea of the sea as well.

The idea shifts in other ways, too. Fishermen of different ages have given me intriguingly different accounts of how one catches really big yellowtail. An older man explained that good days come in late summer, when the big fish gather into schools to migrate southward. Whenever such a school comes into the bay, he said, you can take your panga half-way to Cabeza de Caballo, cast your lines in, and pull out five fish in less than a minute. On the other hand, a younger man answered me with a question of his own: "Where are you fishing for them?" He smiled know-ingly, and the implication, I gathered, was that he had his own secret spots: if you just knew where to go for the big ones, it wouldn't take you too long at all. As I inquired further, pleading gamely and swearing se-crecy, he revealed that you had to drop your baited lines over certain deep rocky reefs—like the ones out behind a certain island, he said, giving me a little wink.

In the older man's mental map, great schools roil the water in front of town. The younger man's assumption, by contrast, is that you take your panga somewhere special—probably secret and surely miles from town—to find big yellowtail. A similar shift is evident in expectations about shellfish. Certainly no one today expects rich beds of pearl oysters, since such things have not been seen for more than a century. What is more surprising, though, is that expectations about species that were re-moved more recently have also shifted. Just a few decades ago, lion's paw scallops formed beds in certain sandy areas around the bay. In other sand flats, pen shells had beds of their own. And in the right season, murex and conch gathered to mate, blanketing expanses of the seafloor with inter-connected shellfish. Young people close to the business of fishing still know these species and consider them worth taking, but they expect them not in large aggregations, but sprinkled sparsely and haphazardly around the bay. This too represents a shift in mental maps of the sea: a benthos that was formerly partitioned into separate patches is now a watery salmagundi.

The dynamics of memory, in which commercial potential persists as notions of size and distribution change, make it all but impossible for a manager—by which I mean, really, the broader community he symbolizes—to leave a population alone for any length of time. Instead of waiting and watching, which is the only way to find out how productive a population will become, the manager permits what you might call "clean-up fishing"—persistent removal of the recovering residuum of virtually every species that has ever been targeted. Once there are few prospective parents, it takes very little fishing to hold a population down. And, therefore, for every case in which an ecological transition has prevented a population from recovering, there are, I believe, many cases in which clean-up fishing holds the population at its lowest possible value.

Between ecological and cultural forms of decay, the stairway of marine resources is perhaps now looking a hopeless shambles behind those who descended it. But I'm not sure it's truly irreparable. An old fisherman like Pablo knows exactly how big mantarraya—*each species* of mantarraya—ought to be. He even knows how large they are at birth. Another old fisherman once told me where lion's paw scallops used to reach their greatest densities. And he referred to a particular inlet south of Punta Roja as "la caleta de los pepinos," the little cove of sea cucumbers. I once met a hookah diver who said he knew where many sea turtles lie dormant, at depth, during the winter; he would not give me the exact location, however, because I might pass the secret along to the wrong sort of people.

To reach further back than the memories of Pablo and his peers, there are documentary sources. Granted, these sources are rife with errors—mantarraya that engulf divers, for instance—but careful historical reading can sift out some truth. What bodes especially well, I think, for the prospects of recovering memory is that its threads are interwoven, and therefore, recounting one story of decline leads you across so many others. Just think of the intersections we hit upon, incidentally, as we endeavored to follow the skein of Bahía's mantarraya back in time: fishermen are not as well equipped as they used to be to take the occasional manta, because their pangas are filled mostly with small fish traps, not with the large harpoons they carried for goliath grouper, giant sea bass, or sea turtles; the manta was greatly feared by divers descending for two distinct and extraordinary species of pearl oyster; devil rays were ignored in the past, because there were more expensive species to pursue. You may set out to trace a single thread, but the weave of ecology and culture will lead you eventually across the entire fabric, threadbare and fraying though it may be.

But what about the ecological form of decay? Is that not truly irreversible? Hard to say, since clean-up fishing has never ceased. But what if the new ecosystem really is stable without the help of fishing? What if the metaphorical marble has rolled into a deep new swale? I spent an afternoon in the seminar room talking about ecological dynamics with Chris,

Allie, and Ace. We talked about the Humboldt squid, and then Chris said something that gave me goose bumps.

"So why not tip it back?" he asked.

"What do you mean?" I asked.

"Stop fishing all the predatory fish, and if that's not enough, fish the hell out of squid."

"You mean overfish it on purpose?"

"Right. Just fish your way back to the other equilibrium."

From one perspective, this might seem positively hubristic. Try to manipulate an ocean's ecosystem? Steer it just where we want it to go? What if something even less desirable than squid rose up to replace it? What if nothing rose up, and we ended up with a sea full of baitfish and plankton? But from another perspective, such a policy might seem to follow naturally from a clear-eyed acceptance of the role we have in fact already assumed. In this light, the idea that nature met its end long ago seems not only a philosophical notion to be mulled by transcendentalists like Thoreau, but also a hard fact with practical implications.

10

One evening at home, I talked with Graham on the phone. He was skeptical of my claims about the fine taxonomic distinctions that the fishermen of Bahía had made among their rays. I told him about Giuseppe's work with the fishermen of Punta Arena. But Graham wasn't convinced. "That was twenty-five years ago," he said. And he argued that the fishermen of Bahía are simply too modern, too immersed in our contemporary world of televisions and air conditioners, too distant from an earlier connection with the natural world, to make such precise identifications. Did I really believe the average fisherman of Bahía knew the difference between *Mobula lucasana* and *Mobula japanica*, cubana de lomo negro and cubana de lomo azul?

"Well, it's not like I told them about it," I said.

"Maybe you prompted them somehow."

Suddenly I was struck with doubt about my own notes and memories. And it did not help that my notes contained this line: *Cubana? Did he say Cubana? What's Cubana about a mantarraya?*

I sent an e-mail to Lane McDonald, asking if he would do me the favor of dropping in on some fishermen and inquiring about the names of rays. Lane happened not to be at the station, so he forwarded my note to a friend who was. The message I received in response confirmed that the fishermen do indeed distinguish among three sorts of cubana: negro, azul, and café. The message closed, however, with this line: *Roberto Ocaña says you should contact John O'Sullivan at the Monterrey Bay Aquarium.*

I had to smile. Roberto must have assumed, reasonably enough, that what I was interested in was rays, and not what he himself knows about rays. So he kindly directed me to just the right expert. And though the referral was not exactly what I'd been looking for, it was in fact helpful, because it gave the lie to an assumption of mine. In the grip, perhaps, of certain clichés from an outdated mode of ethnography, I had imagined a very clear separation between the locals and the visitors, the fishermen and the academics, words like *cubana* and words like *Mobula*. But the two worlds are not so separate as they were when Giuseppe arrived in Punta Arena. So while Graham was incorrect about the resolution of the fishermen's taxonomic knowledge, he was right, I think, to remind me that today's artisanal fishermen are not remote from modernity, which means not just televisions, but science as well. And there is even a possibility that the fine taxonomic knowledge of rays, which I took to be artisanal knowledge, shared for many decades among the fishermen of coastal Baja, from Punta Arena to Bahía de los Ángeles, was in fact scientific knowledge, gleaned decades ago (by Giuseppe) from artisanal fishermen, and then reintroduced (by visiting scientists, like John O'Sullivan) to the more modern fishing community of Bahía. I doubt it, but yes, it's possible.

On another evening, I thought I'd search the Internet for reports of mantas in the bay. I entered the words *manta ray Bahía de los Ángeles*, and when I pressed return, I received several pages of links. For a moment, I thought the manta might be faring better than I'd believed. But as I began clicking through, glancing at snapshots, reading a line or two, it became clear that all the reports were of leaping *Mobula*. None actually involved *Manta birostris*. Most of the photos, however, bore captions that identified the flying creatures as manta rays. What was in the process of happening in Bahía with the word *mantarraya* had already happened on the Internet with the term *manta ray*. And this struck me as a very vivid instance of the quiet stepwise change in our idea of nature: the name of a huge species had been bequeathed to a small one.

Alongside some of the mislabeled photos of *Mobula munkiana* were accounts of nice trips people had taken. Here are some excerpts:

. . . not easy to reach but worth the journey, especially for those who value unspoiled nature more than a mint on the pillow.

If Baja is one of North America's last great wildernesses, then L.A. Bay is Baja's crown jewel of untouched beauty.

Civilization hasn't changed the place much.
The hundreds of miles of coastal cliffs, beaches and coves are still as pristine as when the Seri and Cochimi Indians fished the same shores and islands centuries ago.

We go in search of wilderness, and so it is wilderness we find. And we write home about it. We tell the story of our rediscovery of nature. We compose suitably lyrical prose. We send back photos, and their iconography is just right: we hold up the large fish we've just caught; we recline on a white sandy beach; a winged ray leaps from the water. These are our dispatches from the wilderness. But when those snapshots are held up against images from the past, something is a bit off: Our fish is rather small, or maybe it's not a fish at all, but rather an enormous squid. That beach is hardly pristine. In fact, before the oysters were removed, it was not a beach at all. And that ray, it is two and a half feet across, not twenty-five.

I too am awestruck by the monstrous squid, lured by the beach, thrilled by the leaping ray. But how can we reconcile our sincere wonderment in the face of what natural wonders we still have with our recognition that what we are seeing is miserably dilapidated? On the one hand, a skein of rays, and on the other, the likelihood that they are but fry. We must be careful, I think, to hold on to both. Not to let one wash away the other. For if we permit ourselves to indulge too thoroughly in wonderment, we will forget where we are on this long stairway downward. But if we let our sense of history overwhelm our wonderment, leaving us with nothing but blasé weariness of the unimpressive present, then we won't much care where we are. What we must do, perhaps, is cultivate our craft of seeing more than one thing at a time.

PART VII

Balaenoptera physalus

THE END OF NATURE

1

The storm, like a dream, is gone by day but real and immediate at night. Cold wind comes once again off the channel at an hour that should rather be warm and still. And the scent has returned, too: that iodine breath of a churning sea. But even with the waves exploding against the station terrace, the kitchen feels safe—its masonry walls thick against the wind, the large industrial stove warm still from Alejandrina's evening work, and, on one of the large worktables, her plate of peanut butter cookies so invitingly reminiscent of childhood. All of which may explain why Yukon and I are still here, though we came to make coffee almost an hour ago now. I had just opened the paper sack of coffee beans when Allie walked in from the seminar room, took a cookie, and leaned against a stone pillar, directly across the worktable from me and Yukon. She asked why I had him on a leash.

"He saw lightning," I said, "out past La Guarda. He thinks we should all run for the desert."

"Oh, Yukon," she said sympathetically, walking around the table to crouch beside him. He leaned into her, so she sat on the concrete floor and allowed him to curl on her lap, as if he were something other than a ninety-pound golden retriever.

Looking up at me from her buried position, Allie said, a little awkwardly, "I hope my dad hasn't been annoying you guys with e-mail."

"Annoying us?"

"Veronica said he'd sent a couple messages about the storm."

"He just keeps offering to help. I think he wants to come airlift you out."

She smiled and shook her head. "Sorry," she said.

"No, no. Don't apologize. He's being incredibly considerate. He offered to drive down and bring supplies."

Allie laughed. "He lives in North Carolina."

"He didn't mention that."

"He's probably on the road already—somewhere in Texas, with life-vests stuffed into the backseat."

"Now I see where you get it."

"Uh-oh," she said. "Get what?"

"No—it's a good thing—just how concerned you are for everyone."

"I hope it doesn't get overbearing. Believe me, I know. It can get overbearing."

"Probably just when it's a parent."

"I don't know. I should be careful about it."

"That's true—you should be selfish and inconsiderate every once in a while, just to set everyone at ease."

She buried her face in the scruff of her huge lapdog's neck. "Oh, Yukon. What are we going to do? Maybe a peanut butter cookie would help. Can I give him a cookie?"

I leaned over the table, took a cookie, and handed it to her. She offered it to Yukon, but he wasn't interested.

"Wow," I said. "I've never seen that happen. Must be quite a storm coming."

Someone appeared in the dim passageway from the seminar room. The figure seemed to hesitate there, beside the tall shelves of skeletons, pickled invertebrates, and other curiosities, but it took me only a second to realize that such a small silhouette could only be Isabel. She came forward, perhaps because she saw that I had recognized her, or perhaps because she had spotted Allie and Yukon on the floor and realized she wouldn't have to talk with me on her own.

"Oh, no," she said, appearing in the doorway, "is Yukon sick?"

"No," I said, "just terrified."

"Oh, Yukon," she said, and walked around the large table to join Allie and Yukon on the floor. She nestled in beside Allie, lifting Yukon's rear legs and tail into her own lap, so he was now spread across two gentle caretakers. His head drooped off Allie's lap and his nose smushed against the concrete, making a small dark butterfly of moisture.

"Oh, no," said Haley, whose tall frame now filled the doorway into the dim hall. "What's wrong with Yukon?" Barely visible behind her was the extremely slight, lanky figure of another student—Lucy. And what occurred to me then, as I saw Lucy's towhead and bright wide eyes peeking around Haley's square shoulder, was that I'd really like to talk with her more, but she is always, somehow, just out of view. It's curious. I am often aware of Isabel—what she might be thinking, how she might be feeling—even though she is still so shy with me that we hardly ever talk. But Lucy, on the other hand, who seems nearly bursting with readiness to converse—those eyes always attentive, a grin so big it looks like it might extend right off the sides of her narrow face—of her I am rarely even aware. It's as if she is so entirely good-natured, so reliably interested, so easily engaged that I never feel worried about her, and so, among all the other students, she is the one I let myself look past. But what a loss—for me, that is. I mean, we've certainly never had a student who sings regularly with the San Francisco Opera. And by the simple numerical measures of a transcript, I think Lucy was neck and neck with Anoop for the title of strongest academic applicant. And yet, despite those qualifications, I am always thinking of Ace as the musical genius and Anoop as the brilliant academic. Which they are. They truly are. But what about remarkable Lucy, stuck there in the dim passageway behind Haley?

"Yukon's scared," Allie said to Haley.

"Oh, Yukon," Haley said, stepping into the kitchen and walking around the table to where he lay like a dying patient across his nurses. Lucy followed, flashing me a bright and enthusiastic smile. What might be most remarkable about her, I thought, returning the smile, is that she has yet to reveal even a hint of impatience with Becca. I have seen Haley simply stand up and walk away as soon as Becca started talking; Isabel often looks mortified; Allie stares sadly down at her own feet, tortured, I suppose, that there is nothing she can do to help everyone out of the situation. But Lucy just listens attentively. I would hold it against her—abetting socially corrosive behavior—if I didn't think Becca would be just as difficult, maybe even more so, without Lucy's encouragement.

Haley crouched now beside Yukon's head, gently lifted his drooped muzzle from the concrete, and somehow managed to slide her legs in beneath it. As Lucy arrived, there was no section of dog left to take on her lap—Haley had the head, Allie the midsection, and Isabel the rear—so she simply settled in cross-legged, all knees and elbows, directly across from the other three young women.

That was when Graham arrived from the staff house, walking in through the stone portico to the west. Instead of his usual tank top, sarong, and flip-flops, he was wearing a wool skullcap, fleece jacket, sarong—an all-season garment, evidently—and heavy brown hiking boots, which made the little sections of leg showing below his sarong look exceptionally skinny. At first he didn't see the gathering on the floor; it must have been hidden from his view by the large tables.

"Hirsh," he said abruptly. "What the hell? I thought you were going to make coffee. I have to lecture in thirty—whoa—what are you guys doing to Yukon?" He bent at the waist to get a better view under the tables.

"We're *petting* him." Haley replied. "He's scared."

"Ah," Graham said, "I see." Standing up, so his face would be hidden by the tables from the young women on the floor, he looked at me in a way that seemed to inquire whether I didn't think this was all a little weird—as if the gathering on the floor might have been around a Ouija board instead of my dog. Then he turned around and walked out the way he came. His voice, echoey in the stone portico, hollered, "Bring that coffee when you come up, Hirsh."

Haley looked at me and shook her head. "Jeez," she said. She has not taken a liking to Graham; his intellectual showiness strikes her, I think, as a form of arrogance.

Yukon's head lifted from Haley's lap, perhaps because her pair of the eight hands responsible for petting him had momentarily paused from their ministrations.

"Here," Allie said, "maybe he'll eat the cookie now." She reached behind her to retrieve the peanut butter cookie she'd left on the floor and offered it again to Yukon. He ate it, drooling on Haley's legs. "He's feeling better," Allie said. Cookie gone, Yukon let his head settle back down on Haley.

"I think he might pull through," I said.

Amid further petting, Allie looked up at me and said, "Isabel and I heard from Alejandrina that they're going to sell the whole ejido."

I was taken aback, unsure how to respond, because I had thought of Alejandrina's news about the Texas developer as something of a secret. And maybe it was. Maybe Allie and Isabel have already become members of Alejandrina's large circle of friends and confidants. On several occasions, I have seen the three of them in the kitchen, chatting in Spanish. Allie is fluent, having spent a semester abroad; Isabel, of course, is a native speaker; and both of them, I've noticed, like to linger in the kitchen, keeping Alejandrina entertained. Once, while I stood filling my water bottle at the large cooler against the kitchen wall, I tried to follow their conversation unfolding behind me. I was struck then by how remarkably different Isabel seems in Spanish. She was telling a funny story about her father, the grocer in Los Angeles, and how there are certain trivial chores he simply refuses to perform because they remind him of his younger, harder years, when he was a migrant farmworker.

Alejandrina was laughing heartily. When my bottle was full and I turned to leave, I saw that she had removed her eyeglasses to wipe away tears. With the other students, Alejandrina is always warm and kind. She greets them with a Spanish teacher's pedantically clear "buenos días" or "buenas tardes." And if someone looks a little tired she tends to him as if he had an acute, life-threatening fever. But there is always a certain distance. Even with Chris, who also speaks Spanish perfectly and has a familial connection to Mexico, Alejandrina seems a bit guarded.

As I stood now in silence, looking down at my dog and wondering what to say about the ejido, Haley finally said, "I heard that, too. I heard they were going to get a billion dollars."

"Alejandrina told you that?" I asked.

"No," she said, "I heard it in town, at the basketball court."

While Isabel, and maybe Allie, too, have been welcomed into the circle of Alejandrina's friends, Haley has become the adoptive queen of a small clutch of preteen boys who seem to live at the town's basketball court. In the past, the netless rims served the boys only as the upper limits of their imaginary soccer goals. But lately Haley's got them all practicing three-point shots.

"That's funny," I said now. "I thought the ejido sale was a big secret."

"I think it is," Haley said, smiling. "So big everyone in town is talking about it."

"It's amazing," I said, "that you guys are already in on town rumors."

"You think it's just a rumor?" Allie asked.

"I can't imagine why anyone would want all the ranchland. And the number—the billion dollars—is absurd."

"Alejandrina didn't think so," Isabel said.

I didn't know what to say, and shrugged my shoulders in uncertainty.

"What about Escalera Nautica?" Isabel asked. Her eyes looked suddenly stern, clouded over.

"What do you mean?"

"Veronica told us it was a really huge plan."

"It was," I said. "But I don't think anyone's going through with it. They renamed it El Proyecto Mar de Cortés and scaled it way back. So far it's a Pemex and that silly sculpture."

"I kinda like the sculpture," Haley inserted.

"So you don't believe it?" Isabel persisted. "You don't think Escalera's going to make the town a big deal?"

"I hope not," I said. "And I'm sure no one's buying the ejido for a billion dollars."

"Hirsh." Graham was standing in the doorway again. Mysteriously, he had changed out of one sarong and into another, but the rest of his winter outfit remained unchanged. "Dónde está el café, amigo?"

"I was just about to make it."

"You were. But then you didn't."

"Look," I said, pointing to the paper sack of coffee beans, still standing open from the moment before Allie walked in. "Everything's all ready."

"Oh, man," he said, turning around and walking back out. "Bring it up when you make it, will you?" he hollered from the portico.

"I can't," I yelled.

"Why not?" he said, appearing again in the doorway.

"Because it's so nice in here."

He stared at me, looking puzzled. I said it, I confess, so that the students wouldn't think Graham was pushing me around. And maybe I was feeling Haley's critical gaze in particular. But when he turned around and walked out again, I felt guilty.

"Have you guys finished the reading for Graham's seminar?" I asked the gathering.

Lucy nodded. Allie said, "Um, almost," and smiled winningly.

"You'd better get to it," I said. "I don't want Yukon to be held accountable."

One by one they slid out from under Yukon, whose limp body drooped to rest, section by section, against the concrete floor. I made the coffee, and just as it was gurgling up through the stove-top machine, Graham reappeared in the doorway. Preoccupied with his seminar on the history of whaling, he seemed unperturbed by the rude way I'd refused him. I gave him his coffee. Then I took a cookie and my own mug and sat in Alejandrina's chair to write in my journal while the wind hissed at the windowsills and the waves pounded the terrace.

2

In the morning, Veronica, Graham and I stand beneath one of the tamarisks on the station terrace. Veronica's binoculars are trained on Cabeza and Los Gemelos, but I know she is looking past them, at the channel beyond. Graham, to her right, and I, to her left, are holding bowls of cereal instead of binoculars, and since our unassisted eyes have no chance of perceiving the distant waters she is presently assessing, we have given up and turned our backs on the blazing sun. So while she looks out to sea, we face our own shadows on the station's masonry wall. The feathery shade of the tamarisk hovers uselessly above our shadows, offering our actual bodies no shade at all.

"Well, Vica?" I ask. "What's it gonna be?" I am impatient to move not only because the back of my neck is burning, but also because I feel silly staring at the wall while Veronica scans the horizon like a sea captain. Fortunately, the students have all finished their breakfast and are already inside, packing their daypacks and dive bags. Our schedule calls for us to take the pangas to Punta Pescador, a spectacular reef that we must venture into the channel to reach. But our schedule, of course, was written before we knew about the storm; in crafting it, Veronica thought mainly about the difficulty and depth of each location, about the timing of tides and how they would affect what we would be able to see on the reef beneath us, and about the ways the dive sites might complement certain lectures. Now all those carefully laid plans may have to be jettisoned.

"The swell is pretty big," Veronica says without lowering her binoculars, as if continuing to look might yet change the diagnosis. Here in the bay, the morning has been calm—not as glassy as usual, but by no means dangerous. But if Veronica is confessing, in Graham's presence, that it is somewhat rough in the channel, then I suspect the waves are at least head-high. "Yeah," she says with disappointment. "Pretty big."

"Maybe that's because there's a hurricane out there," Graham says, motioning with his cereal spoon over his shoulder.

"It's three days away," I say, addressing my comment, for some reason, not to Graham, but to his shadow on the wall in front of me.

"Right," Graham says to my shadow, "but it's four days wide."

"We don't have to make up our minds until tonight," I say. "That was our agreement."

"That agreement is null and void."

"Why?"

"We made it before the sea cow went out on us. Without information, we have to act cautiously."

At dawn, when the three of us gathered around the computer and Veronica clicked REFRESH, the NOAA website failed to load. Graham tried his incantation, murmuring, "Sea cow sea cow sea cow," and for a moment, it seemed it might work—the NOAA insignia appeared in the upper left-hand corner of the screen, and then a square of ocean blue appeared near the center. But those small tiles turned out to be only cruel taunts by the sea cow, which declined to give us anything more. Eventually, a message appeared: *Oops, the page you wanted didn't load. Click here to reload.*

"Oops?" Veronica said. "Oops? We're trying to find out where a hurricane is and this fucking thing says 'oops'? It really is a sea cow." Leaning over the laptop, Veronica typed www.amazon.com into the browser and pressed RETURN.

"Wait," Graham said. "What the hell are you doing?"

"She's proving it's a sea cow," I said, just as images of camcorders and televisions materialized on the screen.

"See!" Veronica cried.

"No," Graham said, "it's working. Go back to NOAA."

Veronica typed www.nhc.noaa.gov and pressed return. The browser's little wait-I'm-thinking wheel spun and finally a solitary blue tile appeared at the center of the screen. Veronica groaned and raised her fist as if to smite the keyboard.

Now, lowering her binoculars, she says, "It's not like we have *zero* information."

"That's true," Graham says to my shadow, as if I were the one who had just spoken. "We saw that little blue square. So we're sure there's ocean out there somewhere. And you're disqualified from decision-making anyway."

"Me?" I ask. "Why?"

"Severe sleep deprivation."

It is true that I spent a large part of the night preventing Yukon from fleeing for the desert. Like an obsessive unable to help himself, he stared at the eastern horizon. And whenever lightning flickered in the distance, illuminating a band of sky above the rough contours of Ángel de la Guarda, Yukon reared against his leash, scratched his claws across the sandy ground, and somehow backed his head out of his collar. Taiga, lying in stately vigilance beside Veronica's cot, observed our sudden bouts of wrestling. Veronica slept. Graham slept. The row of cots on the station terrace offered the occasional cough or dreamy mutter. Finally, at my wit's end, I decided to take Yukon somewhere he could not see the horizon. I led him up to the staff house, into a small bathroom at the back, and closed the door behind us. I sat with him on the concrete floor of a small shower and waited for him to calm down. But when the lightning next flickered, it caused a faint quiver of light in the dark bathroom, which has a small window. And since Yukon knew precisely what this quiver of light signified, he lunged for the door and began digging furiously at its lower edge. Exasperated, I pulled my headlamp from my head and, grasping Yukon in a kind of headlock, slipped the elastic band over his muzzle and head. I tightened the band as if it were a chinstrap, so the lamp was stuck on top of his head, projecting its beam directly in front of his face. Remarkably, the strategy seemed to work: everywhere Yukon turned, he saw only a searingly bright spot of white light against the bathroom walls. And when the window flickered again, Yukon did not perceive it, because he was staring into his own private sun. Eventually, he fell asleep.

"No Pescador," Veronica says now.

"How about San Diego?" Graham says.

"Burnett," I say, with a touch of irritation, "come clean. You don't actually want to leave."

"I don't?"

"No. You just want to be the one making the argument we should leave. That way, you're not responsible for keeping all the students here, but as long as V and I vote for staying, you get to have it both ways."

"Both ways?"

"We stay, but you're not in trouble with parents or deans."

"I'm glad you think I'm so courageous, Hirsh. But what gives rise to this hypothesis?"

"If you really thought we should leave, you would have made a more sincere argument by now."

"Okay, for the record, here's my argument: there is a category-five hurricane heading straight for us."

"See?"

"See what?"

"You said 'for the record.' Because off the record, you want to stay."

"It's category four," Veronica says. "And it's supposed to veer into the Pacific."

"See?" Graham says to my shadow. "It's futile anyway."

"We'll check in with Samy," Veronica says. "If he says it's okay, we could try the west side of Coronado."

What Veronica is thinking, I suspect, is that there may be large clouds of krill in the narrow channel between the volcano and the peninsula. In the past, we've noticed that windy nights and rough waters tend to gather up the tiny red shrimp and deposit them there, in the protected stretch of water west of Isla Coronado. And where there are krill, there are likely to be fin whales feeding on them.

"Good," Graham says suddenly. "Coronado it is. Let's go."

Veronica turns to me with a look of amazement. I shake my head. The sudden change of heart has nothing to do with my accusation that his campaign to evacuate has been a sham. He too knows we go to the channel west of Coronado to look for fin whales. Last night he lectured about the origins of cetology, how the science of whales was in fact embedded for decades in the industry of whaling. And the possibility of a moment of intersection—one of those beautiful, fortuitous correspondences between our discussions in the seminar room and our experiences on the water—is just too tempting for him to resist.

"His lecture last night," I say, in response to Veronica's look of surprise. "Whales."

"Ah," she says, nodding.

"So are we chatting," Graham asks, "or are we going?"

3

We find Samy Diaz seated on the tailgate of a pickup truck that is already hitched to the *Sea Eagle*'s trailer, as though he fully expects us to head out this morning. But then, as he sets down his newspaper—a week-old copy of the *Ensenada Gazette*—he asks, in an offhand way, if we carry marine radios. Veronica says we do, and adds that we're not venturing into the channel, just heading north, to Coronado. "You'll be fine," he says, making a little frown that seems to say there's nothing, nothing whatsoever, to worry about. I want to ask him why, then, he was inquiring about radios, but he is already climbing into the cab of the pickup. And so, ten minutes later, our pangas are roaring across the choppy water, flattening with their fiberglass bellies the irregular peaks and valleys of disordered waves. We head northeast from the beach, making for the old lighthouse on the sandspit, which we will just round, observed by the yellow-footed gulls and brown pelicans that rest there on the thumb of sand. Past the lighthouse, our course will be due north, along the shoreline. It would be more direct to start fading eastward, toward the angels, but the water closer to shore is pleasantly smooth, and the seafloor along that coast is very beautiful—brilliantly white, interrupted here and there by irregular patches of olive-green eelgrass.

Graham stands beside me at the center console, his bandana, loose oxford cloth shirt, and sarong flapping in the wind, reaching out now and again to whack various parts of my body. On the bench seat in front of us sit Becca, Rafe, and Anoop. Becca has learned by now that, besides the bow, where I have asked her not to ride, the one position on the boat that does not get hit with sea spray, even when the wind is blowing and the seas are rough, is the leeward side of the front bench seat. She has also learned to avoid Veronica's boat. For Veronica, it seems, cannot help herself from falling silent whenever Becca is around. And she has confessed to me and Graham that she has even caught herself passing up opportunities to point out interesting sights to the students—on one occasion, a frigatebird harassing a booby into giving up its fish; on another, a fuzzy-headed osprey chick peeking its head over the wall of sticks that encircled its nest. She would only have had to point and say, "Look," but she didn't.

"Why don't you want Becca to see an osprey chick?" Graham asked, trying earnestly to understand and help. "I mean, where's the harm?"

"She'd one-up it," Veronica answered. "She'd turn it into her own story about something she saw while she was doing fieldwork in Antarctica for some National Academy of Science muck-a-muck."

Graham nodded, but his expression said he still didn't understand. Veronica went on, "Or she'll just say in her knowing way, 'Oh yeah, we learned all about that in Ecology 499.'"

"But isn't that a good thing?" Graham asked. "I mean, I think I'd feel ecstatic if someone said, 'Oh yeah, I learned about that in HIS 499.'"

Veronica shook her head. She could not explain, or did not want to. And maybe that's one reason she has not yet found a way out, and continues to close herself off in Becca's presence. What is strangest to me, though, is the way Becca has responded. I might have expected her to view Veronica's silence as a kind of victory, in which she had succeeded in pressing a figure of authority into the role of cooperative spectator. But that is not at all how she seems to view Veronica's withdrawal. For she seems determined now to avoid Veronica altogether—in that respect, Veronica's strategy, if you can call it that, has worked. And yet, at the same time, I do not believe Becca has perceived or felt any kind of personal critique, because I see no hint of doubt or hesitation on her part; to the contrary, she has simply redirected the full force of her conversational imperialism toward me and Graham. It is as if she needs a certain setting for her striving—needs the cast of other students and, critically, a figure of some academic authority to engage.

And that's why, just a moment ago, Becca made straight for my boat, staking out the one dry spot in front of me on the left. And then Graham followed Becca, because he is required to. We have determined that he is the most adept at handling her, so it has become his appointed task to board whichever panga Becca does and, should the need arise, practice what has struck me as a kind of spoken judo. Becca will make her usual move, in which she initiates with one of us a dialogue that the other students will have no choice but to observe, but Graham will somehow take the momentum of her question and swing it around into an inclusive conversation on some topic of relevance to the class. Yesterday, while we were eating lunch on the terrace, he took a sally that Becca had launched in my direction, a pointed and intrusive question about the Judaic origins of my own name, and somehow turned it into an opportunity for several other students to offer their views on the relationship between religion and science. It was masterly. And when that sort of move doesn't succeed, he will simply extemporize a long lecture, which can be trying in its own way, but is certainly better than listening to Becca.

Just ahead of us now, Veronica's *Sea Eagle* is drawing a wedge of opaque white foam across the brilliant clear blue and patches of dark green, and the sight of three tall figures on her rear bench seat—Miles,

Haley, and Ace—gives me a pang of jealousy. Those three are the most diligent in their avoidance of Becca, and I can foresee that on each morning the storm allows us to use the pangas, a predictable line of embarkation will stand knee-deep in the shallows, waiting to climb aboard my boat: Becca comes first, Graham follows Becca, and Rafe, interestingly, follows Graham. That last link, which I never would have anticipated, means of course that Becca and Rafe always end up in the same boat—an amusing turn of events, when I think how I've worried about a rivalry between them. And their relationship did indeed seem to be headed in just that direction, but remarkably, Rafe seems to have experienced a sudden shift in worldview.

If I played a role in this shift, I did so unwittingly. A handful of conversations do come to mind—conversations that could have wobbled certain convictions he held about science. But my intention was always to make a case for a certain way of practicing science—not to call science itself into doubt. So when I first encountered the aftereffects of Rafe's personal revolution, I didn't quite realize what I was seeing. The other night, when I introduced von Baer's ideas, I was surprised by how readily Rafe seemed to accommodate another old German's theory—especially since, just moments before, he himself had articulated, and even seemed to embrace, Haeckel's version of recapitulation. But we were in the midst of a seminar, so I did not have time to wonder long about the current whereabouts of Rafe's usual persona—his impatience with uncertainty, his peremptory expectation of the right and true scientific answer. The disappearance of all that was strange enough, though, that afterward, I went looking for the missing scientific triumphalist. Walking into the kitchen, I found Rafe with Anoop and Chris. The three of them were standing beside the tray of granola bars Alejandrina had left out on the table, and they were still chatting about recapitulation.

"So, you guys—" I said, because I couldn't think of a graceful way to single out Rafe.

"Aye, Captain," Rafe said. And since he was the one to respond, I now had an excuse to focus on him in particular as I said, "I have a question for you."

"We're ready," Rafe said.

"Okay," I began, "now that you've seen both Haeckel's and von Baer's versions of recapitulation, do you think they could be synthesized into a theory that would qualify as the kind of law Haeckel aspired to—a theory of recapitulation that would always be true?"

"Always true?" Rafe repeated, hesitantly. He glanced over at Anoop, who was frozen with a granola bar in his hand and small bits of it in his goatee. Anoop looked dumbfounded. By now, I think, he had gotten used to understanding where I was coming from—what I was up to, pedagogically speaking. But in that moment, he clearly had no idea.

"Always true," Rafe said again, more reflectively this time.

"Right," I said, still focusing on him.

"Captain, I'm not thinking about these theories in terms of true or false. They're just internally consistent systems of ideas."

Now I was the one looking dumbfounded. It didn't sound like Rafe. The Aussie accent certainly did; the voice did; but the words—they simply weren't his. It was as if Anoop had just ventriloquized an answer through the young man standing next to him.

"Ah," I said, nodding, because that was all I could muster for an answer. Then I took a granola bar off the tray and made for the stone portico, on my way to the staff house. I could feel the three of them looking quizzically at my back, and their conversation did not resume until I was out the door. Up at the staff house, I found Graham and Veronica laughing about something. I told them, verbatim, what Rafe had just said. As Graham leaned down to take three bottles of cold beer from the knee-high fridge, he said, "The young man's ideas are evolving." The phrase felt cryptic, a little dark, but I let the matter drop.

For a short while, I thought of Rafe's strange answer like this: He had always had a very clear line between real science and other sorts of investigation; what was new now was just that he had found a way to take an interest in all the other sorts of investigation. They did not hold any scientific value—you couldn't call anything in them *true*—but you could at least think of them as "internally consistent systems of ideas." And in that guise, they could be entertaining. It was surprising, certainly, that he had made his way so swiftly to such ecumenical curiosity. But that, I figured, had to be testimony to the influence of his new friend, Anoop—and perhaps of Graham, as well.

An expanded range of interest, however, would not have qualified as a full-blown revolution. Rafe's shift in view went further than that—much further—and this became clear to me yesterday afternoon in the seminar room. Veronica was leading the day's first seminar, which was about a very clever molecular genetic study of the evolutionary history of whales. It had been published by a group in Japan, and it was just the sort of research I would have expected Rafe to adore: current, technical, and basically irrefutable—a lot closer to his ideal of science, in other words, than our discussions of antique embryology could ever be. What's more, Veronica is a more formal, less improvisational, and—it must be admitted—better-prepared seminar leader than either Graham or I. She connects her laptop to a projector and shows photographs of organisms, images of raw data from the lab, and finally diagrams and animations to clarify the scientific ideas.

The first key step in the Japanese group's work was the discovery of two new kinds of SINE residing in the genomes of whales. Never mind what the acronym is for. What's important is that a SINE is a short stretch of DNA that has a remarkable ability. Its genetic sequence—the A,G,T,C of it—gives the cell these instructions: 1. Make a copy of me; 2. Insert that new copy of me elsewhere in the genome (anywhere will do); 3. Patch up any nicks and scrapes this little splicing operation may have left. The cell cooperatively—and, one could say, quite stupidly—fulfills the request, and so, over the course of generations, a SINE can proliferate through the genome, inserting copies of itself in many locations. In the human genome, for example, there are about a million and a half copies of SINEs, composing about fifteen percent of the total DNA. It's the sort of fact that could drive an adaptationist to distraction—*all that useless DNA!*—but that was not the issue Veronica wanted us to focus on.

SINEs come in different varieties; you could think of them like strains of flu virus—detectably different, but clearly related. When the Japanese scientists discovered their two new varieties of SINE in whale genomes, their next step was to look for them in the genomes of a number of other species. They found that they were present in the genomes of cows and hippos, but definitely not in the genomes of pigs, camels, elephants, horses, or people. This is a simple result, but it is also very meaningful, because there has been a long debate about where whales belong in the tree of life. Which species are their closest relatives? What sort of mammal was it that ventured from the land back into the sea? Darwin thought it must have been bears, which seems sort of reasonable when you think of the way polar bears swim great distances. But modern paleontologists, studying the skeletons of ancient whales, showed that the bear idea was definitely wrong. And they more or less agreed that whales' closest relatives were even-toed ungulates, a very large group that includes ruminants (like cows), hippos, pigs, and camels, but not horses (which are odd-toed ungulates), elephants, or, certainly, bears. The Japanese study showed that this assessment was right—but, as Veronica promised to explain, only kind of right.

First, though, it was important for her to clarify something about the way SINEs work, because the strength of the study hinged upon it. Once a SINE gets copied into a certain location in the genome, there is simply no neat and tidy way for the cell to excise it. Occasionally, a mutation called a genomic deletion will erase some random segment of one copy of a SINE, but even when that happens, part of the copy remains. And if a genomic deletion is big enough to take out the entire copy, the odds are small that such a whopping mutation will leave the adjacent segments of the genome perfectly intact. So even in that case, you would not be tricked into thinking the SINE had never been in that location; you would at least see evidence that it *could* have been there before the whole genomic neighborhood got taken out. To put it concisely: each copy of a SINE leaves a mark that's hard to erase; a whole strain, comprising lots of copies, leaves an indelible mess.

So think about what this means for that simple result from the Japanese lab. Their two new strains of SINEs were detectable in whales, cows, and hippos, but *not* in pigs, camels, or anything else. Veronica projected this diagram on the whiteboard:

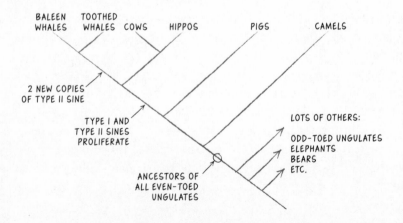

BALEEN WHALES   TOOTHED WHALES   COWS   HIPPOS   PIGS   CAMELS

2 NEW COPIES OF TYPE II SINE

TYPE I AND TYPE II SINES PROLIFERATE

LOTS OF OTHERS:

ODD-TOED UNGULATES
ELEPHANTS
BEARS
ETC.

ANCESTORS OF ALL EVEN-TOED UNGULATES

The label that says *Type I & Type II SINEs proliferate* indicates that the two new strains of SINE must have invaded the genome of the common ancestor of cows, hippos, and whales—and they must have done so *after* that ancestor had already diverged from pigs, camels, and everything else. This contradicts the paleontologists' theory that the whales diverged earlier, from the ancestor of all even-toed ungulates, the larger group that includes pigs and camels. But wait, the paleontologists might object, maybe we're right, maybe the whales did diverge earlier, from the ancestor of all even-toed ungulates, but the SINEs happened to be lost from the genomes of pigs and camels. No, we'd have to say: SINEs leave a permanent mark. There is really no way they could have disappeared without a trace.

There is another nice result here, encapsulated in the label that says *2 new copies of Type II SINE*. The Japanese group had found copies of their Type II SINE at two particular locations in the genomes of both toothed and baleen whales. And since it is extremely unlikely that the SINE would have landed in those exact same locations in two independent lineages, this result demonstrates that a single lineage must have led to both kinds of whales, the baleen and the toothed. In other words, it's not the case that baleen whales shared an ancestor with cows while toothed whales diverged from some other lineage, like hippos. We knew as much already, from other lines of evidence, but it's nice all the same to have it confirmed.

Veronica concluded her presentation by returning to photographs of the mammals we'd been talking about. They were projected on the whiteboard in an arrangement that reflected their evolutionary relationships: Veronica's own photo of a humpback breaching just off Cabeza de Caballo was positioned right next to her photo of the tall dorsal fin of a male orca, also with Cabeza in the background; separated from those two by a bit of empty whiteboard was a brown-and-white cow munching its cud on a lush meadow; farther away, a fat pink pig stood knee-deep in mud; and farther still, nearly off the edge of the whiteboard, a polar bear paddled through an Arctic sea.

When the images appeared, Rafe raised his hand. Veronica finished her summary of the evolutionary relationships, then looked at Rafe to hear his question.

"The Japanese like to eat whales, right?"

Veronica tilted her head inquiringly—an expressive gesture she has picked up, I think, from Taiga. "Some do," she said, awaiting more.

"And the rest of the world tells them to stop, right?"

"Most countries, yes."

"So don't you think it's a little suspicious that they've discovered that the whales' closest relative is beef?"

Veronica laughed, but then realized he was serious.

"Really?" she said to him.

"What do you think their agenda was?"

"Their agenda?" Veronica looked unsure. Most questions the students could have asked she would have thought through in advance. This was not one of them.

From my seat on the bench at the side of the seminar room, I said, "It's a nice conspiracy theory. But this kind of data is really solid."

Graham, from the other side of the room, said, "Aren't PCR-based analyses like this actually quite sensitive to contamination errors?"

Veronica replied, "The basic result isn't PCR-based. They're just blotting genomic DNA with probes for the two new SINEs."

"Ah," Graham said, nodding deferentially.

"And the result," I said, "isn't even really that cows are the closest relatives of whales."

"It's not?" Graham said.

"No. Vica, could you go back to the slide showing the phylogeny?"

Veronica pressed a key on her laptop to move backward through her presentation until she got to the diagram of the evolutionary tree. With that diagram on the whiteboard, I pointed out that the evolutionary distance from whales to cows—essentially, the length of line you had to trace along the tree to get from one to the other—was not actually shorter than the distance from whales to hippos. It was just that Veronica had arbitrarily chosen to write cows on the left and hippos on the right. Then I pointed out that cows are in fact standing in for the entire group Ruminantia. Instead of cow, Veronica could have written moose, or something weird, like water chevrotain.

"Who?" Graham said.

"Water chevrotain. It's a little South Asian animal that looks like a muskrat-deer-pig."

"Ah," Graham said, "a muskrat-deer-pig. And you're saying it would occupy basically the same place on this large tree as the cows do."

"Yes, exactly the same place. Cows were just chosen to represent the larger group called Ruminantia."

"Which species did the Japanese researchers choose to represent that larger group? I mean, did they label their diagram *moose* or *muskrat-deer-pig*—or did they write *cows*?"

"They chose cows," Veronica said, smiling at Graham's lawyerly turn of the argument.

"So the results," Graham continued, "could be entirely right, and yet Rafe's point could still be well taken. They didn't have to present their air-tight results by writing *cow* right next to *whales*. And they might have had some motive for doing so."

"Well done, Burnett," I said.

"Well done, Rafe," Graham replied.

"Indeed," I conceded.

Rafe was looking down at his reader, as if he were still carefully studying some of the Japanese group's data.

If that conversation left me with any uncertainty about the extent or cause of Rafe's gestalt shift, all doubts were answered several hours later, during Graham's evening seminar. The syllabus offered the title "History of Modern Whaling and Cetology," which had sounded to me like two separate topics, but it turned out to be just one, and in fact that was the main point of the seminar. In the 1940s and '50s, cetologists had only one way to pursue the living, breathing object of their research, and that was to ride aboard whaling ships. Many of the seafaring scientists were especially interested in ovaries, and to get their hands on them, they had to stand right there on the flensing deck, plunging their arms into entrails. Graham referred to these hardy, red-armed researchers as "hip-boot scientists," since they often wore waders as they slogged and foraged through the pools of gore.

But why ovaries, of all things? In whales, unlike most mammals, each ovulation event leaves a small but lasting mark on the ovary. Cetologists realized that if they could just calibrate this biological clock—figure out, that is, how many ovulation events an average female whale experienced each year—then they would have, at their bloody fingertips, a reliable method for estimating age. And if they were to apply this method to every female a whaling fleet captured, they would soon get a sense of the overall age distribution, which, you will recall, reveals a lot about a population's welfare or peril.

To calibrate the clock, cetologists simply counted ovulation marks in a number of whales whose ages they knew by other methods—or, rather, *thought* they knew. In actuality, those original, standard-bearing whales were not as old as the cetologists believed, and consequently, the calibration of the clock was mistaken—biased in such a way that the new clock would *overestimate* the age of every whale. And since overfishing is signaled by a dearth of older individuals, the clock's systematic error would obscure the bellwether of decline.

But cetologists caught a lucky break. A scientific second chance. Besides examining ovaries, they sometimes used a method that was basically a version of what contemporary ecologists call mark-recapture. Standing on the decks of whaling ships, they would shoot huge darts into whales that, for one reason or another, the whalemen were not going to kill at that particular moment. The dart contained coded information about when and where it had been administered, as well as a request that anyone who happened to recover it (in the process of butchering a whale) please kindly return it to the scientists who had placed it. Obviously, mark-recapture is a rather chancy method. But if you do your statistics properly, the rate at which you recover your tags can tell you a lot about the overall size of the population. And you don't have to do any statistics at all—you just have to get lucky—for tags to tell you other interesting things, like where animals migrate or how long a certain individual has been alive.

You probably see where this is going. Some scientists were lucky enough to inspect the ovaries of a mature female they had first darted as a very small calf. Their count of ovary marks indicated an age that was obviously far too great, implying that the canonical relationship between ovary marks and age had inflated hundreds of estimates. Many population biologists understood immediately what this meant: the population was faring even more poorly than anyone had known. Remarkably, though, a few influential cetologists hewed to a very different interpretation. They argued that the surge of young individuals testified to high fertility, and thus represented a promising sign, not an ominous one. They were wrong, and they ought to have known it. For while it is true that upswings in a population's growth can generate surges of youth, there are ways to distinguish that scenario from the slide in age distribution that is due to overexploitation; in fact, it was just such analyses that persuaded many population biologists that the rosy interpretation was absurd. And yet the optimists persisted and, tragically, proved an influential minority. Through their caviling, they helped whaling nations—Norway and the

Soviet Union—to postpone by at least five years an international moratorium on killing those species most at risk, the blue whale, *Balaenoptera
musculus*, and its closest relative, the fin whale, *Balaenoptera physalus*.

It was at this point in Graham's story that Rafe raised his hand.
When Graham looked at him, he said, "It's not exactly surprising, is it?"

"What's that?" Graham asked.

"Well, the hip-boot scientists were on ships with whalers. Having
their pints with whalers. Getting bloody with whalers. Even shooting things at whales. So it's not surprising that they would end up siding
with the whalers."

Graham looked a little uncomfortable. I think he wanted to support
Rafe in the general direction of his thinking—his placement of scientists
in their social milieu, his supposition that context, and not just the un-
adulterated operation of method and analysis, must have influenced their
positions—and yet, at the same time, Graham could not quite bring himself to endorse such a ham-fisted thump on the seminar table. He had
already conjured a compelling sense of the scientists' entanglement in the
industry of whaling, but it was not so much an argument as layer upon
layer of imagery and innuendo. So when Rafe capped that delicate accumulation with a such a hard and explicit conclusion, it all seemed in dan-
ger of collapsing. Rafe clearly wanted to join Graham in his perspective
on science, but his deeper habits of thought, his desire for clear-cut propo-
sitions, had not vanished overnight.

After a moment's pause, Graham said, "Yes, very interesting," and
took the problem head-on. Early cetologists, he said, were not only immersed in the world of whaling, they actually depended on it for the continuation of their profession and livelihood. After all, they needed tissue,
ovaries, samples of other sorts, and the only place to get them was on a
whaling ship. And yes, this did influence the positions that some of them
took on early regulatory decisions by the International Whaling Commission. In the ovary debacle, however, we have to attend to certain details of
the story. The most prominent scientist to make the fallacious argument
was a Dutchman who had never actually been a hip-boot scientist. He
was more of a stay-at-home mathematical type—though, it should be
said, one with deep ties to the whaling industry.

As Graham spoke, I watched Rafe, and it dawned on me: he was smit-
ten. And I understood then that while my own conversations with him

had constituted little shoves on his worldview, Graham's influence has worked differently—and far more powerfully—as a gravitational pull from afar. Just a few days ago, Rafe seemed so comfortable and assured in his presumption that what produces scientific conclusions is nothing but good data and rigorous analysis; if cultural, social, and historical circumstances were visible to him at all, they were like a painted canvas backdrop behind the scientist immersed in his work and thought. Darwin's cabin in the *Beagle*, Mendel's monastic pea garden—these were nothing but the spare stage sets for the strivings of great men. But now, evidently, Rafe's new presumption is that a scientific result is primarily an outcome of all those surrounding circumstances, which exert their influences through the inevitable biases of the scientist himself. That flat, mute backdrop has become a living, diverting, and profoundly consequential context. Japanese scientists determine that whales are closest to cows not because that is what their sophisticated molecular tools tell them, but because such a result aligns with their interest in eating whales. Not so differently, whale scientists interpreting demographic data *must* be deeply influenced by their brotherhood with whalers, not to mention their professional reliance on them.

But why? *Why* did his fixation on the foreground (the scientist, brow furrowed in concentration, hands poised on gleaming beakers) shift so suddenly to an equally infatuated fixation on the background (sharing pints, angling for whale meat)? I think that what was truly appealing to Rafe about scientific certainty was that it gave him a sense of epistemological superiority. It let him feel that he, unlike most, really knew what it meant to *know*. But maybe he had never encountered people who brought sciencelike rigor to a critique of science itself—which is exactly what Graham and Anoop would have shown him, and in such persuasive style: Anoop's hyperacademic demeanor; Graham's confident and virtuosic command of conversation; and even—I hesitate to say it—his ability to dominate Becca, who had threatened to become Rafe's academic rival. Maybe once he'd seen all this, Rafe wanted to be in on the discipline that appeared to be even one step more in the know than science.

Still, such aspirations, essentially intellectual in nature, cannot entirely account for the way Rafe was looking at Graham last night, at the end of seminar. There is something more. Rafe aspires to a certain masculine image. He wears a gold hoop and a ponytail. He's into rigor and irrefutability. He is extremely fit. Graham, with his crew cut and sharp features, projects a severe, almost militaristic confidence. In conversation, he can be dashing and powerful. He is also, of course, quite lean. I find myself suddenly and newly aware of these vague parallels and reflections only because, just a short while ago, as I pulled the *Cortez Angel* up to the beach to pick up Graham and the students, I was reminded what deep interweavings there can be among intellect and vanity and attraction. As I hopped over the rail to stand in the shallows and hold the *Angel* off the sand, I watched Graham and the students rise from the beach, where they'd been sitting, and I had to smile then as I noticed the latest addition to Rafe's costume. He was wearing a sarong.

Now, leaning forward over the center console so that Rafe, who is seated at the center of the front bench seat, can hear me over the roaring outboard, I say, "Hey, Rafe, where'd you get the skirt?"

He twists around and looks up with a smile. He's generally game, Rafe is.

"Borrowed it from Lucy," he says. "It's nice and airy. You should try it."

One way to describe the content of Rafe's realization would be this: there is a potentially messy human story behind every intellectual commitment, even those that might appear purely rational or cerebral. And maybe, in this sudden revolution he has experienced, Rafe has been his own object lesson. Perhaps there has been a kind of positive feedback, a cyclical dynamic: *this idea must be true; witness what I am now experiencing.* And so he has gone spiraling from one personal equilibrium into another, very distant one.

4

Halfway between the lighthouse and La Gringa, the coastline bends to the west, away from our northward course, leaving us a mile or so offshore. In the same stretch, Mike's Mountain and the rugged hills that rumple against its shoulder drop off into a sandy yellow flatland of ragged mesquite and creosote. And the change in topography, the opening to the inland desert, is not only visible; it is palpable, too, for we have placed ourselves directly in front of the desert's exhaust pipe. And a moment after I notice that my left cheek is hot and dry, the sea is in furious disarray.

The hurricane swell is coming unabated now from the east, sneaking through the gap between Coronado and the smaller angels, and it is smashing through the desert-blown waves from the west, like one mad crowd running headlong into another. Just ahead of us, the *Eagle*'s bow eases down as Veronica pulls back the throttle. In the *Angel*, I follow her unspoken advice, and trace her slow, foamy path through the hills and troughs of water. It is like we are skiing moguls together.

Coronado is a long island. From the volcano, which is still a good three miles north of us, the island drapes southward like the long train of a black gown. Veronica, making a counterintuitive decision, turns us eastward, putting us on a heading for Coronado's trailing hem. My own instinct might have been to continue very slowly northward, hugging closer to the coast. But she doesn't want our pangas sidelong to the water's main axis of movement, even for a short stretch, so we will head instead into the swell, and away from the desert wind, until we are able to duck into the protective flank of Coronado, at which point we will head northward in peace, along the island's western edge. At least, that is what I take to be the plan.

"Hirsh," Graham says beside me, not loud enough for the students to hear, "what the hell is she doing? I'm feeling slightly uncomfortable about this."

"We're fine," I say.

He says something, but I can't hear him above the crashing of the hull against the waves, so I repeat, more loudly this time, "We're fine."

And then, as quickly as it came, the tumult is gone. The southern edge of Coronado is blocking the swell, and the water has settled. We still have a riffle from the desert's hot wind, but it feels almost quaint after the seascape we've just traversed.

The marine radio crackles for a moment and I hear a fragment of a word from Veronica.

Taking the radio in hand, I say, "Again?"

"I said, 'This west wind makes me nervous.'"

Graham scrunches an eye in skepticism. "*This?*" he says. "She's worried about *this?*" He sweeps his arm at the chaotic water behind us. "What about all that?"

"She knows what she's doing," I say defensively. Then, into the radio, I say the wind is almost always out of the west here, where the coastal hills are low.

"But this is too strong," she says.

"Head back?" I ask.

"*Back*?" Graham says. "We just rounded the bloody cape—to find whales."

"Come on, Captain," Rafe chimes in, from the front bench seat. "We've been talking all about whales. We gotta see one."

"It's true," Graham says, "it's a—"

"Burnett!" I say sharply, because this is a call he should leave to Veronica.

Beside Rafe, Becca says, piously, "We shouldn't take any risks."

I hold the radio up, indicating that we will await Veronica's verdict. Up ahead, in the *Sea Eagle*, she is facing the desert, assessing the breeze, or maybe just watching the inland desert hills for signs of rising dust.

"We should just keep an eye on it," she says, and a little cheer goes up from her panga; her passengers too were waiting on her judgment.

"Good on ya!" Rafe shouts.

"At least it's a warm wind," Allie says behind me. Turning around, I realize that the three students on the rear bench seat—Allie, Lucy, and Chris—are soaked. The stretch of high waves and wind must have swamped them repeatedly, and now their hair is plastered down against their heads and their shirts are dripping. They have been enduring it silently, stoically. Fair-skinned Lucy looks almost translucent. Her lips are blue.

"Why are you guys always stuck in back?" I ask loudly, so that those in front are sure to hear.

Allie shrugs. Chris reaches behind him to rap his knuckles on one of the gas cans.

"I'm Mexican," he says. "I love the smell of gasoline in the morning."

Skirting Coronado's rocky verge, we travel slowly. Fin and Bryde's mothers occasionally deposit their small calves here while they work the deeper channel between Coronado and the peninsula. And the island's coast, which from farther away appears unbroken and monotonous, turns out on closer inspection to be cut with small curving inlets and alleyways, some of which our pangas can fit through. In one, we encounter a pair of snowy egrets, perfectly white and curved, like violins made of porcelain.

Farther north, a line slanting across the water's surface divides blue glare, on our side, from dark shadow on the other. As the bow crosses the line, the sun flares and dies behind the rising slope of the cinder cone. Remarkably, the wind is gone here, the water vitreous and black. I kill the engine and our hull eases to a gurgling, then silent, glide through dark water. Leaning over the rail, I peer past the small bubbles that slide slowly by and perceive there, lingering at depth, a thick, rusty brume. "Krill," I whisper.

Veronica has taken her panga just far enough away to prevent any attempt at conversation between boats, and I'm certain she is whispering, too; if we can just keep everyone quiet, we will be able to hear an exhalation anywhere between the island and the opposite shore. Looking our way, Veronica lifts an arm in the air with a raised thumb: she too has seen clouds of krill beneath her boat.

A gentle hollow gulping sound seems to come whenever a certain section of the panga's hull pulls free of the water's surface. From the island, two shrill whistles—American oystercatcher—then no more. We are suspended, it seems, in a landscape of immense and abstract geometry—our pangas specks of white, like dust motes floating at the equator between the dark cinder cone and its inverted twin; above the barren cone, blank cerulean; beyond the cone's shadow, expanses of aquamarine.

"This reminds me of something," Becca whispers. No one responds. Graham's eyebrows rise above his black hornet eyes. "This one time," she says, "when I was doing fieldwork in Antarctica—"

A short laugh escapes Graham.

"What?" she says.

"Nothing," he says. "Please. Go on." There is a level of sarcasm and ferocity in his voice that I have not heard before. She's getting in the way of his whales.

"We had to wait for these seals—"

"But very, very quietly," Graham says.

"So we set up these lawn chairs—"

"No," he says, "please. Even quieter."

"And we got a huge cooler, but instead—"

"Uh-uh," he grunts, like he's correcting a dog. "It's gotta be quieter. But please, go on."

She tries, and it feels somehow like she has no choice—like giving up on her story would be an embarrassing admission that it's not worth telling anyway. But Graham has forced her into an absurd, inaudible murmur, in which she is speaking to no one but herself. And I'm afraid he might even keep battering her down, until her mouth is moving mutely. This is not the sort of conversational judo for which Graham was appointed Becca's watchman. His more graceful moves have always laid her into silence softly, mercifully. But this feels cruel, and when I can't bear her muttering any longer, I say, in a normal voice, "Graham, maybe while we're waiting, you could take us further back in the history of whaling. It would be good background for my afternoon seminar."

The interruption serves its purpose, gives Becca an excuse to stop. But as I watch Graham step carefully over the forward bench seat, briefly steadying himself on Anoop's shoulder, and into the bow of the panga, where he seats himself facing the rest of us, I am suddenly not so sure this was the best strategy. Did I really intend to initiate a lecture? Is that it for the splendid silence? When he begins speaking, he at least does so in a gentle whisper, just loud enough for the students sitting on the rear bench seat to hear him. And though I cannot deny that I have lately been feeling a little impatient with Graham—in part, perhaps, for this very virtuosity he is about to display, this intellectual dramaturgy that comes all too easily to him—I am no less astonished than I've always been that he is capable of this, devising a lecture on the spot and delivering it in cogent and charismatic fashion.

Graham knows, from the syllabus, that our seminar this afternoon will be about the complicated controversy that presently surrounds a straightforward question: How many fin whales roamed the sea before whalers began hunting them? He therefore begins his story at the moment whalers were about to shift their sights fatefully to fin whales. For much of the nineteenth century, whalers had left the fins alone, because they were just too big and fast for an open boat of whalers to row down and kill. At eighty feet long and 260,000 pounds in weight, *Balaenoptera physalus* is the second-largest animal in the history of the earth, just behind its closest of kin, the blue whale. And fins are built for speed: they are torpedo-shaped, sleek, and lean. On those rare occasions when a fin just happened to surface near a whale boat, and the harpooner moved fast and hurled a strike, the celebration was short-lived. Often, the powerful whale tore free or tipped the whalers into the sea, but even when the boat remained attached and upright until the animal bled to death, the lean corpse would sink before the whalers could haul it somewhere for processing. The only way to profit from a fin whale, therefore, was to stick it with a buoy before it sank, and then return several days later, hoping that rot might have ballooned and floated the animal. But that was an uncertain business, and probably not worth the risk of poking a fin in the first place.

Then, in the second half of the nineteenth century, three technological advances suddenly changed whalers' prospects for taking huge, fast whales. During the 1860s, Scandinavian whaling ships began supplementing their sail rigging with steam engines, enabling large ships to pursue species that had always outpaced the open boats. In the same years, Svend Foyn, a Norwegian whaling captain, was applying himself to a difficult design problem. Since fins and blues often tore free of harpoon lines, it could be helpful, Foyn knew, to kill the whales more quickly, giving them less time to struggle. The obvious means of swiftly dispatching a whale was a bomb, but that created its own problems: the explosion would sever the harpoon line, and since lean whales were quick to sink, a cut line probably meant a lost quarry. Foyn developed a harpoon in which a hinged joint linked the shaft to the head. The head bore a grenade that would explode inward on the whale once the harpoon had penetrated the body, and the hinge prevented the blast from knocking free the line. This two-piece device could be shot from a cannon, which was easily mounted on the deck of one of the new large steamers.

Foyn harpoons could keep a whale attached, but heaving the dense corpse alongside a ship and holding it there for flensing—cutting off the blubber—remained a difficult task. The problem was solved with a clever, if grotesque, device. Alongside the cannons, air compressors were bolted to the deck boards; their hoses fed into giant hypodermic needles, which could be plunged into a freshly killed whale in order to inflate it like a blimp. Flensing was thus rendered as easy with a fin as it had long been with more buoyant whales: cut free the leading edge of a strip of skin and blubber, much as you would initiate the peeling of an orange; clamp that leading edge to the end of a winch line, and pull; as the strip elongates, the inflated whale rotates in the water, like a spindle unwinding its ribbon.

Our panga has drifted, with imperceptible slowness and a stately clockwise rotation, from the wide base of the cinder cone's shadow out toward its narrower, lopped-off mountaintop. On account of the rotation, my view—the landscape behind our seminar leader seated in the bow—

has been very slowly reeling by. I started out looking north, at a little yellow spur of peninsula that juts out just beyond the northern end of Coronado. Then I looked at the volcano itself: mountain of granite and iron, streaked top to bottom with the scars of small rockslides. And now I am staring south, at other angels: guano-white Calaveras, reddish Bota, the distant open archway of Isla Ventana—all of them levitating above the dark, calm water our panga rests upon, separated from it by a thin line of shimmering blue and white, which I know to be the tongue of fierce water we traversed to get here. The *Sea Eagle*, which has drifted on its own away from the island, is somewhere behind me, and that means Graham is now speaking in its direction. Veronica will assume that Graham is speaking softly but continuously because Becca led the students into a discussion that grew too loud or maddening to tolerate, and something had to be done. But that's only half true, and Graham is now making the most of his setting—using the whisper to his advantage, compelling the students to lean in, as if he were telling a ghost story. There was one moment, somewhere around the appearance of the darkly innovative Svend Foyn, when I thought I might have heard the echo of a breath, somewhere out toward the peninsula. But when I turned to look over at the *Sea Eagle*, expecting to see Veronica standing atop her seat and zeroing in with her binoculars, I found her instead seated on the rail alongside Haley and Miles. They were all peering down, dipping their toes in the water. The krill are here, but evidently, the whales are not.

By around 1910, the newly equipped steamships had destroyed virtually every known concentration of whales: the North Sea; the coasts of Alaska and Canada; Iceland and the Faroe Islands. But just as the last populations of the Northern Hemisphere were vanishing, explorers fiddling around the edges of Antarctica came across unimaginable numbers of whales. The narrow Antarctic convergence—a circumpolar band about thirty miles wide, in which cold water flowing northward converged with warmer currents heading south—supported about as many whales as the whalers had found in all other locations combined.

Initially, whaling ships hauled their catch to the shores of islands— South Georgia, Deception, the South Shetlands—where they flensed the whales and boiled down the blubber in great cauldrons. It is hard to imagine what hellish scenes these were. Deception Island is an extinct volcano that has subsided just far enough to leave a ring of land brimming with seawater and accessible only through one narrow passage. It was an ideal site for flensing, and the whalers left the peeled bodies to float ashore in the enclosed bay. A journal from one whaler, written in 1928, reported that you could not leave your ship, because the shore was no longer rock but putrid flesh. If you tried to walk on it, you would sink under and drown.

Graham pauses, looks up at the volcano, and suddenly he is in the sun: we have just drifted across the edge of the cone's shadow, encountering our second sunrise of the day. He has been sitting just inside the bow rail, which passes behind the small of his back. But now he reaches back to place the palms of both hands on the rail, and he is just starting to press himself up, intending to sit on the bow rail itself, when a great blasting explosion of water and mist detonates right behind him. The shock of it propels him forward, onto Becca, Rafe, and Anoop, who are instinctively curling into protective crouches, cowering against one another, and for a second Graham is laid like a blanket over the three balled-up students. Chris, leaping up from his seat, appears to my right, Allie to my left, and I feel Lucy's hands grasp my shoulders as she looks around me. Through the sunstruck mist, a billowing cloud of glinting prisms, we can just perceive the rising island of smooth and shining black, large enough to step upon, walk across, but moving—sliding and rolling at once, as if a great moon of black marble were rolling on the seafloor, rising now just enough to expose this turning knoll of stone. If it keeps rising, it is going to lift our bow. The water at the knoll's edge is agitated—frothy and electric—and now it starts to pull closed, to gather in and press the island back down. Just as the circle of foam collapses shut, and Graham is standing back up from his tangle amid the students and turning now to look past the bow at the site of the blast, there comes another explosion of water and mist, and Graham is back down, not prone this time but sitting on Anoop's lap as Rafe and Becca recoil once more, crouching sideways in their seats.

As the second hillock rolls forward and the water draws closed around it I turn to look at the *Sea Eagle*. Veronica is there, standing on her seat, watching us. She raises her hands high in the air, then grasps her own head. Here, on the *Angel*, it takes us several seconds to recover our voices. Graham stands, saying, "Jesus Christ." Chris emits a sound that is part yell and part laugh, an awestruck and giddy holler. And now others laugh, too: Allie and Lucy beside him, Rafe in front.

"That was *awesome!*" Rafe cries.

"Was that dangerous?" Becca asks.

"I have no idea," I say.

"Did they know we were here?" Anoop asks.

"Yes," I say, then, "I think so," then, "I don't know. I guess I don't really know."

"They totally knew," Rafe says. "They snuck up on us."

"It did seem—" Graham begins, but can't finish.

"Watch," I say, pointing toward the *Sea Eagle*. "They'll come up again." Fin whales tend to breathe in sets of three to five spouts. Wherever they surface, they leave a circular patch of oil, a footprint, and their usual pattern is to put together a short string of them, each one a few hundred feet from the last. I'm hoping this will put them right next to the *Sea Eagle*, but instead they rise halfway between our pangas. Now that they are not so shockingly close, their spouts sound less like explosions and more like the crashings of great waves coming ashore. They are thick geysers of mist that bloom and billow out high above, sparkling in the sun like icy oak trees. From this perspective—the whales are moving directly away from us, toward the *Sea Eagle*—their staggering scale is evident only in the subtle camber of their broad backs: so large are their round, tapering bodies that the part above water appears almost perfectly flat. But if you can just imagine extrapolating that ever-so-subtle arc of the back until it forms a full circle, you ascertain, perhaps, a sense of the cross section of this immense fuselage. But the mind balks: it just can't be; such scale contradicts the fact that this thing belongs in the elementary mental category called animals. As the second whale's geyser ends, there is a split second in which the only sound is the frothing of water around its body, and we can see two large nostrils gape open at the center of its back. And since the blowhole is tilted slightly back, with the leading edge just higher than the trailing one, we are looking, it seems, straight into the darkness of the whale's capacious interior. And now comes inhalation, a sound like someone pressing two keys on a great cathedral organ, one a very low note, the other very high. And at last the hollow, reverberating tone cuts to silence just as the blowhole disappears underwater and the roadway of shining black slides down behind it. Where they were, the water is smooth and opalescent with oil.

They are sixty feet long, at least. And they weigh more than two hundred thousand pounds—whatever that means. And when you behold creatures like this, it seems very much like you *feel* them as much as you see them, the same way you feel a roaring cataract or a frightening clap of thunder. But in a way this is different, because even as you feel the overwhelming scale and power of it, you also sense that this rising mountain is *aware* of you. Here, the sense of immense indifference, the transcendence of surpassing magnitude, the palpable fact that *this thing is immeasurably greater than I*, is joined to something like its very opposite, the acute sense that you are perceived. Strangely, the closest word for it, the only concept that somehow combines these two antitheses, is *mercy*. You feel an immensity which is cognizant of you. *Mercy*. And so it is not surprising to me at all when I glance at Allie and find her looking vaguely confused, distraught, like she might break into tears, for this is a feeling of such strong opposites that it seems almost like it can pull you open. She senses me looking at her, and that quickly, what she's been feeling is gone—she gathers herself, shivers herself awake.

"Did they really do that on purpose?" she asks.

"Don't know," I say, though it feels like a lie. Of course they did.

There is a clamor on the *Sea Eagle*. Everyone is climbing up to high places. Onto the bench seats, onto the bow platform. Veronica has told them to do this, I realize, because the whales' path is likely to pass directly under the panga, and in order to see through the glare on the water's surface, you need to look from higher up. From here we cannot see the whales at all, but their path is now perfectly evident, because everyone on the *Sea Eagle* seems suddenly to stiffen and focus their attention downward. The slow dips and turns of heads, the leanings of bodies, the sudden shift of students from one side of the panga, the side closer to us, to the opposite rail—all these movements track the whales in great detail, and it's almost like we are watching them ourselves, like the people in the other boat are angled mirrors, allowing us to look down into the water. We are looking, in a sense, as Cameron looks, for he too shifts from one rail of the *Sea Eagle* to the other, at exactly the same moment as everyone else, as the whales are sliding silently below.

Because of the way they came, without warning, I expect them to vanish in the same fashion. But they do not. They stay. And while the sun rises from the rim of its cauldron to its apogee and then a short way down the other side, toward Mike's Mountain, we sit in our pangas and watch a pair of sixty-foot fin whales as they lunge-feed on the clouds of krill. The whales move in close parallel and, at exactly the same moment, they roll onto their sides: two pectoral fins cut the water, at first like shark fins, but then, rising higher, and higher still, until they are falcate blades standing a full story from the sea's surface. And then, as the colossal bodies break the surface at an oblique angle, the mouths gape open, revealing plates of baleen like paling fence along their jaws, the pleats of their undersides, pearly white and deeply grooved, swell and billow with strange flaccidity, and we watch the ocean water flow as a river into their great cavernous mouths.

Mostly, we are silent. Quiet reverence—*observation*, in the fullest sense of the word—seems the right posture to assume in their presence. And everyone on the boat—truly, everyone—seems to feel this. At one point, when the whales happen to be working the water farther out from the island—still easy to see and hear, but not quite so close as they've been—Graham concludes his seminar with a single line, spoken quietly: "By the end of it all, the whalers had killed nearly a million fin whales."

If the fin whale itself is in some sense unfathomable, then I am not sure at all how to contemplate one million of them. It feels like a concept that exceeds my capacities. And I have already been stuck for some time now on a far more basic exercise. As I watch the whales, I cannot stop wondering what it would mean to try to kill one. To hurl a harpoon. To pierce it and watch its blood stain the sea. To bring it to a stop. To make it dead. And yet, obviously, this boatload of people watching now in quiet reverence, we are the weird ones, the historical outliers. For decades, centuries, *millennia* before us, humans on the shores and at sea, when they have seen a whale, have tried to slay it. But how have they felt toward that which they destroyed? What have they seen that we do not? Great quantities of meat, I suppose; or barrels of clean-burning oil; or baleen to serve as flexible ribbing in expensive things like corsets or collars or umbrellas. Maybe they even envisioned the animal before them taking its future transmuted form: a large family of fattened children; or a long, curving row of brightly lit streetlamps; or a sunny London sidewalk busy with the bobbing of ladies' parasols. And what do we see that they did not? Did they not sense the same transcendence? They must have, for such awe is as old as history. *He makes the depths churn like a boiling cauldron,* Job's God, El Shaddai, says of Leviathan, *and stirs up the sea like a pot of ointment. Behind him he leaves a glistening wake; one would think the deep had white hair. Nothing on earth is his equal—a creature without fear.* And in one of the few moments the Hebrew God speaks with ironic contempt, he asks Job, *Can you pull in Leviathan with a fishhook or tie down his tongue with a rope? Can you put a cord through his nose or pierce his jaw with a hook?*

Perhaps not, but we can blast his innards with a grenade, inflate him with a hypodermic needle, and peel him like a bloody fruit. So while our sense of the whale's transcendence must indeed be as old as history, then again, men have undertaken, also since the beginning of history, to demean their God in one way or another.

It ends, our observation, with a crackle of my marine radio. It is a jarring sound, and when I pick up the radio I speak curtly: "Didn't hear."

"See that?" Veronica asks. In the *Sea Eagle*, she is standing on her seat, pointing at the sky. High overhead, a frigatebird hovers. Its soaring form is an inky black M stretched out laterally into an elongated, lanky wingspan.

"The frigate?" I say.

"It's not moving."

"Yeah?"

"It's facing west."

"Oh," I say, "I see."

"We gotta go."

"Okay."

"Straight across to the shore, okay? Then we hug it, all the way home."

"Okay."

5

What if we delve even deeper in time? Deeper than even the whalemen can tell. There is, maybe, an alchemical way to do it. Alchemical in the sense that we take one sort of information, a whale's DNA, and we try to transmute it into a very different sort of information, the number of whales in the population. It's not unlike trying to unfurl the evolution of an animal just by watching the growth of its embryo: You wouldn't think you could know the whole novel just by reading its last sentence, but it might just be feasible.

So, we look at the sequence in a particular place in a fin whale's DNA. Here it is:

Fin Whale #1: ACTAAGACTA.

In reality, our sequence would be hundreds of letters long, not ten, but that's not important. We've chosen this particular place in the whale's genome for two reasons. First, we know it is prone to mutation. Some genetic sequences hardly ever change; others change almost every time a new individual is born. This one is highly changeable, and that will prove useful. Second, this particular piece of DNA is always inherited from the mother. Maternal inheritance is not critical here, but it makes things simpler.

Now we look at *exactly* the same place in the DNA of a second fin whale. Lining up the two sequences . . .

Fin Whale #1: ACTAAGACTA
Fin Whale #2: AATAAGACTT

. . . we see two differences (in the second and last letters). These differences are due to mutations that occurred somewhere in the female ancestry of these two whales. Now, how many generations back do these two whales share a female ancestor? Do they share a mother? (That would be one generation back.) A grandmother? (Two.) A great-grandmother? (Three.) We can't say just yet, because we need a critical bit of information: the exact rate at which this piece of DNA undergoes mutation. It changes fast, but how fast? One mutation every generation? (What I mean by that, just to be clear, is one mutation in that sequence of DNA every time a mama whale gives birth to a baby whale.) One mutation every ten generations? More fundamentally, how would we ever know? We'll get to that in a moment, but for now, let's just say this sequence experiences one mutation every two generations. Now, which female ancestor do these two whales share?

A *grandmother*, because, look:

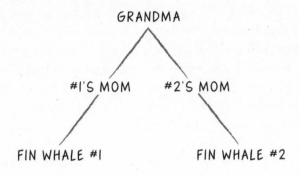

There are *four* acts of reproduction here—just count the line segments. And in every two such reproductions, we said, we expect one mutation. So if fin whale #1 and fin whale #2 share a grandma, they end up differing by two mutations, just as we observed in their sequences. Now, mutation is a chancy process, so just because whales #1 and #2 differ by two mutations doesn't *necessarily* mean they share a grandmother. That's just the most likely scenario. Or, to put that the other way around, if two whales share a grandmother, their sequences will *probably* differ by two mutations, but they could also differ by zero, one, three, four, and so on. (Just as, if you flip a coin four times, you won't necessarily get two heads.) However, if we take enough pairs of whales, and count the number of differences in this particular sequence between every single pair, we can find the *average* number of differences—and therefore, the *most likely* number of generations we have to go back in time before any two whales share a female ancestor.

We were trying to get from DNA sequences to the number of individuals in the population. So far, we've gotten from DNA to the number of generations you have to go back in time before two whales share a female ancestor. But what's *that* got to do with the population size? Say we have a population in which only one female gets to reproduce in every generation. If she makes two offspring, those offspring, of course, must share a mother. So, in a population of one breeding female, we have to go back one generation before any two individuals share a female ancestor. Now, say we have a population of two reproducing females. If we take any two whales, what are the chances they share a mom? Fifty percent, because there were only two possible moms. And if their chance of sharing a female ancestor every generation we go back in time is fifty percent, then the most likely number of generations we have to go back before they share a female ancestor is two.

The pattern is becoming clear: The average number of generations we have to go back in time before two individuals share a female ancestor is (most likely) equal to the number of breeding females in the population. And we've already seen that we can figure out the first part of that sentence—the *number of generations* part—just by comparing the DNA between pairs of whales.

We've almost got it now—a recipe for the alchemy that allows us to look at some DNA and glean from it the size of the population it came from. But we still need a good estimate for the average number of mutations that happen every generation in our particular stretch of DNA. To make this estimate, we will need to compare the sequences from two whales for which we already know—by some other means—how many generations back in time they share a female ancestor. Then we can just count up the number of differences between their sequences and divide by the number of generations separating them; that will give us the number of mutations per generation, which is what we're after.

But where to find two such whales? It would be lovely if we had a vast genealogy for *Balaenoptera physalus*, so we could just pick two whales that we already knew shared, say, a great-great- . . . great-grandmother. But of course we have no such genealogy. What we do have, though, is a phylogeny, a tree that shows the relationships not among individuals, but among species. And we also have a pretty good fossil record of some of the species in that phylogeny, and a fossil record affords all sorts of opportunities for determining dates. One fork in the phylogeny that the fossil record is fairly good on is the one that separates fin whales from humpback whales. Based on geological dating of that separation, and on our best estimate of the number of years in a whale generation, we can estimate the number of generations it's been since fins and humpbacks shared an ancestor. Then we can compare our piece of DNA in a fin whale to the same piece in a humpback whale. There will be a lot of differences, but remember, although our little example contained just ten letters, the actual sequence contains many more, so there will be room to accommodate all the change. Then we divide the number of differences (between humpbacks and fins) by the number of generations since the two species separated, and there it is, our mutation rate.

So here's our recipe: From the number of differences between fin and humpback whales, we estimate the number of mutations per generation.

From the average number of differences between two fin whales, we estimate the number of generations it's been since they last shared a mama of some order. And that number of generations is about equal to the total number of breeding females in the population.

This has been very schematic, and we've glossed over some complications that matter. We've ignored, for example, the possibility that some of the mutations that have arisen in our sequence of DNA—especially in the long lines of descent that separate fin whales from humpbacks— might have been adaptive. The ways this affects our answer are complicated, but suffice it to say, it could make a difference. We've also ignored the possibility of mutations atop mutations: if we spot a difference between fin whale DNA and humpback DNA, we've got no way of seeing whether that particular difference is due to one mutation, or perhaps two, or perhaps even more in the long history of descent. And finally, even if our recipe produces a good estimate of population size, it's not exactly the quantity we set out to find. Our genetic alchemy tells us the total number of breeding females, from which we've still got to extrapolate the total number of whales. And at that point we might pause to wonder, *when*, exactly, were there this many whales in the population? On the day we collected our DNA samples? At some point during the lifetime of those whales?

This last puzzle can actually be turned to our advantage. When we estimate the size of the population by using the recipe we've just laid out, we are in fact estimating a kind of mean population size over the number of generations we are looking back in time. (Never mind exactly which kind of mean.) So, for example, if our recipe tells us that the population includes 10,000 breeding females, then that is the mean size of the population over the past 10,000 generations, which works out to be about 150,000 years, since the average generation time is about fifteen years. What this implies, of course, is that we are estimating how large the population was in an era long before whaling ever began.

A pair of biologists used this approach to estimate the abundance of fin whales that roamed the North Atlantic before whalers began cutting their numbers. Their answer was 360,000 whales. When they tried to account for several sources of uncertainty—generation time, mutations atop mutations, extrapolating from the number of breeding females to the total population size—they came up with a range of 249,000 to 481,000 whales. These numbers caused a stir, because even the lower limit was vastly larger than previous estimates. Those had been in the neighborhood of 60,000 whales, and they had been calculated the old-fashioned way, with some tall stacks of dog-eared, salt-encrusted logbooks from whaling ships, plus a bit of arithmetic. Of course, the old-fashioned methods, too, had their sources of error. How many whaling-vessel logbooks never made it out of a musty cellar in Nantucket or Spitzbergen? More important, perhaps, how many fin whales had been hit with explosives but never retrieved, and therefore never recorded in the logbooks? And how many whaling captains had reasons of their own, ranging from profit schemes to hard drink, for leaving whales off the books?

The scientists tabulating logbooks were certainly aware of such sources of error, and made an effort to adjust for them. And once they had arrived at a final sum for the number of fin whales killed in the history of whaling, they tried to adjust for the fact that not all of those whales were in the ocean at the same moment; some had been born during the whaling years. To make this adjustment, they had to assume a certain relationship between the number of whales and their success in reproducing, and such relationships can be complicated—sometimes a decline in population size means more available resources, which leads to faster reproduction, but sometimes it means fewer available mates, or a breakdown in social structure, or something else that reduces the rate of reproduction. Suffice it to say, this relationship represents another potentially important source of error in previous estimates of the abundance of whales.

But why all the fuss, anyway? Who gets up in arms about guessing the number of whales that swam the North Atlantic before whalers started killing them? The people who want to go whaling do. And the people who want to stop them do, too. The Japanese, Russian, and Norwegian delegations to the International Whaling Commission had liked the previous estimates of historical population size. To see why, recall the theory of fisheries and the idea that a population produces value at the highest rate when it is fished just hard enough to hold it at about half its natural size. The IWC assumes this theory applies to whales, and offers an official guideline consistent with it: populations should be kept above 54 percent of the environment's natural capacity. Recent studies show about 55,000 fin whales presently in the North Atlantic. Prowhaling delegations point out that this is well above 54 percent of the historical estimate of 60,000, and they argue that it is therefore time to resume a sustainable, managed harvest of fin whales.

But this argument stumbles over the new estimate of population size prior to whaling. If the genetic estimate is correct, and there really were about 360,000 fin whales in the North Atlantic, then today's population has a long way to go before it reaches half its former size. Not surprisingly, environmental organizations, including Greenpeace and their allies, have embraced the genetic estimate of population size. Resuming the harvest now, they plead, would knock back a population that has barely begun to recover.

In my opinion, both sides in this debate are wrong—not in their numerical estimates (though those may be off) but in a common premise that their contrary positions actually share. I find the genetic estimates magically clever. And combing through old texts for glimpses of the ocean we've largely forgotten—this too is an undertaking I admire. But in this debate about the resumption of whaling, both estimates, the genetic and the documentary, have been put to work in arguments that are based too simply on the idea that historical population size can tell us how many whales can now be removed in a sustainable form of harvesting. For there is simply no reason to believe that the ocean's former capacity for fin whales is equal to today's, because the ocean itself has been drastically changed. We have removed approximately ninety percent of large predatory fish; we have precipitated headlong ecological shifts from one equilibrium to another; we have changed the atmosphere in ways that significantly alter ocean chemistry and circulation; and the knock-on effects of all this are barely understood. It is unclear whether today's revised ecosystem is less favorable for fin whales, or more so. What is clear, though, is that it would not be wise to stake management decisions on a notion of natural abundance that is essentially obsolete.

Recollection is vital, and reconstruction—perhaps—possible. But mapping from one to the other, from memory into management, will surely require an ecologically broad view. The elements of an ecosystem, after all, are interconnected, and therefore we cannot expect species to return, separately and one by one, to their former levels of abundance. While this may seem perfectly evident, one can also see how a sophisticated and heated debate might be joined over the common assumption that whales left alone would return to their nineteenth-century population size. For such an assumption is recognizable as an echo of that familiar, misplaced dream of wilderness. Thinking of the oceans as wild, forgetting how profoundly we have altered them, we figure we can let the great whales swim out and return to their former state of being. But it is not so.

And there is another, deeper sense in which the estimation of historical abundance seems largely irrelevant to the management of today's whales. Discussion of the environment's carrying capacity, a threshold of fifty-four percent, or the relation between population size and fecundity—it can all obscure what is in truth the more profound question: Do we really want to kill whales? Collectively, it is true, we are their absolute overlords; they are here only, as the phrase goes, at our pleasure. And yet, when you perceive one with your own senses, you will nonetheless shudder, fall down in the bow of your boat, stand and raise your hands in exuberance. More important, perhaps, we know very well that they are long-lived and intelligent creatures, which have lasting bonds and perform all sorts of communicative and cooperative acts. And they are simply too huge and strong to kill in any properly humane or respectful fashion. All of which would suggest that the right grounds for making management decisions about whales lie not in genetics or history or the theory of fisheries, but in more fundamental considerations, about what sort of humanity we want to be.

It is strange to say, but when the whales were greater than us—well and truly greater, and not, in this strange new fashion, both greater and smaller—perhaps it made more sense to slay them. Perhaps it once seemed that when we killed a great whale, we climbed atop its mountainous form, stepping upward in some order. But now it seems only that, when we kill one, we cause it to sink in stature, and somehow we descend with it. In my own view, such a loss cannot possibly be worth the meat and fat gained by it.

6

Eerily, the wind has fallen silent. It's as if the furious harmattan that blasted from the desert all afternoon came up against the cold oceanic gale that has beset us every night, and the adversaries at last annihilated each other. Of the clamorous hordes from opposing sides, desert and sea, the very last has fallen, and we are like the peasants who shuffle now, bewildered, into the strangely quiet aftermath: piles of sand gathered at the door cracks; a colony of small brown bats that have taken refuge in the station rafters; a thick batter of ocean foam and town litter, strewn in long, curving lines along the beach. In the stillness, Veronica and I can resume a ritual we have forgone the past several nights. Before the cold winds came, we would walk down to the beach at night, after the students had unfolded their cots on the station terrace, and bathe in the privacy of dark water. Even the waxing half-moon left us alone, having settled already behind Mike's Mountain. Four of our students are in the habit of staying up late to read, and we could look up from our boundless bath to see four pale glowing rectangles, the pages of their course-readers, illuminated by their headlamps. As we lathered white suds in our hair and rinsed away the layers of sun lotion, we would talk in whispers about the day— the seahorse or cowrie one of us had found, a moment in Graham's seminar that had made us smile, the prospects of a romance developing between Ace and Haley. By explicit agreement, Becca was not to enter the conversation.

Tonight, a gibbous moon is lingering still over the mountain when we approach the water's edge, and we can just make out three silvered figures—Isabel, Lucy, and Allie—standing close together, with their backs to us, and letting only their toes touch the placid water. In the moonlight before a black sea, they appear to be a draftsman's study of contrasts and range in the female form: Isabel so small and her skin dark, the moon making a straight silver line across her glossy black hair; Lucy so long and lanky, pale and towheaded; Allie, a strong-looking figure, in coloration a kind of compromise between the young women to her left.

The three of them are holding hands.

When Veronica says, "Hey, you guys," they startle and then laugh at themselves. They are here, Allie explains, because of something Veronica said to Isabel today, on our way back to the station. It was a hard journey. We traced the shoreline, keeping within a stone's throw of land, but even there, the swell, coming on us from the left, was pushed up by the opposing desert wind into tall walls of seawater, which would crash over the port rail, drenching everyone. It wasn't clear which direction a body overboard would go: to the shore, driven by the swell; or out to sea, driven by the west wind. It seemed to depend on where we were along the shore and just how viciously the west wind happened to be blowing. Evidently, in the midst of those harrowing conditions, Veronica had leaned down to say something to the three students sitting on the bench seat in front of her. "If it ever stops blowing," she said, "the bioluminescence is gonna be totally amazing." Isabel told Allie and Lucy, and the three of them have come down to see it. ·

"But we've just been standing here," Allie says sheepishly. "We're afraid to go in."

"Well, let's go," Veronica says, setting her towel and shampoo down on the beach. She steps forward and takes Isabel's hand. On the other side, I take Allie's, and the five of us shuffle into the water, which is surprisingly, bracingly cold. The stars spark to life around our legs. Every step scares up a swarm of fireflies.

"Dios mío," Isabel says. "How did you know, Veronica?"

"The west winds bring it on, like our reward for the misery. I think it's all the upwelling."

Veronica explained this phenomenon in her first lecture at the station, because it is, in a sense, the fundamental reason we come here. Wind, waves, and tides push along the uppermost layer of the sea, and in places where islands or land disrupt the continuous flow, deeper water must rise to fill in for the displaced volume at the surface. This hoisting of the depths, the process oceanographers call upwelling, is what makes the Sea of Cortez, and especially these waters around us, so remarkably full of life. Over most of its blue expanse, the ocean presents life with an irresolvable dilemma: sunlight, which is necessary, of course, for photosynthesis, falls upon the surface, but most of the critical nutrients, things like nitrogen and phosphorus, have sunk down into the darkness, hundreds of feet below. On land, plants are able to keep their roots in the soil and their leaves in the light. But in the ocean, the tiny green plankton, which play the photosynthetic role and support the entire food chain, find themselves hopelessly stranded between their need for light from above and their hunger for substance from below. It is the planet's most ubiquitous and elemental catch-22, and it is what limits life in most of the ocean. But here, where water is constantly sloshing between the Baja Peninsula on one side and mainland Mexico on the other, and past many islands on the way, the necessities of life are finally unified at the sea's surface, and consequently, the waters teem.

And of course, it's not just nutrients that upwelling draws to the surface, but also any animals too small or passive to fight the rising draft—like the krill we saw this morning, and also the plankton that ignite now, as Veronica sweeps her foot in an arc before her, making a trail of blue milky glow spangled with brighter, yellowish green stars. As we wade in farther, the glow follows us closely, our aura, and the stars swirl into galaxies churned by our movements. When we are in almost up to our waists, we sweep our hands through the water, carving a fleeting calligraphy of light.

"What is it?" Allie asks.

"This is two different things," Veronica says. "The bluish cloud is one thing. The little lights are another. I'd guess one is a dinoflagellate—a little single-celled zooplankton—the other, I don't know, maybe a shrimp larva."

"Shrimp larva?" I say, because I've never heard this.

"*Nyctiphanes*," Veronica says.

"No," I say. "*Nyctiphanes*?"

"Sure," Veronica says, "larval *Nyctiphanes*."

Allie laughs, and this causes Lucy and Isabel to lose their composure, too.

"What?" I ask.

Allie says, confessionally, "You guys have the funniest conversations. We were just talking about it, your . . ." She trails off, unable, it seems, to find a term that will come across as she hopes.

"Lovers' quarrels," Lucy adds, with her exceedingly wide smile promising she means it only warmly.

At my waist, the water is cold. Better to endure it all at once, I think, and I lean forward to dive in headfirst. When I come up, the others dive in, too. I can see their forms, modeled in the substance of the Milky Way, gliding beneath the black water. The students come up beside me, but Veronica stays under, heading out toward the channel like a wayward underwater comet.

When she has been under almost a minute and is visible as a dull glow fifty feet out, I allow myself to say to the students, "I hate it when she does this."

"Maybe she's seen something," Allie offers.

Breaking the surface, Veronica yells back, "The blue stuff is only at the top. The green dots go deeper."

"Thanks for checking that out," I yell, but Veronica doesn't hear me, because she is already swimming back along the surface. She looks like she's riding on a raft of white foam and blue light.

Once we are all treading water together, gathered in a round bowl of milky light, Isabel says, "I think it's getting a lot brighter."

"The moon went down," Veronica says. It has dropped behind Mike's Mountain, and we are suspended now amid two starry skies.

"But what's it *for*?" Lucy asks abruptly and with some urgency; perhaps the apparent equivalency of lights above and below is somehow disorienting, and she wants some grounding, a solid bit of biology to remind her where we actually are. But in truth, no one really knows why these creatures glow.

One hypothesis, though, is uniquely compelling, and it's also a good reminder that we must keep ecology in mind, sometimes thinking two-ply-deep in the ocean's game of chess. The dinoflagellates around us may be offering an important clue when they glow in immediate response to our movements. At first glance, this might seem foolhardy, for dinoflagellates are stalked by copepods, little shrimplike creatures that not only have relatively large, light-sensitive eyes, but also generate turbulence as they swim—just the sort of commotion that would cause a dinoflagellate to flare and thereby give itself away. But this is where we must think one more ecological step away. Much like their evolutionary cousins, the krill, copepods prefer to lurk at depth by day, and then swim upward at night to feed. This affinity for the dark is almost surely driven by the need to elude their own predators, small fish, which are themselves primarily visual hunters. So perhaps, when a copepod is closing in on a dinoflagellate, in the darkness of the deep or on a night like tonight, the predator's turbulence causes its prey to flash, which in turn catches the attention of nearby fish and calls them in—at the last possible second, one imagines—to consume the copepod. In the realpolitik of ecology, the enemy of my enemy is my friend.

Our close circle has migrated back toward shore, just far enough to allow Isabel to stand comfortably. Here, where we need not tread water, the dark collapses in around us. The slightest twitch of a finger still causes a brief localized fire, but now that we have been talking for a while, the water's hypersensitivity to our movements is having a very different effect than it did at first; it is causing us to stand still, and to stare mostly upward, at the stars that are not so ephemeral—nor, I suppose, so intimately connected with the demise of one creature or another.

After a few minutes in which no one says anything, Veronica says we should get out before we get cold.

"So, are we leaving tomorrow?" Allie asks, as if she were assessing the significance of this particular exit from the sea.

"Not sure," Veronica says. "If you'd asked me this afternoon, I would have told you to start packing. But this weather is comforting, you know?"

Earlier this evening, when Veronica, Graham, and I had our moment of truth, clicking REFRESH on the browser to see whether Hurricane John had veered out to sea as predicted, the sea cow again refused to disclose any secrets. Graham chanted, "Sea cow sea cow," Veronica swore like a pirate, and still we saw only a narrow row of the map, pushed up against the top of the screen. It included a bit of the Pacific, Southern California, and some of the Nevada desert.

"Try Amazon again," Graham said.

"Why?" Veronica asked.

"We should order some emergency flares. Maybe some really big pool toys."

We agreed that, if the sea cow remains on strike tomorrow morning, we will walk into town; at last count, there were three other satellite connections in Bahía. But if the sea cow's problems are up in the sky somewhere, and not here, with our own dish, then the town's other connections will be out, as well—a cabal of sea cows. And if we cannot somehow verify that the hurricane has headed harmlessly into the Pacific, we will spend the morning packing and leave by early afternoon, reaching Cataviña well before sunset. It is a ridiculous and ironic possibility that, the morning after the hurricane has turned away from us and the wind has finally settled, we will be driving out of town, fleeing nothing. But we cannot stay to find out if this stillness really is a promising sign, because, as Graham has reminded us, the storm's fringe could easily wash out the only roads home, and one does not want to be trapped somewhere in the desert with a hurricane blowing through.

After we have wrapped ourselves in our towels, we stand on the cool sand, looking out at the islands, which are apparent only as black outcrops against the starry sky.

"Veronica," Allie says, "why were you so anxious to get back to the station today?" Allie was standing right behind me when Veronica came over the marine radio and told me to look up at the frigatebird. She must have heard concern, perhaps even fear, in Veronica's voice.

Veronica answers matter-of-factly, almost as if the interest in what she is saying lies in the natural history or climatology of it. The frigatebird, she explains, was hovering in place and facing straight toward the desert. The motionless suspension was nothing strange—frigatebirds do that all the time; it's how they look for activity on the water. Typically, though, they face out to the channel, since that is where the prevailing wind comes from. The way that bird was hanging there today meant there had to be a firm west wind up at higher altitude. We didn't feel it yet, down below in our pangas, but if it had dropped to water level and caught us too far from shore, it could have been dangerous.

"Wow," Lucy says, with a bit of a shiver, "good thing you saw it."

"We would have been fine," Veronica says quickly. "We would have just pulled into one of the coves on Coronado and waited it out. But it would have been a really unpleasant afternoon."

She pauses, and I expect her to leave it there—a lesson on prevailing winds and nothing worse than an unpleasant afternoon. But she goes on. There is something about this setting, I think, and this hour, and this particular group of young women; something that allows her to speak more freely than she ordinarily would.

7

We were not there. Our understanding of events is fragmentary. We do know, of course, that it was March, when the water here hovers in the low sixties. And we know that three Japanese ecologists—a leading scholar of social insect ecology; a renowned theoretician; and a rising young star who, at the age of thirty-seven, was already a member of Japan's most prestigious research institute—had come to visit Gary Polis at the Vermilion Sea Field Station. Gary was a fixture here, and something of an icon, too. Veronica had met him on her first visit to the station, and had looked up to him ever since. Of the various scientists who worked here, he was the one whose research most clearly transcended this particular place and spoke to the larger questions of ecology. To be sure, he was an obsessively focused student of his particular system—he studied the spare food webs on these desert islands—and yet his thoughts seemed to travel irresistibly from the specific to the general. Shortly after he had published a paper called "Phenology and Life History of the Desert Spider, *Diguetia mojavea*," he published one that bore the title, "Why Are Parts of the World Green?" This was how his science, and his mind, seemed to move—from the bug beneath the rock by his foot to queries that reached as far as ecology possibly could. And his contributions had been significant enough that eminent scientists traveled all the way from Japan just to see where the work had been done.

Gary and the three Japanese scientists were joined by two post-doctoral fellows, two graduate students, one undergraduate field assistant, about a half-dozen volunteers from an organization called Earthwatch, and a local fisherman, whom Gary had hired to ferry part of the group in a second panga. The entire party was out on Cabeza de Caballo when a west wind picked up. The fisherman, we have been told, insisted that the group head home immediately. The Earthwatch volunteers boarded his panga, while the scientists and the students boarded Gary's. Veronica and I both remember Gary's panga very well. It was called the *Alacrán*, the Scorpion, and it was an exceptionally sturdy and beautiful vessel: whereas most panga hulls are fiberglass, the *Alacrán* was fashioned from hardwood; on account of its weight, it sat low in the water, but the bow swept up proudly, to split the waves like a plowshare.

When the boats launched, the fisherman decided not to head directly west, into the rising wind, but rather to cut across to the south, toward Punta la Herradura. From there, he followed the arcing bay all the way around, past El Rincón, La Mona, and Red Mountain. It was a ridiculously long detour, but one with a clear rationale. Mike's Mountain and the high hills on its southern shoulder protect those reaches of the bay from the worst of the west wind. And what's more, if anything had gone wrong with the panga, the boat and its passengers would have been swept safely ashore inside Herradura.

In the hefty *Alacrán*, Gary steered a straight shot across the bay, toward the station. It would have worked, but the engine died. The wind turned the *Alacrán* sidelong to the swell. Water came over the rail. Eventually, the panga capsized.

Gary, the three Japanese scientists, two postdocs, two graduate students, and an undergraduate were cast into cold water around the over-turned *Alacrán*. At this point, the thread of the narrative begins to fray. Perspectives pull apart. Some, surely, are blurred, awash in fear. Some are lost. And so we are left with a jumble of dark images:

Nine people struggle to cling to the inverted hull. But the waves are high and the hull is slick, offering nothing to grasp. One of the older Japanese scientists does not know how to swim, and though he is wearing a life jacket, he cannot manage to keep his face out of the waves. His younger colleague, who is named Shigeru Nakano, is an experienced diver and a strong swimmer. He removes his own life jacket and straps it, for added buoyancy, onto his distressed mentor. At the same time, someone else, someone who can swim, has been overcome by panic, and is hording multiple life jackets, strapping them all on.

The waves are knocking people off the hull, sweeping them away to the west. But each time this happens, Nakano, Gary, and the postdoc named Michael Rose swim out and pull them back in. At some point, Nakano—or maybe it was Michael—manages to swim under the panga with a rope, encircling the hull and thus providing everyone with something to cling to. From the exertion, perhaps, or the hypothermia, Gary· suffers a heart attack. He falls unconscious. Michael holds on to him, keeping him from drifting out to sea.

But the entire panga is drifting out to sea. When the engine died, they were about halfway between Cabeza and shore, but now, several hours later, the wind and waves have driven them almost all the way back to the island. Perhaps for a short while there is hope they might hit Cabeza, but the current is shooting the gap between Cabeza and Los Gemelos. That is where they are heading. There is talk of swimming for the islands. Los Gemelos now look close enough to reach. But Michael will not go, because he is holding Gary, who is perhaps already dead. And Nakano will not go, because he is watching over his colleague.

A postdoc, two graduate students, and an undergraduate decide they will try for the first of the two Gemelos. They release the rope around the panga and swim. One of the graduate students is a strong swimmer, a former lifeguard. But instead of swimming hard and fast for the island, she stays with the undergraduate, who is struggling to continue. She coaches her, urges her on. The pair catches hold of the rocky shore of the island, as does the other graduate student. The postdoc, who has gotten separated from the others, is swept past the island. The others believe he is gone. But the Gemelos are twins, and the second one catches him. All night he is alone on the island, believing he is the only one who has survived.

The next morning, rescue boats find the three who spent the night on the first twin. Then they discover the postdoc alone on the second. The bodies of Michael Rose and the older two Japanese scientists are recovered later that day. Two days later, Gary's body is recovered. Nakano's body is never found.

PART VIII

Galeocerdo cuvier

LIGATURE

1

A strong wind wakes me, and I open my eyes to find that the stars have vanished. Confused, I sit up on my cot. The dark waters before us are a riot of whitecaps and foam. I touch Veronica's arm, and she sits up suddenly beside me, squinting into the wind. "It's the hurricane!" she yells fearfully, still halfway in her dreams. Her blond hair is flying behind her like she's driving a panga.

"Can't be," I say, squeezing her forearm. "It's hundreds of miles away."

"What?" she yells. The sea is very loud and feels frighteningly close.

"It's hundreds of miles away," I yell.

She looks to the other side of her cot to check on Taiga, who is standing there, staring at us: her eyes are round and black in her white wolf's head, and she is intently waiting for the first indication of what to do next, the first twitch of a muscle that will tell her which way to move. Veronica turns back to me, then looks past me, at the other side of my cot.

"Yukon's gone!" she yells over the wind.

"He's in the bathroom," I yell.

"What?"

"In the bathroom! At the staff house. I locked him in the bathroom again."

She rubs her forehead with one hand, visibly gathering her wits, figuring out what I've just said. Yukon is in the staff house bathroom.

"Oh," she says, as the hand drops from her face. "Does he have his headlamp on?"

"He does," I say, though I realize now that since I woke I have not seen the distant lightning that again compelled me to strap my headlamp on Yukon. The horizon must be obscured by dark clouds closer in. And sure enough, the first plump raindrops smack audibly now against our sleeping bags.

"Rain," I say, half to myself; I have seen it so rarely here.

A loud scraping sound makes Taiga turn and crouch for attack.

"Stay!" Veronica shouts.

Twenty feet to our right, Graham is dragging his cot across the rocky earth, toward the pair of concrete steps that lead up to the veranda of Melissa's house. His bluish halogen headlamp shines toward the sea but slides jerkily in the other direction as he shuffles backward, yanking along his cot. As Veronica and I stand and grasp opposite ends of her cot to move it up to the covered porch, I yell, "You okay, Burnett?"

"Hirsh!" he hollers.

"Are you okay?"

"I just want you to know—I blame you for this."

"It's not even the hurricane!" I yell, as if that somehow makes it less my fault that we are still here, awaiting the real storm. We beat him to the steps and place Veronica's cot safely on the porch. As he is pulling his own cot up the steps and I'm going back down them, I take the headlamp off his head.

"What the hell?" he yells.

"I'm going to move the kids inside. Get my cot for me, will you?"

"What am I supposed to do without that?"

Already on my way, I yell back to him, "Go back to sleep."

"It's a typhoon," he hollers after me. "I'm not going back to sleep in a typhoon."

As I pass the whale skeleton, my headlamp shines on the white bone of its skull, revealing that it is dappled with gray spots, like a sort of pox. It takes me a second to realize that it is only the rain. In front of the station, all the folded canvas cots are already migrating slowly, like a line of pale train cars, across the stone terrace and up the steps into the station. Between every pair of cots a student shuffles, holding the foot of the cot in front of him and the head of the one behind. Nobody is talking, and it would seem this method has taken shape on its own. Cameron is in the middle of the train, approaching the steps, but he appears to be having no difficulty at all, as the cot in front of him serves as a guide, and it now tilts upward, indicating that he is about to get to the first stair. One cot behind him, however, Isabel is struggling. At her height, she finds it difficult to keep the cots' wooden legs from hitting the stone terrace and staggering the train's forward movement. She has to hold her hands above her shoulders, by her ears, and in the shine of my headlamp I can see the muscles of her wet shoulders trembling under the strain. At first I try to stand behind her and put one hand on each cot, but this is awkward, so I end up taking over the cot in front as she moves both her hands to the one behind her, and now we've cut the train in half. Glancing back over her wet head, toward the end of what is now the second train, I see Veronica taking over for Becca, who must also have been struggling—even though she was the last one in line, and therefore had only the cot in front of her to worry about. At Veronica's heels, Taiga's white lupine form marks the train's end.

Inside the station, the front train, of which I am the end, moves through the veranda into the seminar room. Then it stops and breaks apart as the students unfold their cots around the large central table. Cameron rolls back into his cot the moment we fold it open. I migrate back through the veranda, where the second train of cots is being arranged into a loose grid, with narrow aisles to walk through. Still no one is speaking—with the roar of wind and waves outside, you would have to holler to be heard. And yet the students are all operating with the same silent purpose, as if someone were shouting out clear directions. Only as they all start climbing back onto their cots and clicking off their headlamps do I realize that the root cause of their magical efficiency has been nothing more than a strong desire to go straight back to sleep, which they will now do with the swift determination of college kids. The last of their headlamps winks out, and I am left standing in the station doorway, with the windblown rain at my back and the bluish beam of my own lamp glancing over their stilled bodies. The image brings to mind some sort of bivouac infirmary, and I have to tell myself that they are all okay. Everyone is fine. I'm not sure where Veronica and Taiga have gone—probably to check on Yukon—but he too will be asleep, with his headlamp shining brightly on the bathroom wall six inches from his nose.

They're all okay, I tell myself again.

As if in retort, Millie, the little white stray with the crooked ear, whimpers loudly from the back corner of the veranda. I cannot see her there, but I can imagine what's going on. She is trying to settle beside Miles's cot, and she suffers horribly every time she has to stand up or lie down. Toward the end of the evening seminar, she humped into the room holding a hind leg off the ground. Her eyes were glazed and she was panting. She limped over to Miles and settled beside his feet, in her usual way, but she wailed as her hip landed on the floor. Miles stood to go to the kitchen for water, but Millie scrambled to follow him and wailed all over again. Miles sat, helped Millie back into her spot, and Haley went for the water. There is no veterinarian here in Bahía. When an animal is injured, it dies—painfully if no one shoots it. And if this little stray continues to suffer, we will be faced with a hard decision when we depart—tomorrow or later—because we cannot leave her in this condition.

2

Puzzlingly, I do not see the dawn sky. I am surely lying on my back. But what I see is wooden planking: brown, water-stained, and utterly unknown to me. Only as I look to my side and see Veronica asleep on her cot do I understand that it is the underside of the covering over Melissa's porch. We have not gone far. Fifteen feet or so. And yet, when I prop myself on my elbows and look out at the bay and its islands, I find myself in an unfamiliar place: the skies are leaden and low; no fire is about to crest the horizon; and strangely, I have no idea what time it is—a feeling I have never had in Bahía. The wind is gone. The water looks like oily metal beneath the dark clouds, and now that I've been staring at it a moment, I can feel its movement—a long, slow undulation, a bass viol waveform rolling slowly in from the channel.

"Feels weird," Veronica says beside me, as she props herself up to look.

Even the gulls on the shoreline rocks seem oppressed. Every now and then one gives a nagging squawk, but no one responds, and their contemptuous fits of laughter can't get going. When a fish makes a plop in the shallows and its ripples expand to the rocks, the gulls seem to shuffle their feet in annoyance, peeved to have their galoshes sloshed. No one goes for the fish. When we first learned of the hurricane, I imagined black storm clouds thundering in with lightning and wind. I had no idea we would receive instead one strange emissary after another—cold wind then hot wind; mute lightning and glowing seas; a sudden cloudburst and now this lightless gray. It has come to feel as though the cavalry is preceded by a traveling carnival.

"Did you know," I ask Veronica quietly, "that so much craziness comes before a hurricane?"

"Don't say before," she says. "John is heading out to Hawaii by now."

Beyond her, at the other end of Melissa's covered porch, Graham is still sleeping. Last night, Veronica found him in the staff house bathroom, sitting on the shower floor and reading *The Faerie Queene* aloud to Yukon. When she came back and met me here, on the dark veranda, she was wearing an expression of real concern. But as soon as she told me what she'd just seen, I broke into laughter.

"What did he say when you walked in?" I asked.

"He said, 'Hey, V,' and then he kept reading."

As I laughed, Veronica added, "Yukon wasn't even awake."

"Would it be better or worse if he had been?"

"I guess I don't know."

I tried to explain that Graham has always had a weird, sort of naïve way of interacting with dogs as if they were people, so reading to one might have made perfect sense to him. And he knew where I've been putting my headlamp at night, so maybe he just went in search of a reading light and then realized that leaving Yukon in the dark might be cruel. "It's kind of sweet," I said.

"He's been saying we should leave," she said, still concerned. "Maybe he really meant it."

"Definitely not," I said.

Now, as I look over at Graham, sleeping late after his hour of reading to my dog through the storm, I am struck with a pang of regret for having felt a bit irritated with him. Already yesterday, when we arrived from our terrible journey back to the station, my impatience was gone. We had just witnessed something extraordinary together—had been shocked by a fin whale spouting a meter off our bow, and humbled by the spectacle of lunge-feeding—and such an experience is much bigger than the trivial things that had been prickling me.

And I think the harrowing ride home might have helped, as well. Strangely, the frightening threats of wild places seem to work very much like their offerings of sublimity. Or at least they do when you make it through unscathed, and come away feeling terrifically grateful. Blessed or merely spared, you are graced with that same abundant sense of what matters in the people around you. And now, waking in this calm, leaden morning after the rainstorm, I can reach a bit further, maybe, than I did yesterday, and admit that my irritation with Graham arose, at least in part, from a certain pettiness of my own. I console myself with the thought that it is hardly an unusual bit of pettiness. Anyone who teaches, teaches anything at all, must feel that there is some vanity tangled up in it, and connected with that, a possessiveness of one's own students. In Graham's company, such vanity can be challenged, for it is hard sometimes not to feel outshined. And so the sudden conversion of a student from the sciences to the humanities—and specifically to the field that examines the sciences critically—could feel to me only like something of a personal judgment. Protecting my pride, I attributed Rafe's epiphany mainly to Graham's theatrics, and not to another cause, which was in truth just as important: his authentic brilliance.

Now, to wake him, I take the empty water bottle that is beside my cot and toss it in his direction. It bounces off his hip and clunks hollowly on the concrete. He sits up, pulling out one foam earplug. Looking out at the bay beneath its strangely somber light, he says only, "Whoa."

"Time to check the sea cow," I say.

"I know a great sushi place in San Diego," he replies dreamily. "They're open late."

Up at the staff house, the three of us position ourselves in front of the laptop. Veronica leans down, retypes the entire URL for the National Hurricane Center—just to be sure—and clicks return. Without delay or complaint, the sea cow revs to life and the image of the map materializes strip by strip from the top of the screen. The motion is that of someone speed-reading—left to right, left to right, zip, zip, zip. It takes a moment longer, however, for us to process the image that has appeared.

"Wait," Graham says. "Can they just do that?"

"Shit," Veronica says.

The small boat propeller, the dingbat that marks the eye of the hurricane, has not followed the bright red line that so confidently marked John's *highest-probability trajectory*. Where the line once swerved left, the propeller has gone straight, ducking into the mouth of the gulf. And now there is no trace whatsoever of that once-promised path into the Pacific—it's as if no one had ever said a word about it—and a fresh red line, cleanly redrawn, runs straight up the gulf to connect with a small white dot, which is labeled *Bahía de los Ángeles*.

"But what happened to the yellow?" I ask, imagining that those widening zones of probability—the yellow for a twenty-five percent chance of impact, the pale red for a fifty percent chance—might at least still offer us some hope, a chance that whoever has drawn this new picture might have been right in the first place, and only now has made an absurd error. Veronica leans down close to the screen.

"The yellow is there," she says grimly. And I see it now: The wide comet tail, that spreading and curving cone of yellow and red, has collapsed in around the line. The fiery path of the hurricane is now without doubt straight up the Sea of Cortez—hemmed in, evidently, by the land on either side. And our little town stands on one wall of the narrow corridor that the great storm is lurching through. What's even worse, though, is that the little boat propeller is actually much farther north than I had anticipated. The storm is already past La Paz. The thick of it will be upon us by tomorrow night. Whether it accelerated unexpectedly, or I simply miscalculated a few days ago, I have no idea. It is certain, though, that these dull pewter clouds hanging over us now are continuous with the hurricane. We are looking no longer at peculiar messengers, but at the storm itself, and the rain could begin anytime now. None of us has to say it, because we all know: it is too late to start packing. Too late to leave.

Graham and I walk out onto the covered porch and sit down, staring at the field station in the dreary light: its ocher stucco walls, thick and sturdy; its windows, reasonably small over here, on the western side, but on the other side, facing the sea, large and vulnerable; the doors—the one to the kitchen, wooden and heavy, the other one, though, much lighter, with a visible gap underneath; and the roof, sadly thin and flimsy-looking, good enough for a decade of shade, evidently, but that's about all.

It is surprising to me how quickly the initial shock of recognition, and the shaky denial that followed thereafter, has given way to an almost calm resignation—albeit with a distinctly nauseated feeling in my stomach. And while my nausea may know better, even now there remains something weirdly unreal about the coming storm. All the strange emissaries, rather than serving to substantiate it, have seemed only like part of the hoax, or maybe the first act in the entertaining show, to be continued when the storm itself arrives.

"I guess we should board up the windows," I say.

"Mmm," Graham says, nodding a little.

"Lane's got some plywood in the garage."

"Mmm."

I am grateful to him for not pointing out that it is my fault, and

Veronica's, that we are still here. But I wish he'd say something, because this steely silence is worrying. Maybe the nocturnal reading to my dog was a more meaningful sign of distress than I'd thought.

From inside the staff house, where Veronica is still kneeling in front of the computer, clicking around and reading things, she says, "Ten thousand residents in shelters in Baja Sur."

A wince passes over Graham's face, but still he doesn't say anything.

"Wow," Veronica says, half to herself. Her voice sounds no different than it does at home, at breakfast, when she can't help herself from reading aloud little bits of the newspaper. In that case, though, we ourselves are not part of the story, so it seems strange that the same tone should preside now. I don't think the storm feels entirely real to her yet, either.

"Four-meter storm surge," she says, impressed.

The rocky embankment between the sea and the station is at least three meters. And then there's the stone terrace, which is another meter or so high, and then six stone steps from there up to the floor of the station. We ought to be safe from the waves. The Diaz cabañas, though, down by the beach—they'll be deep underwater.

"Mud slides," Veronica says. Graham looks at me briefly out of the corner of his eye. I wish she would stop.

"Mike's Mountain is not coming down," I say, and he nods weakly.

"Half a meter of rain," Veronica says.

Staring again at the station roof, I realize that, oddly, the west winds might actually save it. The roof doesn't look like much, but we know, from experience, that it has been built to withstand the west winds, which can gust to a hundred miles per hour. I don't think the hurricane's winds could possibly be much stronger than that. So while the roof will surely let in a lot of water—it is simply not made for rain—I don't think it's going to tear off.

"We could dig a ditch along the back of the station," I say, pointing vaguely.

Graham doesn't reply. He's just gazing straight ahead, at the gray horizon.

"You guys," Veronica says.

"Water's gonna be coming off the mountain," I start to explain.

"You guys!"

"But we could just make a little ditch to send it around the station."

"You guys—it's dying."

"That way," I say, pointing to the left of the kitchen. "What's dying?"

"The storm. I think it's dying."

Graham seems to be very slightly shaking his head. He's had it with our optimism. But I think he's determined to maintain his composure, and by now he's resigned to the nature of his companions.

Veronica goes on: "I'm reading these memos, by the NHC meteorologists. There's a lot I don't get about eyewall replacements and shear and stuff, but I'm getting enough to understand that the storm is actually over the peninsula now, not the gulf, and it's losing a lot of power. Listen to this: 'The initial intensity is reduced to seventy knots . . . John should continue to weaken at a steady rate . . . recurvature into the northern Gulf of California appears less likely than yesterday . . .'"

I stand up to walk inside and look at the screen. Veronica has left out some lines, like, *there is a fair amount of uncertainty* and the hurricane will weaken *as long as it remains over Baja*. But at the moment, concerned as I am about Graham's state, I'm not inclined to focus on those lines, either.

"In twenty-four hours," I say, "the winds are supposed to be down to forty knots."

"Forty knots," Veronica repeats, smiling a great, unrestrained smile. "Forty knots," she yells out to Graham. "We can handle forty knots." She walks to the back of the room, where our bags are, and starts packing her small daypack. Out of habit, she opens the bottle of suntan lotion, but before she squirts any onto her hand, she closes it again and tosses it into the backpack.

"What are you doing?" I ask.

"Getting ready," she says.

"Ready for what?"

"To go out in the pangas. I'm curious to see the bay like this. I've never seen it like this."

I glance anxiously in the direction of the porch, where Graham is sitting. No sound.

"We're still gonna be in a big storm," I say. "That memo didn't say partly cloudy with a chance of showers."

"The ditch is a good idea," she says. "We can dig it this afternoon."

She passes me on her way to the doorway. "Come on," she says. "It's late." As she walks out onto the porch, she glances to her right, where

Graham is still sitting, out of my present view. "Come on," she says to him, and then continues down the stairs, across the dirt lot to the kitchen door.

A few seconds later, Graham appears in the doorway. He looks at me impassively for a moment, then walks past me, to the back of the room, and starts changing into his white oxford-cloth shirt and sarong.

"What are you doing?" I ask.

"I'm getting ready," he says. Then, shaking his head a little, he adds, "It's time to go out on the boats. I'm getting ready."

3

On the southern side of Isla Ventana, a crescent of dark gravel beach is enclosed by vertical slate cliffs. The sheer faces wrap tightly around the sides of the beach and converge toward the back, but they don't quite meet there. They leave a narrow seam, a rocky arroyo that is our usual route on and off the island. Cameron, Ace, Miles, and I have just emerged from that jagged passageway and are now lolling in the cool shallows—a relief from the day's weighty, tropical heat—as others debouch, in twos and threes, onto the beach behind us. Our entire group made it up the island's flank of red rocky scree to the concrete cross that marks the summit. Looking out over the channel, we watched an endless field of slow-rolling swell as it migrated steadily toward us beneath the quilted, low-hanging clouds. Veronica spotted a small bottlenose dolphin stitching its way across the surface, and its white splashes gave scale to the undulation, proving it to be more in the proportion of rolling hills than marching combers. With everyone focused on the dolphin, Veronica crept out in front of us, set her camera down on a rock, and ran back to the group just in time to kneel down facing the camera and say, "Smile—group photo."

"Can we redo it?" Becca asked. "I was looking behind me." But a small group of students was already splitting off to inspect an ocotillo just off the summit. The thorny canes were showing small buds of green—their sudden response to the rare swill of water that came last night. On the way down, I noticed that every little dome of mammillaria cactus, wedged between slabs of rock, wears a tonsure of tiny white blossoms.

Cameron, Ace, Miles, and I were the first ones down, despite a minor mishap in which Cameron's prosthetic came loose from his thigh and tumbled down the slope ahead of us. It could have been an awkward moment, but at this point, there is really no awkwardness left in most of this group. The prosthetic is part of our shared daily life, and it is not unusual to hear Cameron, sitting in the rear of the panga, yell to the people up front, "Hey, toss me my leg, will you? I think it's under those dive bags." So our mishap was nothing more than an opportunity for Miles to chase down the tumbling limb, and for Cameron to tell a story about accidentally dropping his leg from an orange tree, which he was climbing with a friend, into an unsuspecting crowd of Japanese tourists below. He and Miles are still cracking up about that image—a Japanese tour group with a leg landing in their midst—and they keep retelling little pieces of the story and collapsing with hilarity into the shallow water. Ace, meanwhile, is floating lazily on his back, out toward the anchored pangas.

In general, the students are in high spirits. They are pleased we are staying, and even more pleased that the hurricane, or what's left of it, is heading straight for us. Not for a moment, it seems, have they doubted the essential soundness or conservativism of their group leaders' judgment. They have assumed all along that if Veronica, Graham, and I decided to keep the group here, everyone would be perfectly safe, and maybe we would even experience something of a spectacle. We happen to know now that they have misjudged our competence, but they do not know the history of our decision to stay, and we've gotten lucky, so in the important facts of the matter, they are right: everyone will indeed be safe, and the storm may yet provide some excitement. Their mood has helped Graham, too. Such doubtless confidence, misplaced though it may be, is somehow reassuring to him. Or perhaps it's just having an audience that has drawn him out of his stunned resignation; maybe he really did have his own reasons for reading to Yukon during the storm. In any case, as Veronica and I backed our pangas carefully away from the gravel beach, having dropped everyone off before we anchored in the cove, Graham found it in his spirit to joke about our abandoning them all on the island.

"You guys are doing some sort of experiment, aren't you?" he yelled at us from the beach. Turning to the students, he said, "This is just the sort of thing ecologists want to understand: how a small group of humans can quickly deplete local resources, leading to Malthusian crash and—just possibly—some evolution. But the question now is—who is the chosen mutant?"

"I've got my money on Haley and Miles," said Ace.

"Of course," said Becca, "the six-foot Aryans." And, as usual, the conversational game came to a halt as she intercepted the ball. With customary bemusement, Ace merely chuckled to himself and walked off toward the arroyo at the back of the beach.

"I'm gonna catch a chuckwalla," he yelled back to us, referring to the sluggish black iguanas we have told him about. They are endemic to these islands, and terrifically mild-mannered. "Miles," he added, "come on. I need your help—in case it's a chucknorriswalla." Then he chuckled again, and continued on his way to the arroyo.

Now, in the cove, he has just retrieved one of the buoyant seat cushions from the *Sea Eagle*, and he holds it to his chest as he floats on his back, kicking languorously. He is migrating back toward shore, but at an angle, so he's also slowly crossing the cove from right to left. When he is about at its center—with each sheer cliff an equal distance away—he pauses, turns toward us, and emits a curt high note: "Eee." Cameron, just saying something to Miles, falls suddenly silent and asks, "What was that?"

"It was Ace," Miles answers.

"What's he doing?" Cameron asks.

"Not sure."

"He's moving again," Cameron says, hearing the gentle kicking that is now taking Ace farther to our left. Behind him, the anchored pangas look brilliantly white between the dark water and clouds. Periodically, their anchor rodes pull taut, creaking as the hulls pass imperceptibly over the high point of the large, slow swell coming in from the channel. Beyond our boats, I have an unusual view of Cabeza de Caballo—straight up the horse's long nose. To the right of Cabeza, the Gemelos are like two lumps of coal; in this light, the islands have lost their red and the sea has lost its blue; everything is a shade of charcoal. Beyond the islands, the desert mountains of the south bay rise into the clouds.

"Eee," Ace chirps again.

"What the hell is he doing?" Cameron demands.

"Right now," Miles says, "he's looking straight at us, and smiling like he's totally nuts."

"Ooo," Ace says. "Ahh, ahhh, ahhh."

"Miles," Cameron says, "maybe you'd better go—"

"Ahem," Ace says—a theatrical clearing of the throat. "Ahem!" And now it dawns on me: These cliffs form an immense natural amphitheater, and he's been looking for the sweet spot.

Suddenly, powerfully, he begins: "You'll take the high road . . ."

"Whoa," Cameron exhales.

"And I'll take the low road . . ."

The notes are deep, rich, and enveloping—

"And I'll be in Scotland afore ye . . ."

—impossibly loud as they reverberate against the cliffs around us.

"But me and my true love . . ."

And the entire scene, though very beautiful—

"Will never meet again . . ."

—is also unearthly, eerie, because Ace's operatic facial expressions—his huge wide-open mouth and tense throat and unseeing eyes—float on a dark sea, backed by coal-black islands and, beyond them, vertiginous mountains that ascend into closed gray skies. As he reaches the end of his first refrain—

"On the bonnie, bonnie banks of Loch Lomond . . ."

—he holds his last notes virtuosically, and then falls suddenly silent. After a moment of clear sea cove quiet, the others, who have just come down to the shingle behind us, burst into wild applause.

As the cheers die down, Ace calls out to Lucy that she should join him on stage. "You won't believe it," he says. "Awesome acoustics."

But Lucy hesitates—a testament, I think, both to her own humility and to Ace's preternatural gifts; after all, it is Lucy, not Ace, who has sung in major opera houses, and yet she is the one who now says, "No way—you won't even know I'm there. He's a powerhouse."

Hearing the others pleading with Lucy, Ace swims over to the *Angel*, pulls off a second seat cushion, and holds it up in the air as an offering. The hopeful audience cheers raucously, and Lucy really has no choice but to wade in and swim out to float beside him. At first they confer quietly, their two heads close together on the water, though their every whisper is amplified magically around us. Lucy is naming pieces of opera in German—duets, I presume—but Ace is shaking his head.

"What about Mozart?" he says. "I went through a big Mozart phase. *Magic Flute*, maybe, or *Don Giovanni*?"

"Là ci darem la mano?" Lucy ventures.

"Ooo," Ace says. "Yeah. I think I'll get some of it."

Lucy turns to face us, smiles hugely, as opera stars do, and begins. Where Ace's voice was booming and powerful, hers is lovely and precise. And as Ace joins in, making up Italian-sounding words when he comes across sections he doesn't quite know, he constrains his own force, allowing their voices to weave playfully around each other. The music is all the more enchanting for the way it feels profoundly private, offered up on an aqueous stage for a handful of humans, the yellow-footed gulls perched against the cliffs, and the iguanas hidden in the passageway at the back of the cove. And it also feels right that Lucy, who has been so graciously deferential, kind to the point of invisibility, is now, at least momentarily, at the center of our rapt attention.

As Lucy and Ace smile to each other in the fleeting romance of their duet, I find myself glancing at Haley, who has settled into the shallow water beside Miles. She does not look jealous, but she does stare at Ace as though she's in love. Through separate conversations I have had with these two—with Ace one evening on the station terrace, and with Haley one afternoon when I had gone to fetch gas and ended up giving her a ride back from the basketball court—I have begun to understand a bit more of what draws the two of them together. Although Haley grew up in upstate New York and Ace in Southern California, they both come from relatively modest yet very supportive homes, and therefore, on our campus, which can sometimes feel like the country club of princes and heirs, they find themselves surrounded by peers of greater privilege. In both of them, I think, the response has been to repudiate pretension and embrace unapologetically the very unpretentious things they adore—for her, basketball; for him, rock and roll.

What may be just as important is that each of them has a deep but complicated relationship with Christianity. Haley was raised casually Catholic, but on campus, she has leaned into the support offered by a group of Christian athletes and coaches. Their philosophy, in which discipline and individual submission are the path to self-improvement—humility leading to triumph—resonates with her. Their theology is something she's still thinking about. Ace's background offered him a different strain of Christianity. Both of his parents are recovered addicts, who cleaned up in the religiously intense setting of Narcotics Anonymous. His older siblings also struggled with addiction, and when Ace was born, his father gave him his inaugural NA token, commemorating his first clean day. He's been collecting the tokens ever since—they are bestowed to mark important benchmarks, like ninety days or five years—and Ace can report his current tally. But where his parents were saved by a strain of Christianity, Ace himself, it seems, was saved mainly by Christian music, which he first encountered at his evangelical elementary school, and which subsequently gave way to other musical traditions. I could be wrong about this, but I don't think he has a personal relationship with Jesus.

As their voices rise together now toward the end of their duet, Ace and Lucy scare up the birds that have been perched on the cliffs above. Their porcelain white undersides flash before the dark stone wall, then before the gray clouds, as they exit the cove and depart across the dark rolling sea.

"Andiam, andiam, mio bene,
a ristorar le pene
d'un innocente amor."

4

Halfway between Cabeza de Caballo and the station, we cut our engines and float, just to watch the bay in this strange state. In the front of our panga, Graham and I have Rafe, Anoop, and Chris; in the back, Allie, Lucy, and Becca. Our experience on Ventana has somehow left everyone silent. At the close of our short concert, there was wild applause and much swooning. But now it seems that either our deep enchantment or our raucous exaltation emptied everyone out. Or maybe plainspoken words don't feel worthy. Or maybe I'm reading too much into what is really just an effect of the way we are pressed now into the thin layer of space between the dark sea below and the leaden clouds overhead.

A pair of splashes from the other panga makes everyone turn: Cameron and Miles have just plunged into the water. I slip my dive mask around my arm—in case there's anything to see—and follow them, and then the students in my panga follow me. In the cold water, the bay is even more mysterious than it was from the panga. Reflecting the gray sky, the untroubled water resembles molten nickel, and now that I'm in it, I believe I can feel the slow but steady rise and fall, which was somehow lost to me on the boat. And I'm not the only one who senses something uncanny. Beside me, Allie looks down into the water she's treading and says, "*Wooo*, this is kinda—*wooo*." Migrating into the center of the group, I tell them about the theory of selfish herds: sparrows flock, minnows school, and antelope herd for the same unseemly reason; everyone is trying to move toward the middle, because it's those on the edge who are most likely to be snagged by a predator.

With such talk, we kindle apprehension like kids around a campfire, and when something slaps the water far away, Lucy gives a startled soprano shriek, then laughs at herself. I ask Veronica if she can see it from where she's standing, on the rail of her panga. Rising on her tiptoes, hesitating, she says, "It's a big sea lion, and he's sort of—playing or something."

"Can we get closer?" Miles asks.

"Sure," I say, "but stay together, okay? Think selfish herd." The naturalist's conventional wisdom on male sea lions is that, although they're awfully intimidating, they won't behave aggressively unless they feel surrounded. So it is as a single, tight cluster that we move slowly forward. We can't see him over the rolling hills, but we can still hear him splashing, and at one point I glance back at Veronica, thinking she would surely stop us if she thought this dangerous. She's still there, on the rail, and now she has her binoculars trained on the commotion. She offers no sign of concern.

When we're one wavelength away from him—about thirty or forty feet—he rises on a swell and lifts his giant head high above the surface, getting a good look at us. He exhales loudly and barks—a warning—as he disappears once again into a broad swale.

"Let's just wait," I'm starting to say when he explodes through the surface beside us. His head rises high out of the water, jerks to the side, and hurls an enormous, silver yellowtail—a three-foot metallic torpedo, pitching end over end, with bright crimson divots in its side. It splashes down a few feet from Chris, at the edge of our group.

"Fuck," he says, with unexpected calm, and we all pedal instinctively backward, away from the floating meat. The lion hits the fish again, but this time we see only his thick brown back roiling the water—he can't be five feet from Chris—and then, strangely, the silver form is still there, in the same position, but now it is half a torpedo, reddish black in cross section.

"Did he just throw his fish at us?" Miles asks as we're moving farther back, and I hear myself answer, "Let's head toward the boats." Veronica is still up on the rail—I can just see her head above the swell—and I'm confused that she hasn't come for us. Has she not seen? Is she not concerned? I glance at Cameron: Miles is swimming beside him, holding his arm. And then, since the lion hasn't appeared for a moment, I slip on my dive mask and put my face underwater.

For some reason—perhaps just to get my bearings—the first place I look is at the students' bodies, dangling in space beside me. Cameron's prosthetic stands out, anomalous among the pallid, treading legs. Then I look for the lion, and my heart tightens at the sight of a sleek black form, emerging suddenly from the vacant depths, heading up for us at an angle. It's coming incredibly fast, I know, but it also seems impossibly slow, and I can hear myself think, as if someone were speaking to me: *It's the fear. It's the fear making things go so slowly.* And then I hear, *Too small, way too small,* and I realize with a rush of relief that it's a female, not the bull, and surely harmless. And I watch her round black eyes—puppy eyes, edged in white—as she makes a graceful, banking turn to her right and passes inches beneath the feet dangling beside me. Her form disappears in the distance, and I look back at the place I saw her appear and suddenly my heart seizes with fear—real, freezing fear—as I see, exactly where she first materialized, a shark too large to be real. *It's not real,* I think. And then I think of all the blood—*the fish blood in the water*—and then again, *It can't be real.* And yet it is heading up toward us at an angle, up toward my feet, which I watch pull back into my body as the large gray head grows larger and larger—*I couldn't wrap my arms around it*—and then I see its cruelly black eye and its white mouth, the side of its mouth, as it swerves, passes beneath the students—*its tail is as tall as their legs*—and disappears.

I have yelled something into the water, must have, because as I raise my head, Becca is looking at me with wide, fearful eyes. I try to sound calm as I say, "We should head back." But then, when I turn toward the panga, it is gone. The swells must have grown just the slightest bit taller, or the panga has drifted a bit farther away, and now I can no longer see Veronica's head peering over the wave tops. "Let's head back," I say again, this time more loudly, and we start to move. I do not know which direction the boat is in, but the others seem to, and so I follow them.

"What is it?" Becca asks, in front of me, and I can hear she is on the verge of panic.

"Don't worry," I say, stupidly. "Let's just get back on the panga."

The others are silent. Sensing my state, they ask no questions, and we move as a group in the direction of the boats. In my mind's eye is that image—the shark beneath their dangling feet—and I almost expect to see one of them go suddenly under. Miles, who could easily be in the boat by now, is still holding Cameron's arm, moving with the rest of us. When the *Sea Eagle* finally comes into view, Veronica yells, "What? What is it?"

On the other boat, Graham yells, "What's going on, Hirsh?"

Miles shows Cameron's hand the boat. Then he lifts Becca, who is having difficulty pulling herself up onto the rail. As he pulls himself up, he notices that Cameron hasn't moved. "*Come on*, Cameron," he says.

"I was waiting for you," Cameron says, hauling himself onto the rail alongside Miles. "I have a fifty-fifty chance he hits fiberglass."

PART IX

Chelonia mydas

THE SHAPE OF CHANGE

1

All night the rain and the shark came and went and came and went again. The rain thrummed the roof of Melissa's porch, but the shark came in silence from below. It materialized out of the vacant blue, rising toward us, expanding to its actual immensity, swerving at the last second to show me its black thumb-gouge of an eye. And that was when I would wake and stare for a while at the dark wooden slats above, listening to the rain and imagining how it must also be drifting down on the forest of cardón behind La Mona, on the pale beach of Coyote Cove, on the black rolling hills of the Channel of the Whales. But soon the shark was rising again through the sunken blue silence, sweeping to the side, revealing the puckered eye. Only once did the vision continue on. Looking away from its horrible anvil of a head, I saw the students' dangling legs. To my surprise, Cameron was a few feet underwater, his head just even with everyone else's waist. He was looking straight at me, but his eyes were not their usual clear blue-green. They were stony white, like marble. And then I noticed that a few feet in front of his face, a large seahorse was hovering. Its mouth was a cornet; its head and body together formed a treble clef; its tail, segmented and curling forward, recalled a chambered nautilus in cross section. Cameron, I realized then, was looking not at me but at this lovely animal, and I had to yell, had to warn him of what was coming. But I could not, and just then he raised a hand and extended it to touch, with his index finger, the seahorse's protruding belly. The shark would be on him in no time, but still I could not bring myself to cry out. I could only watch, transfixed, as his finger grazed the front of the seahorse, and its abdomen yawned open, like a mouth, and out came dozens and dozens of miniature seahorses, dancing like the baubles on a mobile. I had to warn him, had to shout, and as I opened my mouth the water streamed in and I choked myself awake.

This time, though, my eyes opened not to the dark wooden roof but to a dim gray light, and I have decided now not to return to sleep, because I cannot face that wide gray head yet again. The islands are hidden this morning behind veils of rain. On the black rocks along the shore, the gulls hunker in puffed chagrin. The swell seems smaller than yesterday, but maybe it's just that the tide is slack now and the waves break, therefore, farther away. Careful not to wake Veronica or Graham, I rise from my cot; only Taiga watches me step out into the mizzle. The sandy ground is strewn with shallow puddles, and rivulets are carving fine canyons down to the sea.

The station terrace, wet and empty, feels sadly abandoned without the students sleeping here on their cots; they set up camp indoors last night, expecting the rain to arrive before morning. Coming to the edge of the terrace nearer the kitchen, I see Alejandrina before she sees me. She is standing in the kitchen doorway, looking at the misty bay through the steam of her coffee. The only time she ever looks melancholy is when she doesn't know someone else is looking at her.

"Ah," she exclaims, summoning her smile, and then, in a bit of hurry, as if she might forget to say it if she doesn't say it right away: "Professor— Samy says you must not go out in the pangas this morning."

Turning to look at the bay lost in rain, I tell her I don't think we're planning on it.

"You never know with Veronica," she says, smiling.

"That's true," I say, a bit surprised. I didn't know Alejandrina knew that about Veronica.

As I follow her into the kitchen for my coffee, she asks, as she does every morning, if I slept well.

"I kept seeing the shark," I say, trying to laugh at myself.

She sets down the coffeepot beside the mug she's about to fill, turns to me, and says, quite deliberately, "Professor, it could have been very bad."

Last night, when I first told Alejandrina about the shark, she seemed frightened and disturbed. And she was certain that I was missing the point of the experience—that I had stubbornly refused to grasp its meaning and seriousness. But in truth, when I tried to minimize it, telling her that sharks hardly ever attack people, that people swim with them all the time, that I myself have seen them on several occasions, I was speaking not just to her, but also to myself. Because although I had no doubt about all of those claims, I was having a lot of trouble bringing my psyche into line with the facts of the matter. I had indeed seen sharks before, but this time, reasonably or not, I was shaken. Perhaps Alejandrina sensed as much, and her own expressions of fear and dismay were in fact offers of sympathy for the terror she knew I was sequestering. Or perhaps her own fear of sharks really is so acute that she considered everything I was saying—about rarity of attacks and predatory wariness and specificity of prey and so on—as so much city-kid nonsense. Alejandrina, I was reminded then, has a complicated relationship with the sea. She knows its fringe in intimate detail—knows just where to dig for Venus clams; knows how often, at what hour, and in which season a whale shark comes past this very point on the shore; knows, perhaps even as well as her cousin Samy, when the wind is likely to rise and when it ought to fall—and yet, despite such close knowledge of the edge, she looks on the interior as terra incognita. She cannot swim, because girls in this culture were not taught to swim; they were meant to stay ashore while their men went out to sea. A shark, then, might be all the more terrifying to her for the fact that it lurks beneath the dark regions of her world map, a monster of legends told to her by fishermen, who may have had their own reasons for exaggerating the threat.

"In the future," Alejandrina says to me now, as she resumes pouring my coffee, "you must be more careful."

"More careful how?" I ask, a bit defensively.

"No stealing jurel from lobos," she says, handing me the steaming cup.

"We weren't stealing his yellowtail."

"That's not what I heard," she says, turning away to walk over to the refrigerator—and perhaps also to hide her mischievous smile behind the open refrigerator door; she finds it amusing, I think, that she has a secret source of information on the missteps of el Profesor—Isabel and Allie, probably, or maybe even Veronica.

"We won't steal anymore," I pledge.

"Good," she says, emerging from the refrigerator with two cartons of eggs. "You will sleep better."

"Did you sleep well?" I ask, returning to our daily ritual.

"No, I was too excited."

It is rare that Alejandrina will answer a pleasantry with anything but an even sunnier pleasantry, and rarer still that she will turn the conversation to herself, so I am genuinely intrigued as I ask her what could have been so exciting as to keep her up at night. Her swift and exact hands have already cracked several eggs into a large stainless-steel bowl, but now she pauses and looks up: "Colonet just sold."

What does she mean? Colonet, as I know it, is a windswept, silt-coated town that we are obliged to drive through on our way to and from Bahía. Like most of the settlements along that brown coastal plain, it consists of a strip of crumbling asphalt, two broad margins of dusty hard-pan, where cars and trucks are parked at haphazard angles, and two rows of low-slung shopfronts, each bearing across its brow a large piece of plywood painted with colorful and bubbly lettering: *Abarrotes*, groceries; *Llantera*, where car tires are fixed; *Mariscos*, seafood, which is quite fresh, because the Pacific Ocean is actually just a few miles away; *Birrieria*, where one buys goat meat. In the afternoon, when the kids have just been released from school, the overall effect of the small stalls, the garish signs, and the happy throngs is that of a carnival unpacked by mistake in a dust bowl. Was it even possible for someone to purchase such a thing? The tires, the groceries, even all the goat meat?

"Someone bought the whole town?" I ask.

"No, Professor, not the town. More. The ejido. Ejido Colonet."

The ejido must go by the name of its largest town, Colonet. And Alejandrina, I'm afraid, is excited because she suspects that her own ejido, Tierra y Libertad, will be next.

"Who bought it?" I ask.

"It's a secret," she says, resuming the cracking of eggs: she takes one in each hand, knocks them both on the hard worktable, and then holds them side by side over the bowl as she dexterously parts the shells until the pair of yolks drop. The shells are placed back in the carton, and her hands rise with two more eggs. "People are saying maybe the same investor from Texas. But they also say Ruffo is part of it, and he has investors from Japan. And maybe Carlos Slim, too."

"Who's Ruffo?"

"You know, the governor."

"The ex-governor is buying an ejido?"

"They say Ruffo knows something about Escalera Nautica." She sets down her bowl of twenty-four eggs, lights a match, and holds it close to a burner that hisses increasingly loudly as her other hand turns a dial.

"Bahía's a big part of the Escalera plan," she is saying. "This is where the Land Bridge comes." A small blue explosion makes the hand with the match recoil. She slides a heavy cast-iron skillet onto the flame.

That a particular name is associated with this reported transaction—and it isn't just Carlos Slim, the famous billionaire, or a Texan to be named later—strikes me as an ominous detail, a hint of some reality in the reports.

"How much did Ruffo and his friends pay?" I ask.

"Sixty-five million dollars," Alejandrina says, as her whisk whirs with mechanical speed through the eggs. "But Colonet's much smaller than Tierra y Libertad," she adds, as if the discrepancy between sixty-five million dollars and the previously reported offer of a round billion needs to be explained. From my perspective, sixty-five million is also wildly unrealistic, but an order of magnitude less so than the round billion had been, so in this detail, too, something feels disturbingly factual, and I am suddenly vaguely queasy, uncertain how I'll continue this conversation without saying something that will make Alejandrina think I'm out to dash her hopes.

"Oh!" I say, trying to sound like I've just remembered something. "I've got to let Yukon out!"

"*Pobrecito*," Alejandrina says, setting down her bowl. "He's in the baño again?"

"Sí," I say, heading for the door. "I should fetch him."

"Ah, my Yukon. So many stories about my Yukon. Tell him I'm making extra eggs for him."

2

When someone speaks to you in a whisper, you tend to whisper back. And it turns out the murmurs of soft rain and waves have the same effect, as if we were all in dialogue with the weather, the desert, and the sea. So it is that Veronica and Graham, staring out at the steady rain from the staff house porch, are conversing so quietly I catch only the occasional phrase, though I am sitting close by, at the wooden desk inside. What I keep hearing are the words *green turtle* and their variants—"*Chelonia mydas*" or, more often, just "the greens"—and the refrain's got me wondering:

Are there other species whose histories of demise are inscribed in the very names by which we know them?

I suppose the pearl oysters would count: rainbow-lipped and black-lipped pearl oysters, their names fittingly lusty and disclosing the particular hues of beauty for which they were mined. And then, of course, there is the right whale. The etymology of that name has always had a suspiciously pat ring to it, but I think it's actually true: species of the genus *Eubalaena* were deemed the *right* ones to kill, probably because, unlike fin whales, they were sluggish while alive and conveniently buoyant when dead. Of course, such concerns were rendered quaint by steamships and deck-mounted air pumps, and therefore the fin whale's own smaller cousin, Bryde's whale, also has blood in its name, though you don't catch the scent of it unless you pronounce the name in the original—not "Bride's" whale, but rather "Broota's," as in Johan Bryde, Norwegian consul to South Africa, and the man who brought modern whaling to the coast of Durban.

In the case of the green turtle, which started this morbid exercise, you have to pry a bit further to find the name's epitaphic connotations. The first clue to this effect is that most green turtles are not actually green. Their shells are a gorgeous dark chestnut with brush swipes of yellowish gold; and the same colors mark the head and flippers, where the scales are chestnut and the reticulum between them golden. What is rather green, however, is turtle soup. And the reason for this is that the layer of fat beneath the animal's shell is the color of freshly sprouted grass. But of course you never see that hue unless you tear off the shell, which is fused with the turtle's spinal column. So the name does in fact say a lot about the animal's fate: it tells you which view of the turtle was widely familiar and historically significant (cracked open, that is, and cooked down); and it also tells you one of the main reasons *Chelonia mydas* declined, over the course of several hundred years, from an abundance that today seems unimaginable. Captain Edward Cooke, who sailed the coast of Baja in 1709, wrote that the green turtle, of all species, *is sweetest and best*, and he reported taking a hundred turtles in a single night. Similarly prodigious hauls litter the logbooks of the next few centuries of merchant ships, privateers, and even scientific expeditions. On one of the voyages of the *Albatross*—a series of oceanographic expeditions mounted between 1890 and 1905—Alexander Agassiz, son of the famous anti-Darwinian ichthyologist, wrote of capturing more than a hundred in one seine haul off the coast of Baja.

Delicious green fat would not have been such a tragic flaw had the turtle not also possessed heroic powers of endurance. When sailing ships captured a hundred turtles at a time, they would stack them on their backs, alive, in the ship's hold, where the animals managed to survive, without food or freshwater, for dismayingly long periods of time—six, sometimes even eight weeks. They were therefore the ideal provision for long voyages—delectable and perfectly preserved, albeit in horrific fashion. And the same properties made them a prospective international commodity. In the 1950s, there were enough turtles left in Bahía de los Ángeles to support a newly intensive industry focused on export. By the mid-seventies, they had taken their place in the descending staircase of depleted resources.

In other parts of the world, the impacts were different, but the results basically the same. Green turtles nest on beaches in and near the tropics, and many human populations had come to rely on the eggs as a source of food; as those human populations expanded, the turtles, not surprisingly, disappeared. Coastal construction destroyed entire nesting grounds. And as drift netting grew in scale and ubiquity, so too did the incidental catch of green turtles. By the mid-nineties, the extinction of *Chelonia mydas*, once a globally abundant species, had become a real possibility. Lately, a few disparate populations—in the Hawaiian Islands, for instance, and on Atol das Rocas, off the coast of Brazil—have shown signs of recovery, so perhaps there is cause for hope.

On the other hand, the other species of sea turtle, though perhaps not *the sweetest and best*, are subject to many of the same assaults as the green, and each seems to have fatal flaws of its own, as well: the hawksbill bears a shell that makes especially lovely accessories; the loggerhead travels and nests in subtropical latitudes, where it encounters shrimp trawlers and a lot of beach development; the Kemp's ridley nests in a single location on the eastern coast of Mexico, and feeds amid the highest density of shrimp trawlers on earth; the leatherback has an unfortunate penchant for swallowing plastic trash. On the whole, these species have fared even more poorly than the green. And it is a stupefying but not entirely unrealistic possibility that an entire bough of the tree of life—one that diverges deep in the tree, well over a hundred million years ago—may reach no higher than the next few decades.

Unable to endure my own darkening company any longer, I stand from the desk inside the staff house and walk out to sit with Veronica and Graham. I place my plastic chair next to Veronica's and prop my own feet up beside hers, on the low concrete wall that bounds the porch. A few inches in front of our feet, vertical threads of water, streaming off the edge of the gutterless roof, form a semitransparent wall, like a screen composed of hanging strings of clear blue-gray beads. It feels a little like we are sequestered in a cocoon, spying on the rainy world. Chris, Allie, and Isabel have just come out of the kitchen portico to work on the ditch that now traces a broad arc around the back of the station. I started digging it earlier this morning, possessed by a very clear vision of how it would protect us: the rivulets draining off Mike's Mountain would flow into the ditch, follow it around the station, and drop as a river into the sea. At the moment, though, my ditch is only about six inches deep in a number of places—the desert hardpan, even in a soaking rain, is difficult to penetrate with a pick—and therefore what my dawn labors created was not so much a drainage ditch as a pretty necklace of pools, gradually filling and forming a little standing moat around the station. During our breakfast of tortillas and scrambled eggs, which we all ate on the terrace, beneath a blue plastic tarp that Chris and Allie had rigged between the station roof and the tamarisks, I told the students about my incipient earthworks, and suggested it might provide a good diversion from the exam they were going to be working on all day.

"You'll need all the time you can get," Graham said, smiling wickedly. But surely the students had understood that what I was really offering them was a way of escaping any seminar-room conversations Becca decided to commandeer. Earlier in the trip, I often found myself wishing we could all glance knowingly at one another, and thereby bond in semi-secret agreement: *that is not who we are.* To state it more plainly, for all its ugliness: I found myself wanting her to be ostracized. Fortunately, greater decency prevailed, and for this I credit mostly Lucy, Allie, and Isabel, the three young women who have let slip no more than an occasional sign of their own discomfort.

By this point, though, I find my meaner instincts gone. Where before I wanted our collective irritation at least partly out in the open, now I just hope we can keep it bandaged up for the rest of the trip. And while I'd love to think I've suddenly contracted a bit of Lucy's or Allie's own grace and generosity of spirit, in truth my change of heart has more to do with my experiences with Becca over the past few days. I don't know if she was authentically dumbstruck by those lunge-feeding fin whales, but I do know that she at least had the sense not to stand in the way. She was not probing for a way to interject herself into everyone else's experiences. And then there was the helplessness, the fear-stricken helplessness, that I saw yesterday when I lifted my eyes from the water. It has stuck with me, that look of panic, perhaps because now I see that something like it—a dull and chronic version of what, in that instant, was bright and acute—may be with her all the time, telling her that if she does not have a tight grip on the situation, something horrible will happen to her.

Isabel is swinging the pick, which is nearly as tall as she is, while Chris and Allie jump repeatedly onto their shovels, like kids trying to mount pogo sticks. It does not appear that they are making much progress, though they do seem to be enjoying the exercise. Isabel grunts theatrically every time she heaves up the pick, and then, as its head strikes the hard ground, she jumps straight up, as if she were being catapulted by the sudden impact. For the past few days, these three have been conversing only in Spanish. It began, I believe, when Veronica sat with them at lunch and asked if they wouldn't mind giving her the chance to practice. "Alejandrina won't correct my mistakes," I heard her saying to them. "I need someone to tell me when I goof." But Veronica doesn't goof often, and I suspected then that she had another motive. She too had seen Isabel chatting with Alejandrina, so she must have known how surprisingly different Isabel would be in Spanish—must have known that, for some reason, antics like we're watching now would be possible only in her native language.

Beside me, Veronica and Graham are still talking about *Chelonia mydas*. The reason they're deep into sea turtles this morning is that they've seen some. While I was still hacking determinedly at my ditchlet, they climbed into the pickup truck and drove off over wet sandy earth to find Antonio Reséndiz. He and his family live in a lovely, single-story hacienda, built by his own hands out of heavy wood beams, stucco, and terra-cotta tile, just behind the long beach north of La Gringa—the place on the coast where the seafloor is flat and mottled with dark green grass. Veronica and Graham found only Antonio's enormous German shepherds at home, so they walked with the dogs over to Antonio's turtle tanks, which reside a few hundred feet down the beach in a resourcefully constructed enclosure. Reclaimed boards have been hammered together into the skeleton of a small warehouse or barn. Dark nylon fabric, stretched over the frame's rafters, offers the turtle tanks some shade, and eight-foot chain-link fencing, stapled onto all four sides of the rectangle, makes for airy but strong walls. It is a very valuable stash Antonio keeps in his tanks—a large green turtle will fetch five hundred dollars on the black market in Tijuana—and therefore security is a serious concern, which explains not only the fencing but also the enormous German shepherds. Inside the skeletal warehouse are three circular concrete cisterns, each about four feet high, fifteen feet in diameter, and filled about two feet deep with seawater. In each tank, Antonio usually has two or three sea turtles—greens, loggerheads, and even the occasional hawksbill.

Veronica and Graham didn't find Antonio at the tanks, either. In a town of five hundred inhabitants, it sometimes seems that Antonio has daily appointments with most of them. "Antonio? Ah, he was *just* here" is the phrase we often find ourselves chasing though the sitting rooms, bars, and businesses of Bahía de los Ángeles. Today, though, the enclosure's chain-link gate had been left unlocked, which seemed to indicate that Antonio would not be gone long. So once Veronica had discussed the matter with the suspicious shepherds, she and Graham let themselves in and passed a half hour with their elbows propped on the edge of one of the cisterns, staring down at a pair of green sea turtles flying graceful, if also poignantly futile, clockwise circles. Their discussion of *Chelonia mydas* started then, and now Veronica is telling Graham about a remarkable discovery that is about to be published.

3

When green sea turtles hatch from the eggs their mothers buried in the sand about fifty days earlier, they dig their way out and make a spastic, flailing dash for the sea. They look like tiny rowboats trying, quite insanely, to row their way across the sand and into surf that will topple and crush them. As they reach the breaking waves, the two-inch dinghies tumble like helpless toys. And yet, when the sea foam recedes, there are no toys left broken on the beach. They are gone. Five years later, they

turn up again—or at least, ten-inch juvenile turtles, which biologists estimate to be about five years old, are found along coasts like this one, grazing on seagrass and salicornia. But where have they been in the meantime?
What becomes of a green turtle from the day it first tumbles into the surf
to the time it shows up as a juvenile, chewing seaweed in the shallow
water? That is the mystery that has just been solved—and by a splendidly
sophisticated bit of detective work.

An element, you'll recall, is a kind of atom, placed on the periodic
table according to the number of protons it has in its nucleus. Hydrogen's
got 1 proton, helium 2, and so on, up to carbon (6), nitrogen (7), oxygen
(8), and on further to the exotic—tantalum (73)—and the scary—uranium
(92). An isotope is a particular version of an element, defined by the number of neutrons that join the protons in the nucleus. So a given atom of
carbon might be the most common isotope, carbon-12, in which case
there are 6 protons and 6 neutrons, for a total atomic weight of 12; or it
might be the more unusual isotope carbon-13, which has 7 neutrons; or it
could even be carbon-14, which has, of course, 8 neutrons. Similarly, nitrogen may be nitrogen-14 (7 neutrons) or nitrogen-15 (8 neutrons). Some
isotopes, such as carbon-14 and uranium-238, are prone to radioactive
decay, which shrinks them into different atoms. But many other isotopes,
such as carbon-13 and nitrogen-15, are about as stable as their more common isotopic brethren.

Even so, they do behave a little differently. Carbon-13 and nitrogen-
15 are of course slightly heavier, their nuclei just a bit bulkier, than their
common counterparts, carbon-12 and nitrogen-14. In many contexts—
like most experiments in a chemistry lab, for instance—these differences
in mass and bulk are too subtle to matter. But many biological molecules
are extremely finicky chemists; they do notice the differences. Rubisco,
that most abundant but imperfect protein of photosynthesis, is reluctant
to handle a molecule of CO_2 if the carbon is slightly too plump, weighing
in at 13 instead of the usual 12. Pep-c, the new-and-improved protein
that works in Cam photosynthesis, is by comparison less biased against
heavier carbon. As a result, plants that use pep-c to grab CO_2 from the
atmosphere have more carbon-13 in their tissues than do plants that use
rubisco for the job.

This difference filters its way up the food chain: herbivores that eat
pep-c-reliant plants have more carbon-13 in their tissues than do

herbivores that eat rubisco-reliant plants, and so it goes, through the carnivores that eat the herbivores, all the way up to the top of the food chain. In fact, so robust is the signal, as it moves from plant to prey to predator, that you can decipher, from the amount of carbon-13 in the feather of an eagle, what sort of landscape provided the raptor with prey: Was it a forest, where the plants use the ancient, rubisco-reliant photosynthetic pathway? Or was it instead a dry savanna grassland, where the plants possess a newfangled form of photosynthesis that utilizes pep-c? This seems to me extraordinary: You are tracing a bit of the very air as it seeps into plants, and then into the rodents that eat the plants, and at last back up into the air, albeit in a new way, as the feather of a predator.

Whereas carbon filters its way up the food chain, with the relative amount of the plump variety holding steady all the way, nitrogen behaves differently. Animal bodies tend to hang on to their heavy nitrogen-15 more greedily than they keep their commonplace nitrogen-14. It's not clear which of our biological molecules is responsible for this effect, but the discrimination does not appear to have any functional significance. Rather, it seems to be a physiological accident meaningless to animals— though quite meaningful, it turns out, to ecologists. As every consumer's body hoards a bit more heavy nitrogen, the ratio of heavy to light nitrogen increases with each step up the food chain: the herbivore is richer in nitrogen-15 than the plant, the predator richer than its prey, and the top predator richest of all. And for an ecologist, that is an extremely useful pattern: From a measurement of the ratio of N-15 to N-14 in an animal's body, you can estimate just how far up the food chain the creature resides.

Return now to those juvenile green turtles arriving on the coasts of Baja. Where have they been? And what have they been up to? Two basic facts of sea turtle natural history are essential to understanding the detective work that disclosed where green turtles spend their first half decade. The first is how a turtle's shell grows. It is made of hard keratinized tissue, much like your fingernails or the baleen of whales. Each of the scutes—those hexagonal plates that fit together so neatly to form the gently domed shell—grows outward at the front and on the left and right, but not on the trailing edge, the side of the hexagon closest to the turtle's tail. And as the scutes extend, the turtle's shell also thickens: a thin layer of new shell is deposited uniformly across the underside of the existing shell. The combination of these two processes—extension of scutes and deposition on the underside—creates a pattern in which the oldest shell is thickest, and is located toward the back of each hexagonal scute, while the young shell is thinner, and located near a scute's growing edge.

The second important piece of natural history involves a curious contrast between green sea turtles and loggerheads. Whereas large green turtles like to loll around in the shallows eating seaweed, loggerheads have a varied diet: they spend about ten years out in the open ocean, where they are mostly carnivorous, eating jellyfish, comb jellies, and salps. Later on in life, they consume just about everything, from seaweed to sponges to small fish.

So here's what a team of researchers did to find out where the little green turtles were going. Working at a study site in the Bahamas, they caught sixteen juvenile green turtles that had turned up in the area fairly recently. The researchers were quite familiar with the site, and had been tagging the resident turtles there for many years, so they could be pretty sure that any small turtle without a tag had arrived sometime within the last year or so. From each of these turtles the researchers took two minute cores of shell. One core came from the thin, very new shell at the front of a scute; the other came from the thicker, older shell at the rear of the same scute. The researchers also took small samples of young shell from two other groups of turtles: twenty-eight green turtles that had been residents of the coastal study area for longer than a year (they had tags); and a dozen small loggerheads in their more carnivorous, oceanic stage of life.

The researchers measured the amounts of carbon and nitrogen isotopes in each shell core. In their content of both heavy nitrogen and heavy carbon, the samples of old shell from juvenile green turtles—the turtles that had recently arrived at the study site—closely resembled the shell cores from oceanic loggerhead turtles. What does this mean? The green turtles' old shell was of course produced at a much earlier time in the turtles' lives—during the period of mystery, that is, when no one knew where the little turtles were. With that in mind, take the measurements one isotope at a time. Oceanic loggerhead shell and old shell from green turtles both have a paucity of heavy carbon. Why? Just as Cam plants and old-fashioned rubisco-reliant plants have their own distinctive appetites for heavy carbon, so too do oceanic algae and coastal seaweeds. For a variety of reasons, coastal carbon sources are relatively rich in heavy carbon, while oceanic sources have very little. What the similarity of carbon signatures means, then, is that the baby green turtles were, like oceanic loggerheads, gathering their food far from the coasts, out on the open ocean.

In retrospect, we can perhaps say that this was to be expected; after all, if the little turtles had been anywhere near the coasts, then surely people would have bumped into them now and then. Still, it's nice to have an argument that's more compelling than *We haven't seen 'em 'round here.* And if this result is essentially confirmatory, the story told by the nitrogen isotopes is more surprising. Heavy nitrogen, we've said, accumulates as it works its way up the food chain. A match, therefore, between the heavy nitrogen content of oceanic loggerhead shell, and that of shell laid down during a green turtle's lost years, must mean that young green turtles are eating at roughly the same place in the food chain—the same trophic altitude, we could say—as the oceanic loggerheads. Which is to say, baby green turtles are carnivores. This is extremely weird. I cannot think of another case of animals shifting their lifestyle from pure carnivory to pure herbivory, and there is probably a good functional reason for this: the two diets demand different physiological specializations, not just in the gut, but all over the place. An extraordinary developmental shift takes place in juvenile green turtles when they return to the beach, renounce meat, and begin subsisting on seaweed.

The samples of young shell from juvenile green turtles were consistent with this story. Although those turtles had not been on the coast, living as vegetarians, for very long, their growing shell had already begun to take on the isotopic profile of older green turtles that had been hanging out around the study site for years. Carbon is known to be assimilated to tissues more gradually than nitrogen, so it makes sense that, in the newest shell, the quantity of heavy carbon was still somewhat reminiscent of the open ocean, while heavy nitrogen had already dropped from levels befitting a carnivore to those of a strict vegetarian.

Focusing narrowly on the results of this study, you could say that what this very sophisticated bit of detective work has taught us is really just another basic fact of natural history: green turtles spend their youngest years on the open ocean, eating the animals that drift there. But the more you let your perspective expand, the more intriguing the study's results become. Add some zoological context, and that basic fact of natural history looks very strange indeed: an abrupt switch from oceanic predator to lolling herbivore presents us with what is essentially a new and different way of being a big animal. Bring into view the concerns of conservation and management, and suddenly there are vital implications of the fact that *Chelonia mydas* ventures beyond the coastal shelf: you cannot direct your efforts only to the beaches where the turtles are born or the seagrass beds where they graze as adults; now you must worry also about what's happening hundreds of miles away, in blue open water.

In terms still broader, what the researchers have done can be described like this: they have read, from the atomic nuclei encased in an animal's tissue, a long historical record of where the animal has been and what it has been eating. Granted, the record seems vague: *open ocean* is not quite a treasure map; and *drifting jellyfish* is not a detailed menu. But this is exactly where I start to feel sort of dizzy with the possibilities. The researchers were reading with a scarce vocabulary: they knew the meaning of two terms, carbon-13 and nitrogen-15. But those terms are embedded in a manuscript of many others, for carbon and nitrogen are hardly the only atoms with potentially legible meaning. When you look over the periodic table for elements that are biologically essential, you discover that the majority of them exist as more than one stable isotope: there are three weights of oxygen and three of magnesium; four of sulfur and five of calcium; two of potassium and four of iron . . . It seems almost certain that many of these nuclei could be understood to signify something distinct and interesting about an organism's history. It's as if every creature on earth is busily soaking up its various experiences and filtering them into its own autobiography. And now we must learn to read the atomic language in which such books are written.

Suddenly those sea turtles look like they are carrying around tomes on their backs. And of course it's not just the turtles. You could just as easily read the beaks of birds and even their feathers; the baleen of whales; or, looking back in history, the carapaces of those turtles that Agassiz netted in 1900, or even the feathers of mockingbirds that Darwin shot in the Galápagos. Ultimately, the histories written in nuclei might restore our faded memories.

4

As Veronica's been telling Graham about the mystery of baby green turtles, Miles, Ace, and Haley have come out of the station to join in the ditch-digging. When they first appeared in the kitchen's portico, they offered to take over the shovels and pick, but Chris, Allie, and Isabel were enjoying themselves and declined. Instead of going back inside, though, the three newcomers elected to stand around in the rain, offering encouragement, and now they've gotten down on their knees to scoop wet sand from the sections of ditch Isabel has already softened with her pick. Meanwhile, our hobbled mongrel lies in the dry shelter of the kitchen portico, keeping an eye on Miles, her beloved caretaker.

"I bet Becca's acting up," Veronica says. "Soon they'll all be out here."

"At least we'll have a really good ditch," I say.

Graham declares he's going to march down and tell the seminar room that talking is strictly prohibited. He seems to be waiting for me or Veronica to restrain him—to say, *No, don't be harsh, the students will find their own way.* Neither of us does, but he stays anyway.

"You know what I don't get?" Graham says, speaking more quietly than before.

"What's that?" Veronica asks.

"How you could be so annoyed with someone who basically wants to follow in your footsteps."

"Wow," I say. "I thought you guys were on turtles."

Veronica is nodding her head slowly, her gaze fixed on the ditchdiggers. To me, she appears merely to be taking in what Graham has just said, adjusting to the conversation's abrupt shift from *Chelonia mydas* to her own psyche. But to Graham she must look offended, because he seems hurried as he adds, "I mean, don't get me wrong—I find her tiresome, too. But she doesn't aspire to be me, so I don't have that compensation."

"Rafe, on the other hand," I say.

"Exactly," Graham says. "I adore Rafe. So why doesn't V adore Becca?"

Still Veronica doesn't respond, and Graham sounds nearly apologetic now as he says, "I've been wondering if, maybe—and I don't want to be presumptuous here, so just tell me to shut up if you want—but if it might have something to do with being a woman in science."

Veronica glances at me but doesn't say anything. Graham continues, and his voice sheds its exaggerated deference as his interpretation gathers momentum. Of all our students, he explains, Becca is the one most obviously focused on becoming an academic scientist. And the more difficult aspects of her personality might actually be responses to the challenges that face a young woman trying to make her way in the scientific fraternity: Maybe she thinks she's got to command every conversation simply because if she fails to do so, she will be overlooked or talked over by all the bright young men around her, all the boys who have been told since they were two years old that they're natural-born math whizzes and scientists-to-be. And maybe the obsession with the academic setting—the department, the campus, the who's who of biology—is just a reflection of how she thinks she's got to commit to and master her academic universe if she's going to have a chance of succeeding. And maybe the constant effort to engage one of us and relegate all the other students to the periphery reflects her belief that she's working with a handicap and has to go to extremes to make an impression on her professors. "Am I overstating the challenges?" Graham asks, recovering deference now that his argument has run its course. "I mean, you guys are a lot closer to this issue than I am. But from what I hear, it's real."

Veronica looks at me, then out at the ditchdiggers. Still she says nothing. "It's definitely real," I say. "And I don't know—maybe you're right about Becca. But what's that got to do with Vica?"

Graham hesitates, joins Veronica in watching the ditchdiggers. Miles has finally wrested the pick from Isabel, who has in turn taken Chris's shovel, leaving Chris to wander off and investigate the little barrel cacti that are growing beneath the station's western wall. I don't think Graham will say more unless Veronica gives him some kind of nod, which feels unlikely. As long as I've known her, Veronica has seemed reluctant to talk about this topic, the challenges facing women in science. On a few occasions, I happened to be present when a friend ventured to engage her in a conversation about certain professional slights that had seemed to turn on gender. Each time, Veronica listened attentively and nodded sympathetically, but shared none of her own relevant experiences, though I myself could think of a few she had to choose from.

Chris now has the other five students gathered around a barrel cactus, looking straight down on the dimple at its crown. Maybe there's an insect in there. The pick and two shovels lie derelict beside the newly impressive ditch.

"Anoop is taking swim lessons from Miles," Veronica says suddenly. "Did you know that?"

"No," Graham says, "I did not know that." He sounds a little disappointed that instead of engaging with his question, Veronica has offered a strange non sequitur.

"I don't know if you remember," she goes on, "but Anoop's application said something like, *last time I checked, I was competent in Australian crawl.*"

"I remember that," I say.

"And right away we knew he could barely swim. But we took him anyway—because you loved him for loving Kierkegaard or something—and now, here, he's taking swim lessons from a guy who could race in the Olympics."

A sudden eruption of noise comes from the students around the barrel cactus. Someone has provoked whatever is perched there, and it has startled them all into springing back with laughter.

"That was really close!" Miles says.

Chris is holding up his finger and looking at it.

Allie says something I don't quite hear and grasps Chris's hand to inspect the finger. They all huddle around the digit in question, which Allie is holding in the middle of the circle. The only thing that could really be dangerous, I think, is a small scorpion, which seems impossible in daylight. Still, I wonder if I should go investigate; maybe the dark storm or rain has done something strange to the bugs, brought them out at the wrong hour.

"And Haley," Veronica is saying now, "she knows half the town. She knows more of the kids' names than I do. And they adore her. She has a little rec league going. And Ace found a broken ukulele in that heap of junk in the corner of the equipment room—"

"Is that what that thing is?" Graham asks. "I was wondering where he got it."

"Yeah," Veronica says, "he told me it's a ukulele. And he fixed it, I guess, and figured out how to play it—did you know it has four strings instead of six?—but somehow he figured out the chords, and then, yesterday, I found him sitting by himself, in the veranda room, with the ukulele in his hands and a piece of paper and pencil on the floor by his feet. I asked him what he was doing, and he said he was writing songs about Bahía. He played a few for me, and they were amazing—ironic and funny, you know, the way he is, but truly amazing. One was a rocking blues song called 'Boatyard Dog,' and the other was sort of an eighties heartbroken love song—only it was about Japan, which seduced him with cool stuff like Atari and Kawasaki but then turned out to have this terrible addiction to whale meat and tuna—I guess you'd have to hear it, but I swear, it was incredible. And Rafe. Even Rafe. He's like some like sort of religious convert or something. Anyway, the point is, I could go through every single student, and it seems like every one of them has found their own way into this—into the town and the bay and this whole experience. Every one of them. Except Becca. And it's that, I guess, more than anything else. That's what drives me crazy. She's so determined to achieve whatever it is she thinks she's got to achieve that she can't look up—look up for just one second—and see this place and what everybody else is doing here. A couple days ago I walked into the seminar room, and everybody was sitting there, waiting for seminar to start, and Becca was getting them all to compare their course lineups for next semester. And I guess that sounds like a normal thing to do—I mean, they're all undergrads and they share a lot of classes. But it drove me crazy. I wanted to say, 'Shut up, you're here. You're here now. Stop with all the academics and just get into it.' But she's stuck. Absolutely stuck. And you may be right, Graham. You probably are right—about what makes her this way. But you just can't let it turn you into that. You can't let it do that to you.'"

When she falls silent, she is looking in the direction of the students, who are once again huddled around the barrel cactus, investigating something that may or may not be hazardous, but she does not seem to be watching so much as staring just over them, into the sodden skies. She is close to tears—close enough that neither Graham nor I wants to push any further. And I don't think we have reason to, anyway. I see now the connection that Graham was unable, or perhaps just afraid, to articulate any more explicitly than he did. And he was right, I think, in what he nearly managed to say: that Veronica's immoderate reaction to Becca does in fact stem, at least in part, from her own relationship with the various extra obstacles that face a young woman who has chosen to pursue a scientific discipline. And when Veronica says, *You can't let it turn you into that,* she casts some light not only on her intolerance of a certain student, but also on her determination to avert her own attention from the struggle that seems to be largely to blame for making that student who she is. What Veronica is saying, I think, is that she is no more fearful of the obstacles themselves than she is of her own responses to them—the compensatory measures, the retaliations—all of which, as Becca seems to illustrate, have the potential to bend and distort a person. And so it is safer, Veronica seems to think, to adopt a strategy of simply pushing ahead, working with such chilling concentration that you never pause to consider the fact that you are contending with certain difficulties mainly on account of gender. This strategy has worked for her, at least so far, but still, it also has its drawbacks, one of which seems to be that a young woman coping in her own ways may not be easily understood, or gently tolerated.

5

Antonio Reséndiz has a great blockish head with closely cropped black-and-silver hair and thick black eyebrows shaped like licks of flame. He sometimes holds one cheek tensed in a kind of imminent wink, which has the effect of raising his upper lip into a half snarl, revealing two rows of large and very white teeth. The line where the ivory rows rest against each other is oddly flat, lacking the typical dental pattern of serration. And after you've watched Antonio for a short while, you understand why: He presses the rows hard against each other and grinds them back and forth with the excess energy of a restless pacing animal. As a result, his jaw is broadened by flanking bulges of muscle. And this, you soon realize, is but one of many signs of bristling strength and drive: his neck muscles ripple beneath his skin like a bundle of cables; when he embraces you, which he does freely and often if you are even so much as a friend of a friend, it feels as if you were trying to reciprocate the hug of an enormous oak tree; and when he shakes your hand, which he tends to do, oddly, just after he has embraced you—as if he must vent one last little burst of affectionate greeting—he grasps hard and shakes vigorously. His hands, arms, and torso are built for hauling three-hundred-pound sea turtles, but his legs are different; they are slim and twiggy, like a marathoner's, peculiar beneath his bodily bulk. But they too are signs of implacable energy, for I believe Antonio may run nearly a marathon on some days, as he rushes from one household to the next, from one establishment to another, tirelessly organizing the town's nascent environmental movement and campaigning for his vision of the bay's future.

Today he cut his rounds short. When his topless, doorless jeep skidded to a halt in the rain-soaked lot behind the station, he leaped out, wiped the rain from his eyes, and saw me sitting on the staff house porch.

"Arón," he cried, bounding up the steps and crushing me in a hug. "I heard you guys were looking for me!" After the embrace, he kept a firm grasp on my upper arms and looked me in the eye. "You guys are okay? Veronica is okay?"

Veronica appeared in the doorway. "Veronica!" he cried, releasing me and capturing her. Before he had released her, he was getting straight to work, saying, "You want to bring your kids over? We could talk about the turtles, you know—a little lecture, maybe. And I gotta clean the tanks today and my people, they're all home keeping their houses dry."

"Sure," Veronica said, "we'll put everyone in the truck and drive over."

"Where's the history professor?" Antonio asked.

"Graham?"

"I heard Veronica and the professor in the skirt were looking for me." Antonio smiles mischievously, enjoying his town's assessment of our eminent historian.

"He's in the station. The students are taking an exam and they're all panicking about his questions."

"So you can't come?"

"They can finish the exam later."

"Graham's gonna have a fit," I said.

"They can finish later. They've got the opportunity to touch a turtle."

"Excellent!" Antonio said, turning and bounding back down the steps. "I'll see you soon."

"Antonio," Veronica shouted after him, "you want a cup of coffee?"

"Oh, Veronica, thank you very much—but we gotta catch high tide, you know? So I can pump in water after we scrub."

The jeep skidded out of the lot and raced up the rocky road. And now, less than thirty minutes later, Antonio is standing beside one of his turtle tanks, his massive forearms resting atop the concrete wall as he leans forward to shout instructions over all the noise—a gasoline-powered pump and a torrent of seawater spouting from a pipe and plunging three feet onto the tank's pavement.

"Don't get your feet near their beaks," he yells, "they can bite through bone. Make sure you scrub along the side there; push all the stuff to the drain in the middle; that's right; excellent; perfecto!"

Our first team of four—Anoop, Chris, Rafe, and Becca—wield wide janitors' brooms and a garden hose to scrub away the soft quilt of algae. The turtles, two greens and a loggerhead, seem largely indifferent to the bustle around them; they lie in reptilian torpidity, their chins and flippers resting in the half inch of grubby water that remains on the tank's pavement. Every now and then, though, one of them is struck by some sort of deep turtle idea and starts slapping its flippers powerfully, splashing everyone nearby with a mixture of seawater, algae, and turtle feces. The largest animal, one of the greens, turns out to be more athletic than the others, so when the idea of movement takes hold of him, he actually humps and scrapes his way across the tank until he crashes beak-first into the concrete wall. And for some inscrutable reason—perhaps he believes he can break through, or the neural message is taking a while to travel from his face to his flippers, or he simply doesn't much mind mashing his beak against the wall—he just keeps paddling, renewing his crash with every forward heave of his hard mound of a body.

"Hey!" Antonio shouts. "Don't let him do that. Amigo—grab his rear flippers and pull him back."

Chris drops his broom and leans over behind the sea turtle. Grasping its flippers, he hauls it slowly backward. The movement seems momentarily to distract the turtle, which stops flailing as it slides in reverse.

"That's good." Antonio says. "Excellent. It's good to have a strong guy around. You want a job?"

"Sí," Chris says, still shuffling backward. "En serio."

"Hey," Antonio says, turning to Veronica, who stands to his right with her elbows propped on the tank wall, "he even speaks Spanish."

"He's Mexican," Veronica explains.

"Ah, that's why he's such a good worker. Mexicans know how to work, you know that? People in the U.S. say Mexicans are lazy, but I don't know where they get that idea. Where do they get it, Veronica?"

When Veronica says she doesn't know, he turns to me, on his left. "Why do you think, Arón? Why do people from the U.S. say Mexicans are lazy?"

Antonio's mind is just as kinetic and hard-driving as his body. And it sometimes seems that the mental hunger is so acute, the energy so irrepressible, that every line of conversation is bound to trigger an entirely new topic, on which Antonio is sure to demand your position. And when he does so, you must be very careful, because whatever you say he will commit to memory, and though the matter will not be pursued any further at present—because the next topic, arising by some oblique connection, will seem to him just as irresistible as the current one—nevertheless, whenever conversation happens to return to one of those many issues that have been touched on and swiftly abandoned, Antonio will remind you of your previous position on it and ask you to elaborate.

While I'm still considering what to say about the myth of Mexican laziness, Chris releases the huge green turtle that has, until now, slid so cooperatively backward from the concrete wall. In the instant its rear flippers touch the pavement, the turtle is newly possessed by the desire to flee and immediately starts heaving itself forward, even faster now than before. Its entire humped mass seems to lift up and launch ahead with every stroke of its powerful flippers, and it executes two or three such hurls before I realize that Becca happens to be scrubbing a spot directly in the turtle's path. As she tries to move quickly out of the way, she slips on the algal mat and falls on her butt in the water. Chris lunges to catch the turtle's rear flippers, but it is moving so fast now that they slip from his grasp and he falls on his belly in its wake. I am starting over the wall, aiming for the turtle, which I think I might be able to grasp by the front of its shell, just behind its neck, when I see that Antonio is already in the tank and taking two long strides—not toward the turtle but to Becca, whom he grabs around the waist and lifts straight up into the air, as if she were some sort of cheerleader before a crowd. In the same motion Antonio spreads his legs apart, taking such a wide stance that the turtle passes straight between them, crashing its beak once again into the concrete wall.

Depositing Becca to one side—and on her feet—Antonio turns, grasps the turtle's rear flippers, and rotates it, sending it back across the tank. After three more lunges, the animal suddenly gives up, resting its chin once again in the shallow water.

"I don't know what's wrong with this guy," Antonio says, shaking his head. "He's got something stuck in his magnetite. You guys know about magnetite?" He looks expectantly at the students around the outside of the tank, but they are dumbstruck. What just happened? Or, more horrifyingly, what *almost* happened? Had Antonio said something about beaks *snapping bone?*

"You don't know magnetite?" he is saying now. "It's metallic material in the turtle's brain. Works like a compass needle. You understand?"

No one says anything.

"Anyway," Antonio says, vaulting out of the tank and resuming his position, forearms atop the concrete wall, "where's the history professor? Maybe he knows something about why people say Mexicans are lazy."

"He stayed at the station," Veronica manages to say. "But Anoop knows a lot of history," she offers.

Anoop smiles weakly, adjusting his glasses.

"What kind of a name is Anoop?"

"Um, it's Hindi. I was born in India."

"India!" Antonio exclaims. "Wow, India. Lot of turtles in the Indian Ocean. Hawksbill and leatherback sea turtles—our rarest sea turtles—nest on two beaches in India. You ever see them?"

"Um, no," Anoop says. "India is a big place."

"Yes, very big. Over a thousand million people. World's biggest democracy, right?"

"Yes, I think that's right."

"That's great, right?"

"I guess so, yes."

"Why'd you guys stop working? Even my Mexican worker stopped working. You guys can't talk and work at the same time? How about we finish with the tanks, and then we'll talk about turtles, have a little lecture."

Antonio organizes the students into two more teams to clean the other tanks. As he rushes around—jogging out into the rain to fetch more brooms from a shed, leaping into each tank to open the drain, yanking the cord on another gas-powered pump and turning valves to release two more torrents of seawater—I take a bottle of medicine he has left with me and climb into the tank we've been standing beside. I am supposed to inspect the turtles for cuts or abrasions, giving each wound a dab of bright purple ointment, a prophylactic against bacteria that colonize wounds in seawater.

The large green that just crashed into the wall has a cut between his eyes, just above his beak. Surprisingly, he allows me to approach him, lean down over his head, and paint his wound purple; turtle psychology, or what there is of it, remains obscure to me. His eyes are large and black and forever teary. And his smooth, rounded bowsprit of a head is encased in a luxurious mosaic of chestnut-brown tiles limned with yellow. Such lovely animals really ought to be the easiest ones to protect: the eyes of a sorrowful fawn would seem to be the strongest possible political asset. And all politics aside, truly global species, of which *Chelonia mydas* is one, are supposed to be robust against human depredations—extinction-proof, relatively speaking—for they have more places to hide, more little populations hedging the bets of the species as a whole. And yet here we are: every species of sea turtle threatened; several right on the brink. If we cannot salvage this animal, or its tankmate, the loggerhead, or the hawksbill or the leatherback, then we will have failed at the first and most straightforward task before us, the preservation of a global and charismatic species. We will have flunked out at level one.

The loggerhead has a deep nick out of one of its flippers—a bite, perhaps, by one of the greens. I paint it purple, my lame yet hopeful little offering, and wonder for the hundredth time whether all this hard work of Antonio's really is to the good. Right here before us is the vice of it: a handful of turtles live in a state of some debasement, swimming stupid circles in concrete cisterns, gazing with their huge eyes into water clouded by their own shit. And sometimes they crash furiously into the wall, splitting their gorgeous skin. The virtues, by comparison, seem so much less immediate, harder to describe. When Antonio came here from Mexico City—"Had to get away from the rush," he said to me once, in an explanation that has always seemed inadequate to the magnitude of the transition—he arrived just in time to witness the final crash of the bay's turtle fishery, and he thought he could help. In a characteristically mad scheme, a plan one could never undertake without either grandiose delusions or prodigious reserves of strength, he hauled hundreds of boulders, enclosing a lagoon beside Coronado. He then began talking with fishermen, and somehow he convinced them to change their ways: when a sea turtle became entangled in their nets, they resisted the temptation to sell the animal and took it instead to Antonio. If the turtle was healthy and mature, Antonio would release it in the bay. But otherwise, he would keep it in his man-made lagoon, caring for it until it reached its reproductive prime. And this, as I understand it, is one of the express purposes of Antonio's operation: he is ushering these sea turtles through a dangerous adolescence, helping them to bypass the risks of drift nets and speeding pangas as they mature toward the moment they will be able to contribute to the next generation.

But if it were just that—fostering a handful of turtles—I'm not sure the virtues would outweigh the vices. The number of animals Antonio is able to assist on their way to maturity is surely too small to sway the fate of a population. As a management program, therefore, this operation is quixotic. But Antonio's work has other impacts, and while they may be harder to measure in units of turtles saved, they are probably, in the long run, more significant—at least, that's what I tell myself each time I see an open wound and wonder if it wouldn't actually be better to knock a big hole in these concrete walls. Once, when I arrived here in search of Antonio, I saw small schoolchildren sitting atop the concrete walls of every tank: three circles of kids, their little shoes dangling above the murky water. Antonio was standing on a wooden box beside the middle tank, lecturing loudly and holding up, as a prop, the empty carapace of a huge turtle. Afterward, as we were waving good-bye to a mob of kids pressed against the rear window of a colorful bus, which was heading back now to the town of Guerrero Negro, Antonio said, "Maybe the kids talk tonight at dinner about turtles. Maybe they start the argument."

Maybe so. And maybe one argument leads to another—the conversation, though narrowly focused at first, opens out into a broader discussion about the sea, its resources, and local livelihoods. Something like that may in fact be happening here in Bahía. Over the past few years, Antonio has recruited about a dozen fishermen to join what he calls the Grupo Marino. It is mainly a symbolic entity: they make Grupo Marino T-shirts, featuring pictures of whale sharks; they have a kind of environmental catechism, which begins *Don't poach turtles* and also includes the line *no poopoo peepee on the islands*; and they hold official meetings, where they discuss the future of the town and the bay. Antonio has persuaded these dozen or so fishermen of the great promise of ecotourism: they could all become guides; they could be paid very well by rich gringos eager to see whale sharks; they could—Antonio has said, suggestively, in my presence—work as assistants in classes like our own. How T-shirts, catechisms, and conversations grow into a local movement, which in turn becomes a regional campaign, which finally, somehow, influences policy—this is a cascade I cannot claim to understand. But if it ends up happening here, then Antonio Reséndiz will surely have been a key figure, an essential leader, in the growing movement.

About ten years ago, an enormous industrial fishing vessel parked itself in the bay. I've never been able to figure out what the ship was fishing for, but evidently it was a huge operation, and local fishermen were both terrified and enraged. A number of them went to Antonio. He and three fishermen took a panga out to the vessel and boarded it with their hunting rifles. As the story goes, the captain told Antonio the vessel held a permit to fish these waters, to which Antonio responded that the vessel might well have a permit, but he and the fishermen had guns. The captain must have seen the logic in this argument, for I have heard, from a few different sources, that the moment Antonio and his men were back in their panga, the ship pulled anchor and departed.

Why did the fishermen go to Antonio? He is not himself a fisherman. He held no official political office at the time. And he is certainly no veteran of armed rebellion. They went to him, I think, simply because he is the turtle man: the gorgeous creatures swimming circles in his backyard serve as his totem, giving him a special standing when it comes to protecting the sea. So whether these turtles I am anointing now with dabs of gentian are the beneficiaries of a cause or rather its martyrs—on that I am not quite clear. But I do—I think—believe in the operation.

When I have completed my ritual markings, I join Antonio and Veronica beside the tank with the loggerhead and the pair of greens. Our students have climbed out and moved on to assist in other tanks, because Antonio has just plugged this one's drain, and the water level is now rising around three stolid mounds; in just a moment, they will float up off the pavement and resume their graceful rounds. Over the plunging spate of seawater, Antonio is telling Veronica about the most recent ejido meeting, which took place just before we arrived in town. Antonio had brought two important guests, representatives of a multinational environmental organization. They had been working with Antonio for several months to arrange a transaction: the ejido would place a conservation easement on all coastal property—essentially, a legally binding promise to refrain from large-scale development; in return, the environmental organization would establish an annuity, which would be paid out to all ejido members for decades to come.

"You understand?" Antonio asks. "You understand what we sell? Only our right to build big hotels and that kind of thing. But we still own our land, you see? We can even have small hotels, ecolodges, restaurants, whatever we want. We just can't build no megaresort. And for that, we get a check every year. You understand?"

"Yes," Veronica says, "contracts like that are being used a lot now in the U.S."

"Really? It's a good deal, right?"

"I think it can—"

"It's like we're getting algo por nada, you know? 'Cause we don't want no megaresort here anyway. And we can still have jobs. We gotta have jobs, you know? Good jobs, like ecotourism. And sportfishing. And hunting. You know, bighorn ram license, at the auction in Las Vegas, sells for hundred sixty thousand U.S. dollars. Hundred sixty thousand dollars! That's the kinda tourism we want. Those guys, the hunters, I'm telling you, they don't even take showers! All they wanna do is go out in the desert and find rams. No drinking, no girls, no drugs, none of that stuff. And we got plenty of bighorn rams, you know? We got a rare subspecies of bighorn ram. It's—"

"So what happened at the meeting?" Veronica asks, pulling Antonio back.

"So, pretty bad. I come with these guys, and I'm pretty nervous, you know, 'cause these are important guys. Big environmentalists. And you know what? Someone else from town, I don't wanna say who, shows up—at the same meeting—with a developer. Really rich developer."

"From Texas," I venture.

"Right! So you heard about it. And this Texan, he says his company wants to buy ejido lands for a thousand million dollars."

"A billion dollars," I say.

"Right. A billion dollars."

"It's impossible."

"Yes. I think so. It's impossible, right? But you know, everyone wants to believe it. And I understand, you know. I want to believe him, too. That would be pretty good—pay for private schools in Ensenada; maybe we buy a nice apartment there, you know? Sounds pretty good. So I guess you know what happened. Everybody kicks me and my big environmentalists out the door. You understand?"

I ask if he's heard anything about the sale of the ejido around Colonet.

"There's a big project at Punta Colonet," he says. "They're gonna build a megaport there—for the really big container ships. They say it's gonna be bigger than the port at Long Beach."

"They bought the whole ejido?"

"No. Right now, just a lot of speculating, you know? It's like Escalera Nautica. Someone talks about a big project, and then there's crazy speculating. You know Roberto's wife sold her beachfront lot?"

"No, I didn't know that."

"They got a nice new SUV. Their land, for a nice new SUV."

"But is it really going to happen?" Veronica asks.

"Escalera Nautica?"

"No, the megaport."

"I think it's going to happen," Antonio says. "It's maybe different from Escalera Nautica."

"I heard Ruffo's involved," I say.

"He bought some land—some land right on the coast. And a big hill right nearby."

"A hill?"

"When they start construction, they're gonna need a lot of rock."

"I see."

"He bought all the rock," Antonio says, smiling to himself. "You know, I used to say that to people: put a price on the rocks, and soon they'll be gone, too. But it was just a thing to say, you know? I thought it sounded good. But now it's coming true. Put a price on the rocks, and soon they'll be gone, too."

6

By four in the afternoon, the sky is dark, bruised, and roiling, its surface like that of a bubbling cauldron of wine-dark stew. There is nothing more we can do to prepare the station: our roof will hold, or it won't; our ditch will divert the flood, or it won't. And I am learning now that weather on this scale can foster a deep and peaceful resignation. Without much discussion, most of us have abandoned whatever it was we were doing— digging ditches; writing the answer to one last exam question; reading

The Faerie Queene on the staff house porch—and have drifted, like objects on floodwaters, into the calm eddy of the seminar room. At the center of the table, a pair of small lanterns are serving passably as an ersatz campfire; ordinarily, the windows would provide plenty of afternoon light, but not today, and the town's generator went out a few hours ago. Even the dogs seem to sense that it is time to hunker down: Taiga sits with her huge white head beside Veronica's shoulder; Millie is a compact, cream-colored ball beneath Miles's chair; and Yukon has persuaded Allie and Haley to sit on the floor, so that he can lie across their laps.

Only two students, Becca and Anoop, are missing, still at work on their exams. I'm not sure where Becca has sequestered herself, but Anoop I have just seen in the corner of the museum room, sitting on an overturned white plastic bucket. He was hunched over a sheaf of loose-leaf notebook paper, writing furiously. It is dark in there, but Anoop is wearing a headlamp, and the reflection from his paper casts a pallid and wavering light across the rest of the room. On one wall of his corner, just above his head, there hangs a desiccated hunk of baleen: its laminary plates resemble old parchment pages so rippled with age and moisture that they no longer lie flat, one upon another, but press apart, holding the book slightly open. On the adjacent wall, a bookcase is filled with glass jars, each of which holds a small organism suspended in clear auburn liquid. As I walked by Anoop, on my way from the kitchen into the seminar room, I glanced down, expecting to greet him with a silent nod. But he was too absorbed in his work to notice me, and so instead I looked down on the black top of his head and, just beneath it, a brightly illuminated circle of white page, which was densely lined with Anoop's minute and fastidious script. And then, a step farther on, I noticed that the white plastic bucket serving as his stool had a name penned on it: *POLIS*, it said, in the thick black lines of a permanent marker. Gary would have been quite pleased, I thought, with this particular repurposing of his field equipment.

Around the dimly lit seminar room, several quiet discussions are in play, each intermittently audible above the murmurs of the others and the rain on the roof. The topics of conversation are disparate but share something in common: they all originate in what Antonio called his "little lecture." When the students had finished their scrubbing of algae, Antonio gathered them around a single tank. Shouting over the gas-powered pumps and their small waterfalls plunging into the circular pools, he delivered one of his uniquely wide-ranging, wildly impassioned, and partially comprehensible monologues. Since I suspected that Veronica and I would find ourselves, at some point, struggling to decode this rant for certain curious or puzzled students, I tried to make mental notes of key topics, deciding what we should or should not come back to later on: turtles in the bay and what became of them (yes, we ought to talk about that); the use of hallucinogenic drugs by the man who won the Nobel Prize for his invention of the polymerase chain reaction (better to drop it); how hard it is to make a living as a fisherman (yes); India and its environmental problems (probably not); the migratory route of loggerhead sea turtles (yes); and ecotourism (yes). Antonio closed his lecture, such as it was, with this:

"Ecotourism—you understand?—it can give us good jobs. And we gotta have good jobs here. We don't wanna go pick lettuce in Alta California. And we sure don't wanna go into Mexico's biggest export business, if you know what I mean. Sorry, but we're not interested. We live here. This is our home. And you gotta be local, you understand? You gotta know your home. Sure, maybe you wanna go study biology in Bahía de los Ángeles. That's good. But what's *your* place? Where are *you* local?"

You gotta be local . . . You gotta know your home. Everything else in Antonio's monologue—even his fascination with Kary Mullis, the inventor of PCR—was material I had heard him touch on at least once before. But this was new, and I was not sure what might lie behind it. The line about lettuce and drug trafficking betrayed resentment, but was it directed at us? At Veronica and me, and our students, too? Was Antonio telling us to find our own corner of nature and get out of theirs? But how would such rejection fit with his advocacy of ecotourism? Could he be disappointed, perhaps, in our own particular realization of his favorite idea? In the conversations that are now surfacing, separately and intermittently, from the background drone of rain, most of Antonio's topics are accounted for: Veronica and Graham are back to talking about *Chelonia mydas*; Ace, Miles, and Cameron are making a plan to try LSD on their seventy-fifth birthdays, at which point, they figure, they will have little to lose and much to gain from the experience; Chris and Allie are talking about the loggerhead's migration. But no one seems to be contending with Antonio's closing remarks. No one is discussing whether we were told to go away.

Anoop appears in the doorway from the museum room; an instant later, Becca enters through the doorway to the veranda, and suddenly the separate conversations around the table are falling quiet in quick succession, the way talk at a dinner party drops off when someone starts clinking a glass to make a toast. Anoop and Becca take seats at the table, and still no one talks. Graham glances up at the clock, then Veronica does—it must be battery-powered, because the second hand is in motion, and its time, therefore, still accurate. And now I understand: this sudden and anticipatory silence is testament not to Becca's social domination, but rather to the force of habit. It is four-ten, and we are therefore ten minutes late for our afternoon seminar. Only today, there is no seminar; at least, there's none scheduled on our syllabus; it was supposed to be an afternoon off, or maybe some extra time for the students to work on their exams. But just as twelve-thirty brings hunger, four o'clock has by now come to trigger a different sort of appetite or expectation, and so we are all sitting here, waiting to find out what we'll talk about.

Veronica, Graham, and I exchange looks; they too have realized that we ought to have come prepared.

"Well, V," Graham says, "how about turtle migrations?"

Veronica smiles. Happily, she has heard in Graham's suggestion not an attempt to foist our current predicament of unpreparedness on her, but rather a vote of confidence in the sea turtle tutorial she's been giving him throughout the day. As she begins, it becomes clear that she is even going to adhere to our tacit agreement not to present the same material twice. She's not going to talk about green turtles. Instead, she takes as her starting point something Antonio said as he pointed to the lone loggerhead. "That's Lolita," he shouted. "When we release her, she's going to make one of the longest trips of any animal. Longer than the humpback whales. Longer than the gray whales." Then he moved on to other topics, having delivered the dénouement without the thread. So Veronica now picks up that thread much farther back, asking how we could possibly know where Lolita will go.

The story begins in the late eighties. Christopher Golden, a commercial fisherman, was working a patch of ocean fifty miles southwest of San Diego harbor. As he hauled in one of his small drift nets, he found in it a pair of loggerhead sea turtles, entangled and drowned. One of them bore a tag, pierced through its hind flipper. Christopher had a deep interest in marine science—when he wasn't earning his livelihood at sea, he was working toward a graduate degree at Humboldt State University—and perhaps this accounts for his persistence in following up on the tag. Eventually, he learned that the turtle had been raised in captivity and released in Okinawa, Japan.

What did this mean? At the time, not a great deal. There is a large clockwise gyre in the northern Pacific: it runs from west to east at higher latitude, then circles back, from east to west, at lower, nearly equatorial latitude. If a careless turtle had inadvertently merged into this great clockwise current, it might easily have been swept all the way to California. And the northern half of the gyre, passing through relatively productive temperate waters, would not have been such a hard place for a turtle to survive: there is plenty to eat up there, and it is never severely cold. So when Christopher traced his turtle's tags to Okinawa, the most parsimonious explanation probably would have been that the animal had simply lost its way.

And yet there were a few other clues floating around—just enough to raise some suspicions. At the 1988 International Symposium on Sea Turtles, in Hiwasa, Japan, two Japanese scientists hung a poster reporting the recovery of their tagged turtle by a fisherman in Southern California. They noted that there had been very few reports of loggerheads nesting on Pacific beaches in the Americas—and even those few might well have been cases of mistaken identity, where inexperienced naturalists took an olive ridley to be a loggerhead. Could it be, they wondered, that all the loggerheads in Californian and Mexican waters in fact came from Japan? A few months later, a nine-inch loggerhead was found drifting in a scrap of gill net hundreds of miles north of the Hawaiian Islands. He was either a second wayward soul caught up in the North Pacific Gyre, or the second datapoint disclosing a mass migration; it was not yet clear which.

Remember that stretch of DNA we looked at in fin whales? The one that's inherited always from Mom and mutates extremely quickly? In the early nineties, a team of researchers at the University of Florida began collecting blood and tissue samples from Pacific loggerheads, with the intention of examining that same rapidly changing stretch of their genomes. The researchers reasoned that if the turtles in Japan and California belonged to two separate populations, then that particular stretch of DNA, unstable as it is, ought to be quite different in the two locations. By early

1994, the team had obtained more than a hundred DNA samples from a variety of sources: a nesting beach in Australia; a nesting beach in Japan; drift nets in the northern Pacific, where casualties had grown rapidly since that first one was reported, five years earlier; and finally, the coast of Baja. In fact, a number of the turtles from Baja had passed through Antonio's tanks, where the team from the University of Florida had taken the opportunity to draw their blood. As it happened, in July 1994, Antonio took one of those same turtles—an individual whose blood would figure in the Florida group's study—put tags on its front flippers, drove it across the peninsula to Santa Rosalíita, and released it into the open ocean.

In April 1995, the team from the University of Florida published their results. Across the northern Pacific—from Japan to mid-ocean drift nets to Baja—the loggerheads exhibited virtually zero divergence in that highly mutable stretch of DNA. Clearly, enough turtles were making the trip from Japan to California to erase any trace of genetic separation. The Australian population, by contrast, was genetically quite distinct.

At this point, it appeared that the northern, eastbound segment of the Pacific gyre represented something of a loggerhead highway: turtles born in Japan would merge into the current and ride it all the way to the rich feeding grounds off Baja. But what then? As far as anyone could tell, the immigrant turtles were simply not nesting on this side of the Pacific. So were they all reproductively doomed? Did they compose what ecologists would call a sink population—an aggregation that is stable, but only because it is constantly supplied by a distant source? Certainly it seemed strange that such a large number of turtles would make a decision so unwise by any Darwinian measure. And yet the alternative seemed perhaps even more preposterous. Whereas the eastbound segment of the gyre is full of food, the southern part, where the current flows back toward Japan, is essentially a blue-water desert, utterly unproductive and, what's more, even longer than the northern part. If it was remarkable that turtles were traveling six thousand miles from Japan to California, it seemed altogether impossible that they would survive more than six thousand miles of starvation on their way back.

On the ninth of November, 1995, Takumi Tai, a Japanese fisherman, was checking his net in waters near Gamoda Beach, Tokushima, Japan. A sea turtle had drowned in the net overnight. When he hauled the animal aboard, Takumi noticed that its front flippers bore tags. He could not read the roman letters engraved on the tags, but Takumi, like Christopher Golden, his counterpart across the Pacific, was both curious and concerned about turtles. Several weeks later, Antonio received a package from Japan in his post office box in Ensenada. And that—in conjunction with the genetic work, and subsequent studies that used more sophisticated, radio-trackable tags—is how we know where Antonio's Lolita will go when he decides it's time to release her. It is also why Antonio's boat, the one he uses almost daily to collect seaweed for his turtles, is called the *Takumi Tai*.

7

Only as Veronica comes to the end of her story—with Antonio reaching into his post office box, withdrawing a small package, and grasping instantly all that it implies—do I really register it: the rain on our thin roof, the gusts splattering the veranda window, the surf pounding on the rocks, the darkness—it's all been pressing in on us, closing us into an ever-smaller pocket of space in which we can still see and hear one another. But there is nowhere else to go and—I have to remind myself again—nothing more to be done. The appointed hour of seminar must be over by now—I can no longer make out the clock on the wall, but it must read five-thirty or six—and yet no one seems inclined to move. For the time being, our shrinking capsule of lantern light is our lifeboat on a dark sea. And what I find deeply reassuring is that the boat's mariners look hale: their faces tan and unconcerned, some of them calmly dazed as they stare into the light—drained, I imagine, from the strange combination of exam questions and ditch-digging. Only Becca looks pale and queasy; most tellingly, perhaps, she hasn't made any effort at all to turn the end of Veronica's story into a one-on-one dialogue.

"Have you guys noticed your ears popping?" Cameron asks. He is yawning repeatedly, the way a motionless fish does to draw water over its gills.

"Mine can't pop," Rafe says, "'cause of the holes in my eardrums."

Surprise and concern arise around the table—Allie, Lucy, and, touchingly, Anoop, are all very worried about the lasting effects of Rafe's injury. Veronica glances at me with discreet amusement; we are quite certain Rafe does not have holes in his eardrums.

"It's not that bad," Rafe explains. "Kind of awesome, actually. When we're diving, I never have to equalize."

Veronica looks straight down, so as not to make eye contact with me. The storm mounts a few seconds of bluster—a splash of water against the windows, a wave exploding on the terrace—and as it subsides, Isabel says to Veronica, "So is this the last time you guys will do the class?"

Veronica looks surprised. "The last time?" she asks.

"Sure as hell is," Graham says, looking at the black, rain-spattered windows. "Next year we're getting hotel rooms in Manhattan—doing the whole class in the Museum of Natural History. Whale dioramas. No wind. What do you guys say?"

"Why would it be the last time?" Veronica asks Isabel.

"I just meant, if the deal goes through to sell the ejido—I wasn't sure what would happen to the field station and everything."

"No one's actually buying the whole ejido," I say.

"You know," Isabel says, sitting up, her back seeming to stiffen, "sometimes, when you say these deals aren't going through, it seems like maybe what you really mean is that you don't *want* them to go through."

The room, though filled with the sounds of the storm, seems suddenly silent. "That's true," I say finally. "I mean, you're right that I don't want them to go through. But there are also reasons—a lot of reasons—to be skeptical about them."

"It seems like you guys really care about Alejandrina. And she really wants this. So I guess I don't see why you wouldn't want it, too—just for her."

"I don't think a huge FONATUR development would be the best thing for Alejandrina—or anybody else in town."

Leaning in over the seminar table—almost as if to place herself between me and Isabel—Allie says, "Haley heard in town that the ejido next door already sold."

"This is getting a little confused," Graham says. "I think Isabel is right that we need to be clear about whether we're addressing the facts or our desires."

"What we're addressing," Haley says with some impatience, "is whether someone's gonna buy all the land here."

"And from what I hear," Isabel says, "someone is. Alejandrina says there's a big Texas developer who's really interested. And the governor of Baja Norte is in on it, too."

I look at Veronica, hoping she'll respond. Somehow, all of a sudden, I feel like my words are compromised, as if every statement I could possibly make were already in doubt.

"We talked to Antonio about it today," Veronica says calmly. "He has doubts about the Texan's real intentions. And it wasn't actually a whole ejido that sold north of here. There's a big government plan to develop a port at Punta Colonet. Some speculators—including a former governor—have been buying land there, thinking it will be valuable because of the new port."

"But if you believe a big port's going to happen," Isabel says, "why are you so skeptical about the big tourism project here in Bahía?" Speaking to Veronica, Isabel sounds less accusatory, more plaintive. But when, I wonder, did she become so angry with me in particular? And where has her shyness suddenly gone? It's as if the Spanish-speaking Isabel—not closed off but expansive—is suddenly bursting through in English. "I just don't get it," she is saying now. "I mean, it sounds like Escalera Nautica is just as big a government plan as the new seaport, doesn't it?"

She glances at me, and I'm talking again before I realize I ought to be leaving all of this to Veronica. "Escalera Nautica doesn't exist anymore," I say, "not even as a plan. The government changed the name and scaled it way back. And I'm not saying I know what the government—or FONATUR—is going to do. I don't have any idea. But I'm pretty sure no one's buying the ejido for hundreds of millions of dollars."

"So everyone in town is just being totally dumb about it?"

She knows I can't respond to that question, and I don't, so Graham steps in, again playing the moderator. "Okay," he says, "so we've established that we can't be sure about land deals in the works. Maybe we should move on to the next question: Would some form of large-scale development be a good thing for the town?"

"I don't think that's the next question," Isabel says.

Graham looks at her, awaiting more.

"I think the next question is whether it's really any of our business. I mean, it's like Antonio said: you gotta be local. And if they need the jobs, then they can invite whichever developers they want to come here and give them jobs. I mean, it's their home—"

Becca interrupts, talking directly to me: "Do you know of any economic—"

"Hang on, Becca," Isabel says. "Just let me finish. It's their home. And they know better than we do what to do with it. So we should just get out of their way. If they want to invite us here as turistas, then we'll say muchas gracias and come as turistas." As she finishes, she looks at Becca and bows her head slightly to cede the floor, but Chris speaks before Becca can. "Isabel," he says, "haven't you been paying attention? Haven't you heard what everyone's been saying about this place? Even Antonio? I mean, this place should have tons of fish and turtles and all sorts of stuff, so why would the townspeople be better off cleaning hotel rooms for gringos?"

"They need jobs," Isabel says. "They need salaries."

"But what kind of jobs?" I ask.

Isabel's eyes harden. Looking at me, and then at Chris, and then back at me, she says, "You want to know what I think? I think you guys are *ecocentrists*. Rich, privileged ecocentrists. And seeing all your favorite animals is more important to you than the people who live here. You don't really care about them, as long as you can get what you want here."

"Water," says Miles.

"What?" Chris says sharply, like a combatant turning blindly on a bystander.

"Water," Miles says again, pointing at the doorway—not the one on the seaward side, but the one across from it, the one that leads into the library room and then, from there, out to the dirt lot, and our ditch, and the staff house, and the immense, steep slope up to Mike's Mountain—and there, in that doorway, sliding like a lazy river, is a dark, gleaming plane of water.

8

Even as we stood from our chairs the flow touched our feet, and by the time we had reached the doorway, the water was up above our ankles. The sandy lot behind the station was lost in darkness, but I could see just enough to understand that the ground was moving: it was a tracery of small rivers making for the sea, like a great fanning delta viewed from high above. Our ditch, ten feet in front of me, was hardly visible anymore, and the rivers were eating away at it, overrunning it even as I watched. I took one of the shovels that had been left against the station wall and stepped into the moat, started digging, trying to clear the silt and divert the flow once more around the station. Veronica appeared beside me with an enormous plastic trash bag, which I began to fill with mud. For many hours we—meaning all of us, every last one of us—worked furiously to deepen our ditch, build up our plastic wall of silt-laden bags, and sweep the floodwaters out the station doors. So feverish were our efforts that I thought hardly at all about what Isabel had said. And when we stopped, having outlasted the worst of the rain, we retreated to our cots—the students in the veranda room, we on Melissa's porch—and though I still felt Isabel's accusations there in my chest, like a pulled muscle or a foreign body lodged under the skin, I was so exhausted that I slept and dreamed of cutting it out with a sharp knife.

In the morning, when I sat up on my cot and looked out at the water and sky—silver and ash; no longer lead and coal—I felt the most profound relief: the storm was reaching its end. It took a moment before the echo of Isabel's voice came back, making me cringe and turn to look for Veronica. She was there on her cot beside me, but still asleep, too deeply asleep to disturb. Beyond her, though, at the far end of the covered porch, Graham was sitting up on his cot, staring out at the pale gray morning. Rising quietly, I walked over and sat on the low concrete wall in front of him.

Isabel's words were not on his mind at all, and he seemed sincerely surprised to hear they were on mine.

"Hirsh," he pleaded, "*come on*. That's exactly what this course is about—conversations just like that. It was fantastic. Isabel had a seminar coming-out moment."

"Easy for you to say. You're not the rich and privileged ecocentrist."

"We're all rich and privileged, right? I mean, relatively speaking."

"And?"

"I'm just saying this isn't the first time that indictment—that *fact*—has occurred to you."

"I'm not sure previous exposure makes me immune."

"She was trying to make a real argument, Hirsh. She was trying to say conservation is too costly at this stage in development."

Patiently, gradually, Graham drew me out of my aching preoccupation with Isabel's judgment, getting me to focus instead on her argument. While we talked, we watched the islands in the bay drift in and out of misty veils, which the morning was slowly tearing into muslin tatters. I was surprised to learn that Graham actually thought Isabel's claim seemed right: the development that could bring basic improvements to life in Bahía—a steady supply of freshwater; twenty-four-hour electricity; eventually, perhaps, a high school—would unavoidably entail some degradation of the town's magnificent setting. And on balance, the ejido members might come out on top: yes, a few beaches get wrecked by big hotels; there's more boat traffic, pollution, and sewage; but there are some new jobs in town, and at least you can turn a water tap or flip a light switch when you want to.

But to me, this argument seems partly dubious, and partly plain backward. What I find doubtful is the notion that conventional tourism could ever thrive here. Summer days of a hundred thirty Fahrenheit; winter days of fifty; west winds that blow so hard they seem to scour the desert and drive the scurf straight through town—in so many ways, the environment of Bahía de los Ángeles is at odds with the image of a pleasant and leisurely resort. The most darkly absurd outcome would be an immense, government-subsidized development that destroys what treasures Bahía does possess but fails, in the long term, to offer any compensatory economic rewards.

What Bahía has but Cancún does not—at least, not anymore—are the precious seeds of natural resources. And that is why Isabel's argument, as it was taken up the next morning by Graham, seems not just dubious but backward: The people of Bahía don't have to exchange the bay for economic improvement, because their greatest potential source of wealth *is* the bay. And although resources here have certainly been depleted, and will take time to recover, their potential value far exceeds the wages that a big resort would bring—even if it were to succeed. This is one reason it is so important to remember what was here—the missing species, their former abundance, all those leaves torn from Thoreau's figurative poem. If we allow ourselves to forget the long history of depredation—if we take this place for wilderness—then, when we tally up the bay's resources and weigh their value against the dividends of something like Escalera Nautica, our calculations will be mistaken, and momentously so.

It is wrong, then, to claim that objecting to certain kinds of development amounts to a privileged disregard for the locals' prosperity. Still, there is no blunting the sharp truth in Antonio's dictum: *You gotta be local.* Yet the dictum is surely too simple, and must be prized apart to find what's right within it. To that end, I want to tell one last story about a fish. I have seen many old photos of *Totoaba macdonaldi*, and the grainy black-and-white images look like they've been digitally doctored: a fisherman stands grinning beside his catch, which appears to be a common croaker, except that it is laughably enlarged—longer than the fisherman is tall, wider than the portly fellow's belly. A few of the older fishermen in town have told me that when they were children they watched great shoals of such fish pass through every summer. The totoaba frothed the sea from the coast to the horizon—a boy's horizon, at least—and they swam so thick their dark dorsa were like the backs of tightly herded cattle. One of the men said he had watched his father wade into the seething shallows and drive his harpoon into the humped back of a passing fish, the way a bullfighter stabs his bull.

Today, totoaba are on the CITES list of critically endangered species. You can fish around the islands for ten years without seeing a single one. How we got from then to now—from vast shoals to simply gone—is a story in which several far-flung narrative tributaries, each of which we've touched on before, flow together . . .

Tributary one: San Diego sprawl and the Colorado River. Between San Diego and the border, our convoy traveled briefly alongside a broad concrete spillway, like a spare interstate recessed beside our own. A thin black creek of water trickled down the spillway's middle. Curiously, that thin creek bespeaks an elemental connection between the burgeoning suburban developments south of San Diego and their uncannily similar clones outside Denver, Colorado: the distant expansions of housing and mini-malls not only look alike; they also drink from the same river. Although Colorado's sprawl is mostly on the eastern slope of the great divide, most of the water it consumes is drawn from the other side of the Rockies. A huge tunnel runs beneath the mountains, drawing water from the Fraser and Williams Fork river basins, and delivering it to reservoirs on the eastern slope. In fact, the Denver Water Utility, anticipating a serious shortfall as soon as 2016, is about to expand one of those reservoirs from 1.8 to 5.1 billion cubic feet. But whatever comes out of the Fraser or Williams Fork is water that would have otherwise ended up in the Colorado River. And the Colorado River, hundreds of miles farther downstream, on the other side of thirsty neighborhoods and farms in Utah, New Mexico, and Nevada, supplies the sprawl outside San Diego, and departs, finally, as no more than a trickle down a concrete canal.

Tributary two: Hernán Cortés and La Isla de California. The first sunrise I ever saw from the field station terrace was astonishing—the bay a pool of molten glass—and from that morning on, I assumed I knew why the Gulf of California is sometimes called the Vermilion Sea. But then, when I read the journals of the conquistadors, I encountered the name in an entirely different context. Hernán Cortés had sent Francisco Ulloa northward, to circumnavigate La Isla de California. And when Ulloa's ship was approximately fifty miles from what we now know to be the northern end of the gulf, he recorded in his log that the sea around him had suddenly turned ruddy. When I read this, I was mystified, because I'd seen those very waters, and they were not even faintly reddish. Could it have been a red tide that Ulloa witnessed? A tremendous bloom of dinoflagellates? I figured probably not, because the season was wrong, and besides, Ulloa made no mention of bioluminescence at night—something that surely would have caught his attention and merited a note. And then it occurred to me: It was the river, the Colorado River, delivering such volumes of freshwater into its estuary that the entire northern gulf was ferruginous with silt. In effect, the Grand Canyon was staining the Sea of Cortez. And the reason I had never witnessed El Mar Bermejo—in its original sense, I mean—was simply that the Colorado River no longer has any water to give the gulf. By the time it reaches the sea, south of the concrete canal we drove beside, the great Colorado is nothing but a wide swath of sand, an avenue for salty tides to run upland.

Tributary three: history written in isotopes. Most fish, it turns out, have pebbles in their heads. They are bumpy little rocks made up of many fine layers of calcium carbonate—$CaCO_3$, the stuff of oyster shell—laminated together with sticky protein glue. They are called otoliths, ear-stones, and they provide a fish with data about orientation, acceleration, and sound. Most fish have six otoliths, two large and four small, and each one resides in its own cozy cavern, which is lined with a velvet of hairlike sensors. When a fish darts forward, the pebbles roll back, giving the fish an accurate sensation of darting. When the fish swims upside down, the pebbles roll to the roofs of their caverns, and the fish feels inverted. And finally, when sound waves wash over the fish, the dense stones do not vibrate as the wet velvet does, and the relative motion of hairs against pebble generates a neurological transcription of sound.

Taking a pair of relatively large otoliths in the palm of your hand, you might find them pretty and mildly mysterious—they have a pearlescent sheen and their shapes are finely sculpted and absorbingly involuted—but you would never imagine the sheer density of information reposing there. They don't seem quite weighty enough for all they contain. To start, they hold very specific information about whose head they once inhabited, because nearly every kind of fish has its own particular form of otolith. And since ear stones really are mostly stone, they are durable, left behind long after the rest of the fish they traveled with. This combination of species specificity and general durability has made otoliths valuable in several fields of research. In marine ecology, for instance, otoliths recovered from the feces or stomachs of all sorts of piscivorous fish and marine mammals have been used to reconstruct the predators' diets. In archaeology, ear stones found in prehistoric middens have revealed what certain tribes fished for and ate.

But the truly surprising depth of the otolithic library—the extent of its holdings—is revealed only when you take an ear stone and carefully slice a thin cross section. There, on the delicate disk you've obtained, you will see concentric rings contracting toward the center of the stone and alternating between translucent gray and opaque white. Translucency increases with the rate at which material accretes. That rate, in turn, is paced by the fish's overall metabolism. And because fish are generally busier in summer than they are in winter, the seasons of a fish's life can be seen and counted in the concentric circles of its ear stone.

But each little white record contains still more information, and this is where the library starts to get positively Borgesian. The deposition of thin laminae of calcium carbonate and gelatinous glue follows a daily alternating cycle: a day, a layer of stone; a night, a layer of glue; and so on, throughout the fish's life. With a good microscope, therefore, you can see not just how many seasons a fish swam the ocean, but how many *days*. And as we know from those layers of turtle scute, biological deposits record more than the passage of time. The letters of the elemental alphabet—N-14 and N-15; C-12 and C-13; the isotopes of iron, potassium, and others—spell out many messages about the lives of animals: what they ate, where they lived, when they lived there. And in the onion-skin layers of otoliths, such messages will be finely stratified, written one atop the other, such that the history of each diurnal cycle is inscribed in one layer of glue and one of stone. The glue is rich in the various biological elements we considered deciphering before. The calcium carbonate, for its part, forms a crystal lattice, a rectilinear cage that happens to trap many rare elements—exotic things such as strontium and molybdenum, which record, in their own isotopic scripts, daily memos on such matters as the temperature of water or the distance to the nearest nuclear test.

The element that is relevant here, however, in the particular tributary we are now tracing, is nothing so rare or exotic, but rather a common constituent of stone and glue alike: it is oxygen, the O of $CaCO_3$ and a building block of every protein and sugar. The most common isotope of oxygen, O-16, is a bit lighter than its sibling O-18. As a result, while H_2O-16—water containing light oxygen—evaporates from the ocean, jumping into the air to join clouds, H_2O-18 has a sluggard's tendency to stay behind. Therefore, water that has more recently traveled through clouds—rainwater, that is, and the river water descended from it—contains less O-18 than seawater does.

And here our tributaries coalesce, around *Totoaba macdonaldi*. A group of researchers examined ten totoaba otoliths: five recovered from prehistoric middens, and five extracted from fish caught in recent years. In the growth ring corresponding to the first year of life, the prehistoric otoliths had strikingly low concentrations of heavy oxygen, O-18—impossibly low, in fact, for fish living in seawater. By contrast, the first growth ring of the modern totoaba otoliths had about as much O-18 as pure seawater. Moving outward across growth rings—forward, that is, in the lives of totoaba—the prehistoric otoliths began to look more like modern ones: in their second annulus, the concentration of heavy oxygen was still somewhat low; but by the third, the prehistoric otoliths had taken on a typical, seawater-like concentration of O-18.

In the context of what older fishermen remember about totoaba—and what we know about the Colorado River—this pattern makes sense. The oldest fishermen of Bahía were young boys when they last saw it happen, but nevertheless, they remember with certainty that the great herds of totoaba came out of the north, arriving in early summer. By September, the fish had departed, continuing on their way southward. In Guaymas, which is south of Bahía and across the gulf, fishermen remember that the season for totoaba was not summer, but rather from autumn into winter. When the totoaba left there, at winter's end, it looked like they were headed back to the north. And indeed, the old fishermen of Puerto Peñasco, at the northernmost edge of the gulf, will tell you that the best time for totoaba is spring. So it seems that a circle of memories traces the movement of totoaba around the gulf: the herd was describing an annual migration, circling counterclockwise and returning each spring, the season of spawning, to the mouth of the Colorado River. And what those ancient otoliths are telling us is that the fry spent their entire first year of life in the presence of abundant freshwater. In their second year, too, they stayed close to the estuary. But by their third year, it seems, they had joined the great migration down to Bahía de los Ángeles.

Of course we must ask—before we point fingers at thirsty farms and developments from Colorado to California—how much the freshwater really mattered to young totoaba, and by extension, how important the river's depletion really was in the decline of the species. After all, as the memories that encircle the northern gulf suggest, there was plenty of fishing going on. As early as 1915, a commercial fishery was catching totoaba for export. Bizarrely, the fish were taken not for their meat, but only for their swim bladders, which were dried and exported for use in a certain Chinese soup. In a number of respects—the grotesque inefficiency; the mounds of carcasses rotting on the beach; the straw mats arrayed with hundreds of replicas of a certain fish part, drying in the sun; even the particular geographic destinations of the desiccated commodity—the business prefigured the shark-fin fishery that would arise, in the very same villages, seventy-five years later.

In the early 1920s, American entrepreneurs finally realized what a spectacular, untapped resource lay close at hand, and by the early forties, more than two million pounds of totoaba filets were being shipped to the United States each year. Over the same stretch of two decades, from the twenties to the forties, the shrimping business was also expanding rapidly in the northern gulf, and since shrimping gear is extremely indiscriminate, the growing fleet surely killed significant numbers of juvenile totoaba. What's more, further expansions of the shrimp fleet, in the fifties and sixties, exacted a double toll on totoaba populations: the shrimpers not only took juveniles as bycatch, but also retooled their boats and shifted their efforts to focus on totoaba fishing every year in the early spring. This practice ended in 1975, but shrimping, with its tremendous bycatch, has continued. So one must wonder: With almost a century of such obvious impacts to look back upon, must we really contemplate the Colorado River as well?

Two facts make me believe we must. The first is a simple matter of timing. The volume of freshwater exiting the Colorado River fell to zero, once and for all, in 1964. Fishery data from '65 and '66 show no immediate evidence of any serious impact: total catch hovered around four million pounds per year. Beginning in '67, however, catches declined swiftly and relentlessly. And the two-year delay makes perfect sense when we recall that the otoliths showed totoaba lingering in the northern gulf for two seasons before they joined the migration southward: If the complete absence of freshwater somehow stunted growth or disrupted spawning, 1967 would have been the first year that the full impact presented itself in the catch. From that point on, it appears, the fish population simply couldn't keep pace with the fishing pressure. By 1975, the last season before totoaba was protected under CITES, the total annual catch had fallen to two percent of what it had been a decade before.

The second piece of evidence I find compelling resides in those ten totoaba ear stones. Because the pace of otolith accretion is set by overall metabolic rate, a fish's bodily growth in a given year can be inferred from the width of the corresponding annulus of its otoliths. Tellingly, the first two growth rings are significantly thicker in the five prehistoric otoliths than they are in their modern counterparts. By age two, the prehistoric totoaba were topping four feet, and this number represents an important milestone: it is the length at which totoaba reach sexual maturity. Modern totoaba don't make it there until age five. Such a dramatic shift in the age of first reproduction—from two to five—must have had an enormous impact on the population's ability to sustain itself. And since the difference in growth rates is confined to those first few annuli, deposited while the fish were lingering at the river mouth, the most plausible culprit is the complete collapse of the Colorado.

If there is any silver lining here, it is perhaps the indication that not so much water would be needed to let the totoaba population recover. The river's flow first hit zero in 1965, but it wasn't exactly gushing up until then. Over the preceding decade, the average annual flow was somewhere around 70 billion cubic feet, which sounds like a lot until you consider that it's roughly a tenth of what the river was supplying to the gulf before any dams went up. And yet the totoaba were hanging on, and their final collapse didn't begin until that last ten percent of the river had been sucked out. This seems to suggest that if the farms and residents from

Denver to San Diego could find a way to use just ten percent less water—a tenth less water in our tubs and toilets; a tenth less for our golf courses and lettuce fields—the totoaba might come back.

Be local, says Antonio. *You gotta be local.* It is certainly true that the old fishermen of Bahía can tell you as much about the potential value of their bay as anyone on earth. They remember the mantarraya and the totoaba, the green turtles and the tiger sharks, the lion's paw scallops and the black murex. In another sense, however, local knowledge, even at its most profound, is critically insufficient. As Antonio's own mailbox helped to show, many ecological problems are not local at all. To the contrary, their shape is dendritic, extensive, less like a dot or an area on a map and more like some sort of web or network. They arc from the Sea of Japan to the Sea of Cortez, and ramify from Denver to Las Vegas to Bahía de los Ángeles. To put it differently, there is no single locality where one can contend successfully with an ecological crisis that connects Christopher Golden to Takumi Tai to Antonio Reséndiz.

And while I like the ring of it as much as the next guy stuck in traffic, *Think Globally, Act Locally*, the well-worn mantra from the early years of America's environmental movement, suffers from the same shortcoming. Oftentimes, acting locally just won't do, because the shape of any successful solution will have to mirror that of the problem. You could pour infinite resources into protecting loggerheads off the coast of Baja, but unless you do something about the drift nets in the northern Pacific, thousands of miles away, your efforts will be absolutely in vain. You could halt totoaba fishing—they *did* halt totoaba fishing—but unless you restore a trickle of water from Denver and San Diego, you'll have no fish to show for it. In many cases, therefore, we must sacrifice the bumper-sticker brevity of the mantra in exchange for an approach that is more prosaic in description and more daunting in execution. It would go something like this: integrate information on a vast geographic and historical scale; act as an extensive collective to formulate and apply policies on a scale commensurate with the problem. True, if we put it that way, it's a lot harder to feel sanguine. But we'd better be honest with ourselves about the tasks at hand.

And yet there is one important sense in which Antonio's dictum—and, by extension, Isabel's accusation—is irrefutable. As Isabel suggested, my own relationship with Bahía de los Ángeles has been essentially exploitative. Not in any grand way, of course. Not like conquistadors or whaling ships or pearling vessels. But in a small, individual way. Our classes come; we find it interesting and beautiful; we enjoy the snorkeling and the whale watching; and then we leave. No great harm done, really. But no help offered, either. No measurable steps toward protection or management or plans for the right kind of development. We've taken a bit, and we've done basically nothing in return, so the balance, I reckon, is a small negative number. And I think the main reason my personal account here has come out in the red is that Veronica and I are not local. We have not poured our time and energy into one place, our home. By contrast, Antonio has, and he's gotten something done. But here too in the final analysis, *Be Local* seems roughly half right. Antonio surely qualifies as a local, but he came from Mexico City; he was well educated and savvy about relations with the federal bureaucracy. In these respects, Antonio was better equipped than any native Bahíans to contend with the bay's environmental problems. Inspired by his biography, and also dumbfounded by the various ways our students, these brilliant kids, are forever surprising us, I have sometimes imagined that one of them—could it be Chris? or Allie? or Isabel?—would one day return here and do what we, Veronica and I, have never done: help to mend the place we love.

PART X

Pseudorca crassidens and *Homo sapiens*

TELLING OUR WAY TO THE SEA

1

At daybreak, fiery light pierced the folds of gray—the faded silver sky, the drifting tatters of fog, the mist rising off the bay—and struck our faces for the first time in four days. The desert steamed like a newborn landscape. The cardón and cirio looked clean and alive; every spine pricked a tiny droplet quivering with light. It was as if we'd blinked awake just in time to catch the sea being separated from the sky, the earth from the sea.

Our group is giddy with the world's freshness. As Veronica and I swing the pangas from Samy's boat ramp to the station beach, we carve needless curves in the placid water, as if to announce, in giant calligraphy, our presence after the flood: *the hurricane is over, and we are still here!* The students spill from the terrace down to the beach like a clan of wild and hooting street urchins. Rafe is wearing his sarong as a cape. Anoop has a pair of Miles's mirrored race goggles around his forehead like a headband. Cameron and Allie are both coated in wet sand, because the two of them have been heaping up an enormous pile, which they say they are going to sculpt into a totem of the fin whale.

As Ace, Miles, and Cameron climb into the bow of my boat, I notice that Haley has already joined Veronica. It's the first day she's not in the same panga as Ace, though I have no idea if the move is meaningful. Graham, standing on the beach, looks back and forth between the pangas, as though something's amiss. And then I realize: Becca's not here. When I ask the students about it, Ace replies, "She said she's too tired." He manages to say it without sounding joyful, but of course the group's uncorked spirits may have something to do with her absence. I glance at Allie and Lucy, who have seated themselves behind me, and Allie shrugs and nods, as if to say it's true, that's all there is to it, she's tired.

"She's had a rough couple days," I say, but stop there, feeling that the enumeration of details—the shark; the fall in the turtle tank; digging ditches with the rest of us—will somehow come across as mean-spirited.

Allie nods. Even in her, I think, there is a hint of relief, if also a sad resignation: at last a day without Becca's stories, but yes, we have failed as a group. Across the short span of shoreline between the *Sea Eagle* and the *Cortez Angel*, Veronica and I look at each other, and I make a gesture like I'm counting off the students in my boat. She winces and nods a grim acknowledgment. Then I nod, too. We both know we should go check on her, talk with her before we embark, and we've both decided against it.

"Hop on, Graham," Veronica says.

Throttles open, hulls planing weightlessly, we make for the sand spit. From there we'll veer north, toward the volcano, because we've decided to search for the fin whales. On this first, steaming morning after the flood, the image of leviathan lunge-feeding beneath a cinder cone seems irresistibly primeval; it's as if the lifting of the mist has awakened in us some atavistic urge to pursue our greatest of quarries. Our plan, then, is for Veronica to take the *Sea Eagle* clockwise around Coronado, while I take the *Angel* counterclockwise; we'll meet somewhere on the northern side, and hopefully, one boat will have located the whales.

We round the sand spit in parallel, my own boat closer to the verge, where the still water looks greenish over the plunging slope of sand. The beach appears exceptionally pristine this morning, scoured by the rain and the surf, and for some reason abandoned by its usual row of malingering pelicans and gulls. I throttle down and swerve right, crossing Veronica's wake, because from here our panga paths will diverge. Ordinarily, we wouldn't separate, but the day is blissfully calm, and we are eager to cover more territory, determined to find our whales.

Having cleared Ventana reef, I steer for a cloud of birds hovering in front of Piojo—only slightly off our appointed course—and as we approach, I ease the bow down and kill the engine, so we can watch and hear the feeding frenzy: a patch of water seethes where a shoal of silvery grunion are pressed to the surface by a murderous pack of predators—I've already seen brief flashes of yellowtail and skipjack—and the birds have gathered overhead to feast on the panicked throng. Attacked from above and below, the grunion swarm and writhe like eels, catching the sunlight on their sides and frothing the sea into an emulsion of white foam, shattered mirror, and fish blood. The ceaseless roar of a waterfall is punctuated by the percussion of a shooting range—*pop, pop-pop, pop*—as boobies divebomb the shoal from high above: folding their wings, they plunge fast, puncture the mayhem heedlessly. Meanwhile, the pelicans belly flop clumsily into the thick of things and come up with their gullets full of saltwater. One smacks down face-first right beside our boat. As he lifts his head the sea drools from the sides of his beak and the membranous double chin deflates, revealing within it the wriggling shape of a fish; for a second, the bird looks disconcerted, stricken with a swollen and spastic epiglottis; but then he tilts his head gluttonously back and ends the struggle with a single hard swallow.

Behind the crisscrossing paths of bird-missiles, through the fishy mist that rises from the froth, I see now a distant row of quick misty spouts peppering the surface: bottlenose, the first pod we've seen. As we head toward them, they rocket from the sea in boisterous spirals; the students cheer loudly, and the dolphin respond with even wilder antics. For them too, it appears, the morning feels bracingly new. Behind their acrobatics—as if on cue—a pair of whale spouts rise: one towering, the other low—a mother and her calf. We circle round the bottlenose to head for the spouts, and the dolphin fall in behind our boat, as if they regret losing our attention.

"Are you there?" says Veronica's voice. And before I can pick up the radio she repeats herself—"Are you there?"—and my heart tightens, because her timing is off—too quick—and something must be wrong.

"You okay?" I ask.

"We're with a cetacean I've never seen before."

"What do you mean?"

"Much bigger than a bottlenose. High black dorsal. Rounded head, but not as round as a pilot whale. And there are about thirty of them." She's trying to be calm and clear, but there's a slight tremble in her voice. In fifteen years on these waters, Veronica has seen many species, even transient orcas, but never something she could not identify—never a creature whose diagnostic traits and Latin name were not memorized, years ago, by a young Czech girl doting over the soft, worn pages of the Audubon Society's *Guide to Marine Mammals of the World*.

We peel away from the whale spouts and fly north: Veronica must have already made it around to the eastern side of the volcano; otherwise the radio wouldn't work. And sure enough, seconds later, there is her panga—a small white dot on the horizon—and between it and us, something is breaking the water: a wide, traveling front. If I had not just heard from Veronica, I might have thought common dolphin, just because of the way the pod is arranged in a broad phalanx, or I might not have, because the black backs stitching the surface—rising, gleaming, descending, and rising again—are just too big to be dolphin. Now that we are coming closer I can see that the front is moving in the same direction we are—toward Veronica's distant panga. But why, I wonder, is she so far away? Why isn't she with them? Running full bore, veering slightly eastward so as not to cut their phalanx in its middle, we pull even with the traveling front. Then I back off the throttle, and we peer down the length of their row. They are just as Veronica described: twice the size of bottlenose; coal-black and fusiform; powerful and fast. Moving with them like this, I feel almost like we've joined their pack, like we're one among the sea wolves. I pull a quarter mile ahead and cut my engine, so that we will be able to listen to them breathing as they pass.

"We're with them," I say quietly into my radio.

"We are, too," Veronica replies. So the pack must be traveling in ranks up the gulf. "My best guess," Veronica says, tentatively, "is false killer whales—*Pseudorca crassidens*."

Just as they are about to pass us, the entire phalanx disappears. Their wake dissipates as it rolls toward us, and once its slight gurgle against our hull is gone, all is mysteriously quiet. We are looking only at glassy blue water, with the granite gown of Isla Coronado behind it.

"Where'd they go?" Cameron whispers.

"I guess they went down," Ace is whispering in reply when a sudden piercing whistle slices the silence. It's the sound a bullet would make as it streaked by your ear. And then a second one screams by. And still more. And the sounds seem to be growing louder and coming closer and closer together—the whistling bullets from a machine gun now.

"What is it?" Lucy says anxiously, and everyone is now looking in a different direction—up into the sky; at the island; at each other. Cameron is still staring at the water where the phalanx would have been if it hadn't just disappeared; he has one finger pressed against the bone just behind his ear, as if he were a musician doing a sound test or an athlete taking his own pulse. I stare where he is staring, and that is in fact where a whale breaks the water like a torpedo, heading straight for the side of our boat, straight for the center of our hull. He is coming so fast that the water parts in white wings to either side of his black bullet-shaped head.

"Hang on!" I yell. "Hold something!"

But he suddenly halts, his head about a foot from the side of our panga, and just as he stops, the whistle breaks apart, shatters into a cascade of notes—a string quartet playing madly—and it's like we are listening to music that is much too full for human ears. "He's singing," someone says. But it doesn't sound like singing. Too fast. Too urgent. It's more like he's urgently demanding something—asking something—and I wish terribly that I could reply. Then, as the quartet plays furiously on, he lets his long black body swing slowly around: keeping his head in place, a foot from the sideboard, he swings around until he's perfectly parallel with us, so close we could reach him if only we would lean a bit farther over the rail. But we do not. We keep our hands on the rail and on each other's arms, because there is something both fearful and untouchable about him. And then, the torrent of notes still coming, he lets himself slide sideways, beneath us, so that he rests motionless in the shadow of our panga, and his sounds reverberate up through our hull and along our spines and into the base of our skulls. He moves forward, slowly at first, and suddenly he is gone. No one moves. Only when she releases her grip do I realize that Allie has been clutching my arm. On my other side, Cameron leans over the rail, reaches out, and ever so gently touches the surface of the water. I rest my forehead against the rail and let the tears run down my cheeks.

2

It is the strangest day I have ever witnessed here: the hurricane seems to have driven the gulf's whole menagerie northward, and the angels, from La Guarda in the east to little Borrego in the west, have received the motley fugitives as they've stumbled in from the storm. It is an accident of powerful weather, but it has produced what must be a close approximation—at least, the closest any of us shall ever see—to the sea before humanity exacted its toll. In the course of a single morning, we have watched the water boil with frantic shoals of grunion and herring; our gazes have raced after the bubble trails of skipjack, yellowtail, and a fleet of Pacific barracuda as they streaked pell-mell through the clouds of baitfish; we have come upon another skein of rays, allowing Miles once more to swim among them—a pale, elongate member of their flock; we have seen common and bottlenose dolphin, short-finned pilot whales, Bryde's whales, fin whales, and, of course, the false killer whales. There has been something providential about it.

By midday, we felt overwhelmed, incapable of witnessing more, and so we retreated to watch the postdiluvian eruptions of nature from the calm and quiet distance of our field station terrace. Because the sun has already regained its triumphant and cruel intensity, our group has fragmented into the separate islands of hazy tamarisk shade available to us here on the hot stone masonry. And by an accident of unoccupied plastic chairs and the timing of people walking out of the kitchen with their lunch plates, Graham and I find ourselves sitting now, in a row facing the sea, with Ace, Becca, Lucy, and Chris—an unusual grouping of students, or at least, not one of the typical teams.

Out in front of Cabeza, a new feeding frenzy is just gathering strength, and what was violent and electrifying when we were close enough to hear and smell it is, from here, rather picturesque, strangely mute, like a pitched battle viewed from a distant hilltop. With the boobies, pelicans, and gulls flying beelines from all over the bay, swerving suddenly as they merge into the tight cylinder of circling birds, the frenzy resembles a funnel cloud materializing out of a storm, reeling in fresh energy to fuel its rotation, and at last touching down with a furious, sucking appetite. Just to the left of the frenzy, between Cabeza and Ventana, a pair of small whales—Bryde's, probably—are lolling at the surface. And the air is so calm, so painfully and wonderfully windless, that larger spouts are visible farther out in the channel.

"I don't really get it," Becca is saying to me. "I mean, why would you *weep?*" Ace has just mentioned my reaction to our brief and obscure colloquy with *Pseudorca crassidens*. I hadn't known that anyone had seen me bow my head to the rail and let myself go—it was only for a few seconds—and so I was at first a bit taken aback, abashed to have such a gesture disclosed here, on our terrace, where it can only sound overblown and theatrical. Still, it is clear to me that Ace's intention was not to point out my melodrama, but only to convey how significant the encounter felt—only to communicate something like, *He's seen a lot of whales, and even he was overwhelmed by it.* And whatever unease or embarrassment I might have felt has just been canceled by Becca's reaction, which somehow has an effect exactly opposite to the one she intends: instead of making me feel like I need to explain myself or justify my sentimentality, her question reassures me that I do not. So I merely nod and shrug my shoulders, as if to acknowledge the mysteriousness of sentiment. But Becca won't give up so easily: she leans forward now, completely blocking my view of both Lucy and Chris, and says, "Don't you think you're being sort of anthropomorphic?"

She is especially determined this afternoon, domineering even by her standards, and I have to wonder if this might be in retaliation for our departure without her—the way it seemed maybe a little too easy, a bit too willing on everyone's part. And maybe she senses, too, that what we experienced today on the bay has drawn the group even closer together, and she—sadly, but also, somehow, inevitably—finds herself outside of the newly girded friendships. Whatever its cause, her intensified campaign seems especially pitiable just now, because I think she'd been loosening her grip over these past few days, relenting ever so slightly. Maybe it was just that she was exhausted. Or maybe—and this is a troubling thought— she was finally opening up to her experiences. And then, just as she was starting to relax her hold on everyone, we abandoned her completely, which might be why she's become more ferocious than ever. Still, even as she escalates further, saying now, "You guys sound sort of like kids at SeaWorld," no one rises to her provocation; the generosity of the bay has kindled in us, I think, both humility and our own small gestures of kindness.

And surprisingly, Becca tires. When each of her sallies is met with nods, patience, and a contented willingness simply to sit longer in silence and watch the whale spouts erupting all over the bay, she seems at last to give up. She leans back in her chair, draws her skinny legs up to her chest, and hugs her knees. At the end of our longest silence yet, Graham turns to Becca and says, "You know, Becca, what's right under what you were asking before—*why would you weep?*—is a really big question. In some ways the biggest. Why does one care? Why does one care about any of it?"

For a moment, we're all silent, leaving room for Becca to respond, but she remains quietly curled in her chair, and finally Chris says, "I think that's what we've been talking about, isn't it? Sort of indirectly?"

"I'm not sure I follow," Graham says.

Chris explains that, for the past few days, we've been talking about all the ways the sea here can be valuable: there are the various fisheries we've mentioned, and then there are also all the species that tourists will pay to see or catch or shoot. "With the right kind of visitors," Chris says, "those species are as good as gold to the people here."

"That's true," Graham says. "But is that really why *we* care? I mean, it's not *only* that, is it?" When Chris does not respond, Graham goes on: "I think maybe you're giving in to a more purely economic calculus than you'd really want to. It's certainly true that the sea can mean money for the people here, but what I'm trying to get at is an answer that doesn't outsource the value, if you know what I mean."

"Not really," says Chris.

"Why is it valuable directly to *us*? To *you*?"

Chris smiles. After what happened with Isabel, he's just not going to expose himself. "What would be your answer to that?" he asks Graham.

"I think, at bottom, it can only be a matter of aesthetics. We want this place to stay as it is—maybe even return to what it once was—because we value its beauty. And I'd even make a more general claim—that environmentalism itself is, ultimately, an aesthetic movement."

"I think that's incomplete," I say. I'm not sure it would seem so clear to me always and everywhere, but today, after what we've seen, and here, where we sit, overlooking a sea that for a brief moment resembles its primeval grandeur, the intuition that this place should be saved—our response when we imagine or remember the destruction of all this teeming life—seems to me an intuition that belongs not only to the realm of aesthetics, but also quite firmly to that of ethics. In other words, our reaction to the bay's destruction feels not only like a lament over beauty lost, but also like a pang of contrition.

"Contrition toward *whom*?" Graham asks. "Toward the people who would depend on it? Isn't that a restatement of Chris's position?"

"Yes," I say. "But not only toward them. I mean it more broadly—a broader form of contrition."

Graham looks at me queerly: I need to make more sense. And I can do so, I realize now, only if I stop trying to hide or obscure the actual origin of the position I am now taking. So I go ahead and ask: Why is a moment of glancing, near connection with an individual whale so overwhelmingly powerful? It is not merely that we find the whale so beautiful. It is, rather, our sense that we are eye to eye with another *being*: a being that commands our acknowledgment, our respect, and perhaps even—I know I'm pushing it here—our apology.

"*Apology*?" Graham says. "Hirsh—"

"I know," I say, "I know." And I try to explain that I'm not suggesting we say sorry to the whales—or that doing so would do anyone any good at all. I'm just trying to describe the sense of moral obligation toward a creature that is conscious and worthy of reverence, but outside the sphere of human communication. This morning, when we encountered *Pseudorca crassidens*, I didn't actually know the details of its status or history, but I might have surmised that its rarity has something to do with humanity. As it turns out, the pod we saw today represents about half the false killer whales in the Sea of Cortez. And there would be many more, I suspect, were they not competing with people for fish populations that have been annihilated over the past century. Certainly there would be many more in the Pacific had they not been systematically slaughtered by Japanese fishermen, who considered *Pseudorca crassidens* an especially clever predator, with which they would rather not compete. In short, *Homo sapiens* has been awfully hard on *Pseudorca crassidens*. And yet, this morning, a member of the defeated species approached a boatful of the triumphant one, and it tried to communicate. By aesthetic measure, *Pseudorca crassidens* is indeed valuable—it is a rare, magnificent, and thrilling creature to be near—and so, in approaching us, that individual whale conferred a kind of gift. It is a benefaction that the history of *Homo sapiens* suggests we do not merit. And yet I have some vague hope, or wish, that we might, in the future, prove ourselves more deserving.

3

"Well, how did it go?" Veronica asks. Above the onyx water, her shoulders and face are silvered by the moon. It is full tonight—so pale and radiant that, to anyone standing on the station terrace, Veronica, Graham, and I would resemble remnant columns of white coral or stone, just now appearing out of the ebbing tide. At the moment, it's a pleasing fantasy—that one could remain here for as long as it would take the saltwater and waves to wear away the features and eventually the entire form of a telamon; that one could stare stonily out at the horizon and the islands when the tide was low, and just as stonily at the passing fish when the tide came in; that one's skin would become prickly with the tiny cinder cones of barnacles, the upright tubes of fanworms, the layer upon layer of residents. These final evenings make me melancholy, and to become a stone column dressed and undressed each day by the sea, slowly converted to a small tower of homes, feels like an escape from my premature nostalgia.

Each of our pale busts casts two shadows on the black water—one in the moonlight, the other in the fluorescent light that washes down from the large windows of the veranda. We can see the students bustling about in there. They're up to something—a mural on one of the walls, I think, though what it depicts we don't yet know, because the students are intent on hiding it from us until it's finished. Their conversation right now is amusingly hushed yet loud, like the talk of young children excited to be keeping a secret. Between the rising fragments of their animated exchange—*No, no, put the whale spout there . . . yeah, above him*—I can just barely hear a tune being picked out on the old ukulele, and Ace, singing quietly along.

"You mean tonight?" I ask. "Or all of it?"

"Tonight," Veronica says. "How did it go tonight?"

"Really well," Graham says, and I agree with him. Veronica led our final seminar, which our syllabus calls, somewhat grandiosely, the Course Synthesis. And she took a risk. Instead of standing at the lectern at the front of the room and delivering the impassioned valedictory—*here's what you've learned, now go forth, change the world*—which we know the students love to hear, Veronica courageously yielded the floor. She asked the students to call upon what they had learned about this place, and to describe its future—an optimistic future, but not a wild dream. "You know some of the challenges," she said. "So let's imagine how they're met."

At first, the group felt painfully hesitant. Rafe redid his ponytail. Miles scratched Millie's crooked ear with his toe while, beside him, Haley leaned over in her little desk to scratch Yukon's belly, which he was patiently presenting to anyone who might take an interest. Ace looked down at the blank page of his notebook, smiling at one of his own thoughts, while Lucy smiled widely at everyone whose eye she could catch. I was mystified. Our group has had some taut moments over the past several days, but I had not felt this—what can only be described as awkwardness—since much earlier in our trip, and I did not have any idea what could make everyone feel so suddenly that every move or word available to them would somehow be embarrassing or uncomfortable. As the silence stretched on, congealing around us—any remark offered now would sound forced—I began to feel very worried for Veronica.

But she just waited them out. She rested her elbows on the lectern and leaned forward. And though her face was flushed and shining already with the evening's airless heat, she looked as composed as always. I don't know whether she was calling on some profound faith she has developed in this group and the experiences we've had, or if she understood all along—knew not only the nature of the hesitation, but also that a certain visitor could change everything.

Becca was the first to break the silence. "The government should promote tourism," she said, "but only ecotourism. Definitely not other kinds." She glanced at me conspiratorially, letting me know, I gathered, that she was taking my side in the argument I had had with Isabel. For the first time in days, I felt my irritation with her flooding suddenly back.

Fortunately, Anoop stepped in before I could respond. He offered a comment that was so characteristically academic-sounding, so carefully articulated, as to be charmingly his own, though it was clearly excerpted from one of our previous seminars. Raising his hand politely—a habit he just can't seem to relinquish—he said briskly, "Catch limits should be set to hold population sizes well above fifty percent of carrying capacity. It's too risky to hold populations close to their theoretical point of maximal yield." He then carefully enumerated the sources of risk: the uncertain relationship between population size and fecundity; the difficulty of measuring population size; unpredictable ecological feedbacks.

As Allie responded, she tried, I thought, to sound as if she were picking up spontaneously and with great interest where Anoop had just left off. But in truth she too was parroting previous seminars. For many species, she said, it would be better not to set catch limits at all, but rather to protect them in another way, by designating certain areas where fishing is not allowed. Such reserves, she explained, would have the advantage of protecting not only the species being fished, but also the habitat or prey they depend on. "Sometimes," she concluded, doing her best to sound invested in what she was saying, "you have to think about the whole system."

Rafe, Lucy, and Chris followed up with comments of their own, and these too were quite right and well put, but all too obviously clipped from a reading assignment or a seminar. Everyone was trying bravely to push through the awkwardness, and they all somehow seemed very much themselves even as they said things that were not quite their own. But there was just no avoiding it: their comments felt canned. Like the pickled animals in the museum room adjacent to our conversation, the ideas were no longer alive; they were gray and still, there to be pulled off the shelf and demonstrate perfectly preserved anatomical details.

But then something happened. Alejandrina entered the lecture room.

In one hand, she held a stack of small plates, and in the other, a beautiful, caramel-colored flan, a treat she had prepared for our final evening at the station. As she offered profuse apologies for interrupting class, she spooned the smooth custard onto plates, and the students passed them around the room. Then Alejandrina said, "Buenas noches, see you early for breakfast," and she turned to leave. But before she stepped through the doorway leading back to the kitchen, she turned around again and addressed the students. "You are a beautiful group," she said, smiling affectionately. "Thank you for being such a good group."

Then she left, and somehow, with that, the conversation was changed. It was as if, mysteriously and right before our eyes, the ideas stirred to life, cracked the preserving jars they'd been crammed inside, and shook themselves awake, spattering drips of formaldehyde all over the room; their natural colors returned; they took flight or fin; and suddenly our seminar room was a reef teeming with creatures, an oyster-shell beach, a bay and town for the future. And for all its dazzling flourish, the place was not altogether fanciful, because the students really have come to know a bit about Bahía de los Ángeles. As they offered their thoughts and plans, I realized that a number of them had begun—days ago, in their own reflections or conversations—to answer the very question Veronica asked them tonight.

So what changed? What transformed our conversation? Was it just that the sweet, homey custard made everyone feel more comfortable? Did it dispel an anxious hunch that they might be engaged in a final exam in disguise? Maybe, but I think there was more to it. I think what had been holding the students back was a sense that they would be presumptuous to answer Veronica's question with ideas of their own. A few days ago, Antonio reminded us all that we are definitely not local. Then my run-in with Isabel warned us how fraught with tension is this great difference in privilege between us, the visiting academics, and them, the townspeople. But when Alejandrina turned to the students and reassured them, so warmly, that they are welcome and appreciated, she somehow gave them permission to speak to the future of her town.

It was Miles who started it. Pausing from his custard to gesture with his spoon, he said, "You know what I could see?" Curled at his feet, Millie tracked the spoon's small movements intently. "I could see solar panels back there," Miles said, pointing west, to the hillside behind town. "I mean, this place has more sun than anywhere—not this week, but usually. And then they could shut off their damn generators."

"Amen," said Cameron. "Nix the generators." And then he described how different the bay had seemed to him on those nights when the generators had been silenced. It was like the animals suddenly came out of hiding. "And you know," he said, "every afternoon, the wind totally howls. So maybe you could have wind turbines too—like up on the ridge of Mike's Mountain. And with solar panels and wind turbines, you'd have this total energy plant."

And from there, a village by the sea began to come clear. Aquaculture would thrive on the bay's edges: to the south of Punta Roja, brown sea cucumbers; to its east, the black and rainbow pearl oysters that so crazed the conquistadors; in the bay west of Punta Don Juan, lion's paw, the largest and most valuable scallops ever harvested; east of Don Juan, edible oysters; in parts of La Gringa, perhaps, Venus clams like the ones we dug there for our own lunch.

Allie wondered whether bulldozing all the half-built and deserted buildings into the sea might provide a substrate on which the oyster reefs could found their recovery. "But not Alejandrina's restaurant," she said, with a lighthearted but knowing glance at Veronica and me.

Cameron suggested a pharmaceutical company might be interested in the compounds hidden in the bay's diverse fauna. "Just look at holothurin," he said, "and cancer treatment. Maybe that's a way to make cucumbers even more valuable."

"And what about the bighorns?" Rafe said, recalling one of Antonio's ranting detours, when he had described hunting safaris for wealthy gringos. "Let the blokes shoot a ram every year. Right there's a hundred grand for the ejido, right?"

"We came up with this kind of crazy plan," Chris said, and from the way he looked over at Isabel it was clear that *we* meant the two of them. "It takes the idea of catch shares a step further."

Catch shares, like the marine reserves Allie mentioned, had come up in one of our seminars on managing fisheries. And they, like reserves, offer an alternative to more traditional systems in which managers sell licenses and then declare the fishery open for a designated season or until a certain amount of fish has arrived at the dock. Such systems, though they are still common, have corrosive effects. They create a race to fish, because there is a limited quantity of fish to take and many fishermen competing to get as much of it as they can. The race motivates fishermen to buy bigger boats and gear, to head out to sea even when conditions are dangerous or the fishing not particularly good, and to bring all their fish to market at the same time, early in the season—all of which makes fishing a lot less profitable than it would otherwise be. A strategy involving catch shares avoids all of this by granting individual fishermen real ownership, in perpetuity, of a defined fraction of every year's total permitted catch. Roberto, for instance, would have the right to one percent of the total catch of halibut that the fishery manager permits every year. Such a system allows each stakeholder to decide when he wants to catch his fish, and gives him the whole year long to do it. The race to fish, with all of its perverse incentives, is averted.

What may be more important about catch shares, though, is simply that they allow fishermen to take the long view: if the fish population thrives, a fisherman's share increases in value; if it declines, so too does the value of his share. This makes a fisherman less likely to sneak off with more than his quota, because he knows that doing so might compromise his share's value in seasons to come. And the stakeholder is not only a more honest participant, but also a de facto enforcer, since he≈doesn't want the value of his share diminished by anyone else's cheating. The fishermen around here—men like Roberto and Pablo— already have a long-term interest in the bay's fisheries, for the simple reason they have nowhere else to go. But the same cannot be said for the larger vessels that come from Oaxaca or even Japan. And the threat of such vessels might even impel Roberto and Pablo to fish as fast as they can.

In view of the basic difference between those who take a long-term stake and those who do not, Chris and Isabel had invented a new system for managing the waters around Bahía. "The federal and state governments should have nothing to do with it," Chris said, "except for defending the ejido's ownership. These waters—from here out to Ángel de la Guarda and, I don't know, a hundred miles, say, north and south—they should just be turned over to the ejido. And then the ejido members could do whatever they wanted with it."

"You realize," I said, "that you're effectively privatizing the ocean."

"Yeah," Chris replied, "I suppose we are. But privatizing it only to a point. We're saying it has to stay in the hands of the whole ejido—the people who live here. If they decide ecotourism would be best for them, they go that route. If they decide industrial fishing is the way to go, the bay is theirs to destroy."

"But they won't," Isabel interjected. "Of course they won't. Because it *belongs* to them. And it will belong to their children. And they can be sure no big ships will come from Japan, so they don't have to pull out everything before the other guys do."

"Just look at the Seri," Chris said. "Why can't that work for all of Bahía de los Ángeles?"

The Seri are an indigenous tribe that live just across the gulf from our field station. Since the late 1970s, the Seri have held exclusive rights to fish the waters around Isla Tiburón, including Infiernillo Channel, which runs between the island and the mainland. The Seri focus on fan clams, which taste like giant sea scallops and fetch a similar price on international markets. There are about five hundred Seri, and all of them have the same legal rights to harvest fan clams. You might therefore expect competition among them to drive the hurried extraction of such a valuable resource—yet another iteration, in short, of the familiar story of fisheries. But this has not come to pass. Infiernillo Channel today harbors the richest shellfish beds in the entire gulf.

In an article Veronica assigned for one of our seminars—the article to which Chris was now referring—an anthropologist studied how the Seri have averted the typical dynamic of open fisheries. It seems they possess a kind of collective awareness of their resource. Members of the tribe are constantly gauging the state of shellfish populations simply by wading out onto certain sand flats and digging around. They talk about their observations, and older members of the tribe, who have a longer view of historical levels of abundance, have more sway. Whenever their conversations begin to converge on the conclusion that fan clam abundance is low, they start to scale back their harvests. What is most extraordinary, and also a bit amusing, about this collective management system is how their decisions are enforced. If anyone is deemed responsible for declines in fan clam abundance, he faces a severe punishment: Groups of women berate and humiliate him in public.

Responding to Chris and Isabel's proposal, Anoop pointed out that here in Bahía, assessing the size of most populations is not so easy. You can't just wade into the shallows and dig around a bit. "How will you know how many grouper there are?" he asked. "Or tiger sharks?"

"We'll float in bloody water," Rafe said, smiling. "We're awesome at that."

"And then there's enforcement," Anoop said.

"Right," Rafe added, "I don't think everybody in town is going to listen to some grandmother."

"Well, it's not going to be *exactly* the same," Chris said. "But I—we—just meant that if you really hand the place over to the ejido, they'll make good decisions. Because it'll pay off for them."

"And couldn't they get help figuring out population sizes?" Allie asked. "Isn't that something the government could do? Or maybe universities?" As it occurred to her, the idea seemed to fill her with excitement: "We could actually make ourselves useful!"

"Sure," Chris said. "But the management decisions would be theirs—the very same people fish it, own it, and manage it. That's what's important."

"And like Antonio was saying," Isabel said, "nobody knows what was here any better than the old locals. Maybe they'd need technical advice—the computer models and stuff—but they know the place better than anybody."

It's hard to say there's a bright side to fishing all the way down the stairway of marine resources. But if there is one, it is this: potential value is now evident where before it was unseen. From the rainbow pearl oysters in the sixteenth century to the murex and cucumber in the twentieth, there is a long succession of newly discovered resources. If those resources were managed wisely, they could all yield income in parallel, and not, as has been the case for five centuries, in succession. When I made this point to the group, Rafe said, "Anoop and I were thinking about a kind of WikiPlace."

"A what?" I asked.

"We called it a WikiPlace," Anoop said. "It would be a common repository for everything everybody remembers about a location's natural resources and environment."

"It would have Antonio's stories of turtles," Rafe explained, "and all the fishermen's stories of giant sea bass, totoaba, sharks, whatever. And historians could go back even further—like to the Spaniards you guys have talked about, Cortés and missionaries and all."

Veronica added that ecologists could contribute with their own tricks of reconstructing history.

"We didn't figure out how it would all be represented," Anoop explained, "but we thought it would be best to have it all laid out somehow on an actual map. So you could click on a location and see what used to be there."

"And that could be their best management tool," Allie said, connecting the idea back to Chris and Isabel's proposal.

"Right." Anoop nodded. "I suppose it could."

Graham, clearly taken with the idea, summarized it for them: "A vast, geographically structured memory. A memory map of nature."

"Precisely," said Anoop.

"Awesome," said Rafe.

Was it all so much utopian dreaming? There are certainly reasons to be skeptical. For one, the members of the ejido are evidently quite excited to sell their stake in the land. So why would they feel any differently about a stake in the ocean? Why wouldn't they turn around and sell it to some multinational corporation? One could think through such problems— and the group would have gotten right to work if I'd mentioned them. But maybe the more general reason to restrain our optimism is simply that every success would invite, in equal measure, sudden failure: any step in local development would make this forbidding corner of desert a bit more livable, a bit less physically daunting, and, therefore, that much more inviting to the wrong sort of developers. Imagine Miles's solar panels or Cameron's wind turbines, and then imagine the number of hotel rooms they could air-condition.

Still, as the conversation drew to its end, we all saw, spread out in our imagination, a seaside village abounding in natural wealth: that pair of ten-year-old boys, once again trudging along the beach with a yellowtail as tall as they; the turtle boat, riding at anchor on the blue water, only now it's a boat that takes the tourists out to see the turtles, and the whales, and the manta rays; the pangas returning to the beach, their hulls piled high with sea cucumbers or pearl oysters from the coves across the bay, where the creatures are steadily plentiful, because they are cultivated there; and, maybe the strangest part of the entire vision, a new kind of map of the bay and its angels and the Channel of the Whales, a map like the topographic one I know well, only so much deeper, because it holds layer upon layer of time. As all of this was folded up for the night, I felt, for the first time in years, somehow hopeful for Bahía de los Ángeles.

"Veronica," I say now, looking up at the station's bright windows, where there's just been an outburst of laughter, "did you know Alejandrina was going to talk to the students like that?"

"When I paid her last night, she told me the students had been so polite and helpful, she wanted to thank them somehow. I told her tonight's lecture might be a good time."

A good time indeed. And I'm left wondering, as I sometimes do, just how precisely Veronica has planned things.

"It did go well tonight," I say again, "though we let ourselves off the hook—skipped over some of the hard problems." And I mention now my deepest reservation about the visionary scheme: the better it works, the more likely it is that someone comes along and buys it all for their own purposes.

With a note of playful disputation, Graham says, "So you're saying the prosperity of the people and the bay aren't so well aligned, after all."

"They could be in the end. I'm just not sure there's a path that leads to the right end."

"How about the whole thing?" Veronica asks. "How did it all go?" She seems to sense that this is perhaps not the moment to engage the question we've discussed, in different forms and various constellations of people, for a few days now. It is a moment, she thinks, for our own private course synthesis.

"I think we did well in the storm," Graham says.

"We did," Veronica agrees. "We kept it together." After a moment's pause, she says, "I regret that things got worse with Becca. I never managed that very well."

"Not clear what you could have done," Graham says.

"I could have asked her about herself," Veronica replies. "I never did that, you know? I never just said, 'So, Becca, where'd you grow up?' Or, 'Tell me about your parents.' Or, 'Why'd you go into ecology?' She was so busy talking, we never got her story."

4

A storm dwarfs us; a panga can be tipped and its passengers lost; you can bob in a boat beside a leviathan, and feel a reverence that long pre-dates our dominance of the natural world—these too are reasons it's hard to remember that the era of nature is over. Measured by the magnitude of our collective impacts, we are far greater than ever, but individually, we are just about as small as ever—and this is the scale at which we perceive the world. We do not see the ocean as a fishbowl, or sense the century as a passing minute. And therefore we do not possess the perceptual equipment needed to apprehend the world on the scale at which we alter it.

Or do we? Cortés was writing to his king, and this affected not only what he said about the New World, but also, probably, what he saw of it. And his perception was also influenced by Pliny, by the Spanish novelist Montalvo, and by his Mexica mistress Malinche. In a way, then, we do see the world collectively: each of us is forever pushing and pulling on what the other perceives. Seeing is in no sense a solitary act, but rather, to bend a worn phrase to a new and almost contrary meaning: *We see through each other*.

Granted, this operates most palpably in very close groups. It is mostly as intimates that we truly merge our views into one, or, pushing out just a bit further, as small schools that we perceive the world together. And yet, as the letters of Hernán Cortés suggest, the threads of perceptual influence may be more extensive and dendritic than we regularly sense. Anthony Pagden, the historian and translator, writes, *Cortés is the only conquistador to have been aware of the power of the relatively new medium of print. He went to great lengths to ensure that each letter was carried back to Spain as soon as it was ready, and it is likely that his father was responsible for arranging for their immediate publication.*

The extraordinary possibility that his own perception might well become that of an entire people must have figured powerfully in Cortés's vision of the New World. Insofar as he really did see Tlaxcala as Granada or Tenochtitlan as Seville; if he saw Mexica wearing yashmaks or felt his boot heels on the hard parquet of the palace floor; if he found himself seated on a kingly throne and heard Montezuma cede his own empire to a god arriving from the east—he perceived such things in part for himself, in part for his king, and in part for his people.

It is an example of extreme distortion, but as such, it vividly displays how the perception of large groups can be—sometimes, inevitably is—knit together. And if this phenomenon has a dark side, illustrated in the

Old World's view of the New, perhaps it also has more promising aspects. Perhaps it is the means by which we can expand the scale at which we perceive the natural world, bringing our apprehension into common measure with the scale of our destruction. Lest this idea sound wildly strange, I'd note that it resides within the intersection of a broader and older set of ideas. The German Idealists, with their appetite for active and meaningful parallels between the small and the vast, the fleeting and the enduring, the individual and the group, had a word for the emergence of a single social and cultural mind. They called it *Geist*, or "spirit," as in Hegel's *Phenomenology of Spirit*. And for similar concepts, other groups of thinkers have used the terms *collective consciousness* and *hive mind*. What I am talking about is just the perceptual and mnemonic equipment that seems to be common to all of these concepts of collective entities.

A certain image comes to mind: the original frontispiece for *Leviathan*, by Thomas Hobbes. It shows the towering figure of a king, whose entire form is made up of the comparatively tiny figures of his countless subjects. On his head he wears a crown. In one hand he holds a scepter. In the other, he wields an enormous sword. And he looms huge over a miniaturized European landscape of hills and villages. In our collective relationship with the natural world, the scepter and crown are not in doubt, and the sword has been swinging freely to devastating effect. But what we require now are the eyes and ears of that Leviathan, and also his memory—a perspective expansive enough, and remembrance deep enough, to apprehend our impacts. And as Antonio's turtles and the totoaba demonstrate, we will need his reach, as well.

Recognizing the striking difference in scale between the individuals who compose Leviathan and the great figure himself, between those who inhabit the miniature landscape and he who looms over it, we are compelled also to acknowledge that our ethical relation to nature has changed. It may be right for a man to harpoon a leviathan, but not for Leviathan to strike down a whale.

5

The whole conversation started with a sea cucumber, said Cameron. And at first the statement struck me, stayed with me, mainly because our stories—scientific, historical, and personal stories—seemed a way to perceive the meaning and significance, otherwise terribly elusive, of the humble cucumber. For we face profound challenges in learning merely to see the natural world accurately: How are we to preserve our wonder and reverence, even as we recognize degradation? How do we glimpse what remains of the wild, even as we extend our influences deeper into its very heart? How do we recognize that the wilderness has become our garden, when we are still, individually, so very small? Wilderness and the garden; great and small; beautiful and broken—again and again, we are charged with the vital but difficult task of simultaneously seeing opposites.

The whole conversation started with a sea cucumber, said Cameron. And stories seemed the way not just into meaning, but also into memory. Through stories of evolution, we recall millennia; through ecology and history, centuries; through personal tales, decades. Most important, though, the stories interweave, and in their fabric we discover the many rips and holes. Thoreau said he hadn't even realized that some of the best stars were missing, plucked by the demigods who had come before him. And for us too a span of darkness is easy not to see. But the stories that connect organisms form something like a map of the zodiac, and if one piece of an animal's form is missing—a leg of the crab or a fin of the fish—you notice.

The whole conversation started with a sea cucumber, said Cameron. And at first, I had my eye on the animal. But the notion is important in another way, too. For while it is certainly true that our stories, scientific and otherwise, are a way in, a way to start caring about the creatures and the place, it is equally true that the creatures and the place can be our way back to one another. After all, we tell these stories together. I remember coming to the surface with Cameron and listening to the others—Isabel, Ace, and Haley—as they described a cowrie for him; I think of Chris, Anoop, Rafe, and Allie, asking one another incisive questions about recapitulation. And my view shifts then, from the cucumber to the people talking about it. To repurpose a certain neologism: ecocentrism, I believe, is a humanism.

The whole conversation started with a sea cucumber, said Cameron. And I have to wonder: with years of experiences to show me just that, why did the notion feel so new? Perhaps because I have long held a romantic vision of the solitary individual heading out into nature. And embedded in that vision, of course, is the sense of escape from other people. I've inherited this, I think, from writers I've read and greatly admired—nature writers, mostly, but also the prophets and eremites who preceded them into the deserts, the mountains, and the sea. But maybe, sometimes, we come to places like this not to be away from one another, but to meet in ways that only places like this allow. I think of Ace and Lucy, singing their duet at Isla Ventana; or of Miles, lifting terrified Becca onto the boat, while Cameron waits for both of them to get on before him. I think of six people gripping one another in wonder, leaning over the rail of our panga to peer down at a curious and conversive whale.

The whole conversation started with a sea cucumber, said Cameron. And to add yet another turn in the cycle—not the spiral of decay, with which this book began, but one that moves, perhaps, in the opposite direction—our stories take us into the place; the place leads us back to one another; and, maybe, in our expanded conversation lies the very best hope for the place. I do not know if a collective consciousness of the environment can be scaled up from an island tribe to a town and beyond. If it can, it seems certain that it will require some form of seeing through, as well as thinking with, one another. Perhaps Anoop and Rafe's strange dream of a virtual place, a map layered with memory, a vast cortex storing ecology, is an organ of the Leviathan we require. At the dawn of animal phyla, near the base of the Cambrian, a new level of interaction spawned biological diversity. Perhaps now, in the twilight of that diversity, yet another will salvage what remains.

ACKNOWLEDGMENTS

First, I would like to thank the students and teachers of the Baja class. Graham Burnett, Josh Maximon, and Dmitri Petrov have taught many classes with Veronica and me; Josh and Dmitri were with us in Bahía for some of the incidents reported here, and further books could be written about the experiences we've shared with them. The Baja class began, of course, with Veronica's own mentor, Lane McDonald, and I am grateful to him for welcoming us into the tradition he created. I would also like to thank our gracious and wise hosts in Bahía, in particular Alejandrina Diaz (whose beautiful new restaurant is now open), Antonio Reséndiz, the Grupo Marino fishermen, Samy and Octavio Diaz, and Rubén Ocaña.

Eric Chinski, my phenomenal editor at FSG, chiseled bravely at the rather largish hunks of rock I gave him. And though sometimes I couldn't watch, when I finally brought myself to peek, I would invariably find that what he'd made emerge was much better than what had been there before. Another extraordinary editor, Anne Fadiman, stands in some sense at the origin of this work: she edited the essay that later led to the book, and she encouraged me, at a critical juncture, to keep writing. Tina Bennett, my agent, offered guidance, insight, and reassurance; she is the champion and adviser every writer hopes for.

A succession of intellectual mentors have made this book possible: Richard Jones, David Humes, Steve Johnson, William Howarth, Stanley Corngold, Simon Levin, Alex von Gabain, Dmitri Petrov, and Marc Feldman. And my parents, besides being my parents, also belong in that list. Graham Burnett, Jeff Dolven, Josh Maximon, Len Nalencz, Ben Phelan, and Veronica Volny read the manuscript—valiantly, some of them did so several times over—and they all provided insightful and valuable commentary. Elaine Lauterbach, Yan Linhart, and Mike Macpherson provided important feedback on certain sections of the book.

Acknowledgments

I would also like to thank several institutions for supporting the Baja class or me as I wrote this book: the Agouron Institute; the Vermilion Sea Institute; the Department of Ecology and Evolutionary Biology, and particularly Jeff Mitton, at the University of Colorado–Boulder; and, at Stanford University, the Department of Biological Sciences and the Bing Overseas Studies Program.

My family has offered enduring patience, encouragement, and support, for which I will always be grateful.

Most of all, I thank Veronica.